Ergebnisse der Mathematik   Volume 50
und ihrer Grenzgebiete

3. Folge

A Series of Modern Surveys
in Mathematics

Editorial Board
M. Gromov, Bures-sur-Yvette
J. Jost, Leipzig
J. Kollár, Princeton
G. Laumon, Orsay
H. W. Lenstra, Jr., Leiden
J. Tits, Paris
D. B. Zagier, Bonn/Paris
G. M. Ziegler, Berlin
Managing Editor R. Remmert, Münster

Sergey Neshveyev
Erling Størmer

Dynamical Entropy
in Operator Algebras

Sergey Neshveyev
Erling Størmer

Department of Mathematics
University of Oslo
P. B. 1053 Blindern
0316 Oslo, Norway
e-mail: sergeyn@math.uio.no
erlings@math.uio.no

Library of Congress Control Number: 2006928835

Mathematics Subject Classification (2000): 46L55, 28D20

ISSN 0071-1136

ISBN-10 3-540-34670-8 Springer Berlin Heidelberg New York
ISBN-13 978-3-540-34670-8 Springer Berlin Heidelberg New York

This work is subject to copyright. All rights are reserved, whether the whole or part of the material is concerned, specifically the rights of translation, reprinting, reuse of illustrations, recitation, broadcasting, reproduction on microfilm or in any other way, and storage in data banks. Duplication of this publication or parts thereof is permitted only under the provisions of the German Copyright Law of September 9, 1965, in its current version, and permission for use must always be obtained from Springer. Violations are liable for prosecution under the German Copyright Law.

Springer is a part of Springer Science+Business Media

springer.com

© Springer-Verlag Berlin Heidelberg 2006
Printed in Germany

The use of general descriptive names, registered names, trademarks, etc. in this publication does not imply, even in the absence of a specific statement, that such names are exempt from the relevant protective laws and regulations and therefore free for general use.

Typesetting: by the authors using a Springer LATEX macro package
Production: LE-TEX Jelonek, Schmidt & Vöckler GbR, Leipzig
Cover design: Erich Kirchner, Heidelberg

Printed on acid-free paper   46/3100YL - 5 4 3 2 1 0

Preface

When the algebraic formalism of quantum statistical mechanics and quantum field theory gained momentum in the 1960's, it started a very fruitful interplay between mathematical physics and operator algebras. The study of automorphisms and their invariant states became a blooming discipline, and a subject of noncommutative ergodic theory evolved. With the great success of entropy in classical (abelian) ergodic theory it was natural to extend that theory to operator algebras. In some cases, like that of quantum spin lattice systems, that is rather straightforward, since in those models the mean entropy definition in the classical case can be extended to C^*-algebras by replacing partitions by local algebras. But in more general cases the C^*-algebras generated by finite dimensional C^*-algebras can easily be infinite dimensional, so the mean entropy cannot be used as a definition. In order to define dynamical entropy for automorphisms of C^*-algebras one has to rewrite the classical definition in a form independent of the join of partitions and use that as the basis for a definition. This was done by Connes and Størmer in 1975 for finite von Neumann algebras, and a useful definition was accomplished, giving in particular the expected entropy for noncommutative Bernoulli shifts [51]. The theory evolved slowly; it took 10 years before Connes extended the definition to normal states of von Neumann algebras [49], and a little later he and Narnhofer and Thirring [50] extended the theory to states of C^*-algebras. Several other attempts have been made to define dynamical entropy for C^*-algebras, see Notes to Chaps. 3 and 6, the most useful of which being those of Voiculescu [227]. His idea was to consider finite dimensional C^*-algebras which approximately contain finite sets of operators instead of the algebras they generate. In particular he obtained a definition of topological entropy which is an extension of topological entropy in the classical case.

In the present book we shall develop the basic theory for the dynamical and topological entropies alluded to above. Then we shall discuss the special situations which have attracted most attention. We start with a chapter on the classical case, mainly for motivation and background. Then we develop in Chap. 2 the necessary theory of relative entropy for states, which is indispensable for noncommutative entropy. In Chap. 3 we give the definition of dynamical entropy and show its main properties, and follow this up in Chap. 5 with a definition, due to Sauvageot and Thouvenot [188], inspired by the classical concept of joinings. Topological entropy is treated in Chap. 6. The rest of the contents of the book depends heavily on the above chapters, while the other chapters are more loosely connected. The book is divided into two parts; the first contains chapters of general nature, while we in the second part consider dynamical systems in more special settings. At this stage it should also be remarked that parts of the theory have been treated in the books by Benatti [13], Ohya and Petz [147] and the survey article by Størmer [211].

We are grateful to our colleagues N. Brown, M. Choda, D. Kerr, A. Ocneanu, and Y. Ueda for useful comments concerning the preparations of the manuscript.

S. Neshveyev, E. Størmer

Contents

Part I General Theory

1 Classical Dynamical Systems 3
   1.1 Measure Entropy 3
   1.2 Topological Entropy 9
   1.3 Entropy via Partitions of Unity 11
   1.4 Notes 14

2 Relative Entropy 15
   2.1 Relative Entropy for Matrix Algebras 15
   2.2 Von Neumann Entropy 21
   2.3 Relative Entropy for General C*-algebras 26
   2.4 Notes 30

3 Dynamical Entropy 33
   3.1 Mutual Entropy 33
   3.2 Entropy of Dynamical Systems 48
   3.3 Type I Algebras 54
   3.4 Notes 57

4 Maximality of Entropy and Commutativity 61
   4.1 Maximal Entropy of Subalgebras 61
   4.2 Independent Algebras 66
   4.3 Entropic K-systems 69
   4.4 Notes 74

5 Dynamical Abelian Models 77
   5.1 Entropy via Stationary Couplings 77
   5.2 Zero Entropy Systems 86
   5.3 Notes 91

6 Topological Entropy 93
- 6.1 Rank of a Completely Positive Approximation 93
- 6.2 Topological Entropy 96
- 6.3 Notes 103

7 Dynamics on the State Space 107
- 7.1 Measure Entropy 107
- 7.2 Topological Entropy 110
- 7.3 Notes 119

8 Crossed Products 121
- 8.1 Crossed Products by Discrete Amenable Groups 121
- 8.2 Generalizations 125
- 8.3 Notes 130

9 Variational Principle 133
- 9.1 Pressure 133
- 9.2 The variational principle 138
- 9.3 KMS-states 144
- 9.4 Notes 151

Part II Special Topics

10 Relative Entropy and Subfactors 157
- 10.1 Relative Entropy 157
- 10.2 Index of Subfactors 164
- 10.3 Generators and Relative Entropy 167
- 10.4 The Canonical Shift 171
- 10.5 Shifts on Temperley-Lieb Algebras 180
- 10.6 Notes 183

11 Systems of Algebraic Origin 185
- 11.1 Twisted Group C*-algebras 185
- 11.2 Estimates of Topological Entropy 187
- 11.3 K-systems 194
- 11.4 Zero Entropy Systems 201
- 11.5 Automorphisms of Noncommutative Tori 206
- 11.6 Notes 209

12 Binary Shifts 211
- 12.1 The C*-algebra of a Bitstream 211
- 12.2 Entropy of Binary Shifts 217
- 12.3 Notes 224

13 Bogoliubov Automorphisms 227
13.1 Canonical Anticommutation Relations 227
13.2 Topological Entropy 229
13.3 Classical Bernoullian Subsystems 233
13.4 Dynamical Entropy 238
13.5 Notes 248

14 Free Products 251
14.1 Free Products of Algebras and Maps 251
14.2 Free Shifts 254
14.3 Free Product Automorphisms 257
14.4 Notes 263

A Completely Positive Maps 265

B Operator Inequalities 271

C Direct Integrals 275

References 279

List of Symbols 291

Index 295

Part I

General Theory

1

Classical Dynamical Systems

In this chapter we include some background material on entropy in the classical commutative case. We essentially include all the results from the classical theory which will be needed later, proving those which will be used repeatedly and leaving more difficult and rarely used ones without proof. In the last section we show that the classical definition of entropy can be reformulated to a form which has a natural noncommutative extension and which will be the basis for the general theory later on.

1.1 Measure Entropy

Let $(X, \mathcal{B}, \mu)$ be a Lebesgue space, that is, after removing a subset of measure zero X can be given the topology of a complete separable metric space such that μ is a regular probability measure on X, and $\mathcal{B}$ is the algebra of measurable subsets of X. We shall usually suppress $\mathcal{B}$ in the notation. Let $\xi = \{X_1, \ldots, X_n\}$ be a finite measurable partition of X. The **entropy** of ξ is

$$H(\xi) = \eta(\mu(X_1)) + \ldots + \eta(\mu(X_n)),$$

where η is the function defined by

$$\eta(t) = \begin{cases} -t \log t, & \text{if } t > 0, \\ 0, & \text{if } t = 0, \end{cases}$$

the logarithm being considered with base e[1]. We shall write $H_\mu(\xi)$ instead of $H(\xi)$ when we want to stress that our measure is μ.

More generally, assume we also have a measurable partition $\zeta = \{Y_i\}_{i \in I}$ of X. If I is uncountable, this means by definition that up to a set of measure zero the points of the quotient space X/ζ are separated by a countable

---
[1] In the probabilistic literature the logarithm with base 2 is often used.

number of measurable sets. Then X/ζ is again a Lebesgue space. The algebra $L^\infty(X/\zeta,\mu)$ can be identified with the von Neumann subalgebra of $L^\infty(X,\mu)$ consisting of functions measurable with respect to a complete σ-algebra $\mathcal{B}_\zeta$. This way we get one-to-one correspondences between measurable partitions (considered up to a set of measure zero), von Neumann subalgebras of $L^\infty(X,\mu)$ and complete σ-subalgebras of $\mathcal{B}$. Let $E_\zeta\colon L^\infty(X) \to L^\infty(X/\zeta)$ be the μ-preserving conditional expectation, that is, E_ζ is determined by the condition $\int_X E_\zeta(f) g\, d\mu = \int_X fg\, d\mu$ for $f \in L^\infty(X)$ and $g \in L^\infty(X/\zeta)$. The **conditional entropy** is defined as

$$H(\xi|\zeta) = \sum_{k=1}^n \int_X \eta(E_\zeta(\mathbb{1}_{X_k})(x)) d\mu(x) = -\sum_{k=1}^n \int_{X_k} \log E_\zeta(\mathbb{1}_{X_k})(x) d\mu(x),$$

where $\mathbb{1}_Y$ denotes the characteristic function of a subset $Y \subset X$. We shall often write $H_\mu(\xi|\zeta)$ instead of $H(\xi|\zeta)$. If $\zeta = \{Y_l\}_{l=1}^m$ then

$$E_\zeta(f) = \sum_{l=1}^m \frac{\mu(f\mathbb{1}_{Y_l})}{\mu(Y_l)} \mathbb{1}_{Y_l},$$

where $\mu(g) = \int_X g\, d\mu$ for $g \in L^1(X,\mu)$, and therefore

$$H(\xi|\zeta) = \sum_{k,l} \eta\left(\frac{\mu(X_k \cap Y_l)}{\mu(Y_l)}\right) \mu(Y_l).$$

In particular, if $\nu = \{X\}$ is the trivial partition then $H(\xi|\nu) = H(\xi)$.

For measurable partitions ξ and ζ we write $\xi \prec \zeta$ if $L^\infty(X/\xi) \subset L^\infty(X/\zeta)$. If $\xi = \{X_k\}_{k=1}^n$ and $\zeta = \{Y_l\}_{l=1}^m$ this means that up to a set of measure zero each X_k is the union of some of the Y_l's.

If $\{\xi_i\}_{i \in I}$ is a collection of measurable partitions then we denote by $\vee_{i \in I} \xi_i$ the measurable partition (defined up to a set of measure zero) corresponding to the von Neumann algebra generated by $L^\infty(X/\xi_i)$, $i \in I$. If $\xi = \{X_k\}_{k=1}^n$ and $\zeta = \{Y_l\}_{l=1}^m$ then the partition $\xi \vee \zeta$ consists of the sets $X_k \cap Y_l$.

Proposition 1.1.1. *Conditional entropy has the following properties:*
(i) $H(\xi|\zeta) \geq 0$, *with equality if and only if* $\xi \prec \zeta$;
(ii) $H(\xi \vee \vartheta|\zeta) = H(\xi|\zeta) + H(\vartheta|\xi \vee \zeta)$;
(iii) $H(\xi|\zeta)$ *is increasing in* ξ *and decreasing in* ζ.

Proof. It is clear that $H(\xi|\zeta) \geq 0$. Assume $H(\xi|\zeta) = 0$. Let $\xi = \{X_k\}_{k=1}^n$. Since the function η takes the value 0 only at 0 and 1, it follows that $E_\zeta(\mathbb{1}_{X_k}) = \mathbb{1}_{Y_k}$ for some $Y_k \subset X$. We have

$$\mu(Y_k) = \mu(E_\zeta(\mathbb{1}_{X_k})) = \mu(X_k),$$

and

$$\mu(Y_k \cap X_k) = \mu(1_{Y_k} 1_{X_k}) = \mu(1_{Y_k} E_\zeta(1_{X_k})) = \mu(Y_k).$$

Hence $1_{X_k} = 1_{Y_k} \in L^\infty(X/\zeta)$, so that $\xi \prec \zeta$. Conversely, if $\xi \prec \zeta$ then $E_\zeta(1_{X_k}) = 1_{X_k}$, so that $\eta(E_\zeta(1_{X_k})) = 0$ and hence $H(\xi|\zeta) = 0$. Thus (i) is proved.

To prove (ii) assume $\xi = \{X_k\}_{k=1}^n$. One can easily check that for any $f \in L^\infty(X)$ we have

$$E_{\xi \vee \zeta}(f) = \sum_{k=1}^n \frac{E_\zeta(f 1_{X_k})}{E_\zeta(1_{X_k})} 1_{X_k}.$$

Hence

$$\log E_{\xi \vee \zeta}(f) = \sum_{k=1}^n (\log E_\zeta(f 1_{X_k}) - \log E_\zeta(1_{X_k})) 1_{X_k},$$

whence for $\vartheta = \{Y_l\}_{l=1}^m$ we get

$$\sum_l 1_{Y_l} \log E_{\xi \vee \zeta}(1_{Y_l}) = \sum_{k,l} 1_{X_k \cap Y_l} \log E_\zeta(1_{X_k \cap Y_l}) - \sum_k 1_{X_k} \log E_\zeta(1_{X_k}).$$

Integrating over X we get (ii).

To show (iii) first note that if $\xi \prec \vartheta$ then $\xi \vee \vartheta = \vartheta$, so it follows from (ii) that $H(\xi|\zeta) \leq H(\vartheta|\zeta)$.

Assume now that $\zeta \prec \vartheta$. Then $E_\zeta = E_\zeta \circ E_\vartheta$. Since the function η is concave, by Jensen's inequality for any measurable $Y \subset X$ we get

$$E_\zeta(\eta(E_\vartheta(1_Y))) \leq \eta((E_\zeta \circ E_\vartheta)(1_Y)) = \eta(E_\zeta(1_Y)),$$

whence

$$\int_X \eta(E_\vartheta(1_Y)) d\mu \leq \int_X \eta(E_\zeta(1_Y)) d\mu.$$

This implies that $H(\xi|\vartheta) \leq H(\xi|\zeta)$. $\square$

From parts (ii) and (iii) in the above proposition we see that if ξ, ϑ, ζ and ε are measurable partitions then

$$H(\xi \vee \vartheta | \zeta \vee \varepsilon) \leq H(\xi|\zeta) + H(\vartheta|\varepsilon). \tag{1.1}$$

In particular, entropy is subadditive, i.e., $H(\xi \vee \vartheta) \leq H(\xi) + H(\vartheta)$.

Let now T be a measure-preserving invertible transformation of a Lebesgue space (X, μ). In such a case we say that (X, μ, T) is a **dynamical system**. It will sometimes be convenient to assume that T is a homeomorphism of a compact metric space. More precisely, if (X, μ, T) is a dynamical system then by a **topological model** for it we mean a homeomorphism S of a compact metric space Y together with an S-invariant probability measure ν such that (X, μ, T) and (Y, ν, S) are isomorphic. Topological models always exist: take the spectrum of any weakly operator dense invariant separable C*-subalgebra of $L^\infty(X,\mu)$ for Y.

Let (X,μ,T) be a dynamical system. For a finite measurable partition $\xi = \{X_1,\ldots,X_m\}$ set

$$h(\xi;T) = \lim_{n\to\infty} \frac{1}{n} H(\xi \vee T\xi \vee \ldots \vee T^{n-1}\xi),$$

where $T\xi = \{TX_1,\ldots,TX_m\}$. Note that the limit exists by virtue of subadditivity of entropy and the following standard result.

Lemma 1.1.2. *Let $\{a_n\}_{n=1}^\infty$ be a sequence of real numbers such that $a_{n+m} \le a_n + a_m$ for any $n,m \in \mathbb{N}$. Then the limit of the sequence $\{a_n/n\}_n$ exists and coincides with the infimum.*

Proof. Fix $m \in \mathbb{N}$. Any $n \in \mathbb{N}$ can be written as $n = km + l$ for some k and l, $0 \le l < m$. By assumption

$$a_n \le k a_m + a_l.$$

It follows that

$$\limsup_{n\to\infty} \frac{a_n}{n} \le \frac{a_m}{m}.$$

Since this is true for every m, we get the result. $\square$

In fact with a bit more work one can show that $\{H(\vee_{k=0}^{n-1} T^k\xi)/n\}_{n=1}^\infty$ is monotonically decreasing. Indeed, we have

$$H(\vee_{k=0}^{n-1} T^k\xi) = H(\xi) + H(\vee_{k=1}^{n-1} T^k\xi | \xi)$$
$$= H(\xi) + H(\vee_{k=0}^{n-2} T^k\xi | T^{-1}\xi)$$
$$= \ldots = \sum_{k=0}^{n-1} H(\xi | \vee_{l=-k}^{-1} T^l\xi). \qquad (1.2)$$

Since the sequence $\{H(\xi | \vee_{l=-k}^{-1} T^l\xi)\}_k$ is monotonically decreasing by Proposition 1.1.1(iii), the sequence $\{H(\vee_{k=0}^{n-1} T^k\xi)/n\}_n$ is also monotonically decreasing. Furthermore, by the martingale convergence theorem for any $f \in L^1(X,\mu)$ we have $E_{\vee_{k=-n}^{-1} T^k\xi}(f) \to E_{\vee_{k=-\infty}^{-1} T^k\xi}(f)$ a.e. So denoting the partition $\vee_{k=-\infty}^{-1} T^k\xi$ by ξ^- we get

$$h(\xi;T) = \lim_{n\to\infty} H(\xi | \vee_{k=-n}^{-1} T^k\xi) = H(\xi|\xi^-). \qquad (1.3)$$

The **entropy** of T is now defined as

$$h(T) = \sup_\xi h(\xi;T),$$

where the supremum is taken over all finite measurable partitions of (X,μ). We shall write $h_\mu(\xi;T)$ and $h_\mu(T)$ instead of $h(\xi;T)$ and $h(T)$ when we want to stress that we deal with measure μ.

We need a more practical way of computing entropy than taking supremum over all finite partitions.

Lemma 1.1.3. *Let ξ and ζ be finite measurable partitions. Denote by $\zeta^{\pm}$ the partition $\vee_{k \in \mathbb{Z}} T^k \xi$. Then*

$$h(\xi; T) \leq h(\zeta; T) + H(\xi | \zeta^{\pm}).$$

Proof. We have

$$H(\vee_{k=0}^{n-1} T^k \xi) \leq H(\vee_{k=0}^{n-1} T^k(\xi \vee \zeta)) = H(\vee_{k=0}^{n-1} T^k \zeta) + H(\vee_{k=0}^{n-1} T^k \xi | \vee_{k=0}^{n-1} T^k \zeta)$$

and by (1.1)

$$H(\vee_{k=0}^{n-1} T^k \xi | \vee_{k=0}^{n-1} T^k \zeta) \leq \sum_{k=0}^{n-1} H(T^k \xi | T^k \zeta) = n H(\xi | \zeta).$$

Therefore

$$h(\xi; T) \leq h(\zeta; T) + H(\xi | \zeta).$$

Since also $h(\vee_{k=-n}^{n} T^k \zeta; T) = h(\zeta; T)$ for any n, we then get

$$h(\xi; T) \leq h(\zeta; T) + H(\xi | \vee_{k=-n}^{n} T^k \zeta).$$

Letting $n \to \infty$ we obtain the result. □

A measurable partition ξ is called **generating** if $\xi^{\pm}$ is the partition of X into points. From the above lemma we then get the following result.

Theorem 1.1.4. (Kolmogorov-Sinai) *Let ξ be a finite generating partition for a dynamical system (X, μ, T). Then $h(T) = h(\xi; T)$.* □

The following example was motivating for the whole theory.

Example 1.1.5. Let ν be a probability measure on $\{1, \ldots, n\}$, $\nu(\{k\}) = \lambda_k$. Let $X = \{1, \ldots, n\}^{\mathbb{Z}}$, μ the product measure on X and T the shift to the right. The system (X, μ, T) is called the **Bernoulli shift** with weights $\lambda_1, \ldots, \lambda_n$.

Consider the partition $\xi = \{X_1, \ldots, X_n\}$, where X_k is the cylindrical set consisting of $\underline{x} = \{x_n\}_{n \in \mathbb{Z}} \in X$ such that $x_0 = k$. Then ξ is generating and we get $h(T) = h(\xi; T) = \sum_k \eta(\lambda_k)$. ♦

The following two theorems are fundamental results on Bernoulli shifts.

Theorem 1.1.6. (Ornstein) *Bernoulli shifts with the same entropy are isomorphic.* □

Recall that a system (X, μ, T) is called **ergodic** if any T-invariant measurable subset of X has measure either zero or one. If ξ is a T-invariant measurable partition then the system $(X/\xi, \mu, T)$ is called a **factor system** of (X, μ, T).

Theorem 1.1.7. (Sinai) *Let (X, μ, T) be an ergodic dynamical system with entropy $h(T) > 0$. Then for any finite h, $0 < h \leq h(T)$, there exists a factor system which is isomorphic to the Bernoulli shift with entropy h.* □

Let us now list the basic properties of entropy.

Theorem 1.1.8. *For any dynamical system (X, μ, T) we have:*
(i) $h(T^n) = |n| h(T)$ for $n \in \mathbb{Z}$;
(ii) *if (Y, ν, S) is a factor system then $h(T) \geq h(S)$;*
(iii) *if $Y \subset X$ is T-invariant then $h(T) = \mu(Y) h(T|_Y) + \mu(Y^c) h(T|_{Y^c})$, where we consider the measures $\mu(Y)^{-1} \mu|_Y$ and $\mu(Y^c)^{-1} \mu|_{Y^c}$ on Y and $Y^c = X \setminus Y$, respectively;*
(iv) *if (Y, ν, S) is another dynamical system then $h(T \times S) = h(T) + h(S)$.* □

We shall next show properties of $H(\xi)$ which will be useful later.

Proposition 1.1.9. *For finite measurable partitions we have:*
(i) *if $\xi = \{X_1, \ldots, X_n\}$ then $H(\xi) \leq \log n$, and equality holds if and only if $\mu(X_k) = 1/n$ for all k;*
(ii) $H(\xi \vee \zeta \vee \vartheta) + H(\zeta) \leq H(\xi \vee \zeta) + H(\zeta \vee \vartheta)$;
(iii) $H(\xi \vee \zeta) \leq H(\xi) + H(\zeta)$, *and equality holds if and only if ξ and ζ are independent, that is, $\mu(fg) = \mu(f)\mu(g)$ for $f \in L^\infty(X/\xi)$ and $g \in L^\infty(X/\zeta)$;*
(iv) $\sum_{k=1}^n H(\xi_k) - H(\vee_{k=1}^n \xi_k) \leq \sum_{k=1}^n H(\zeta_k) - H(\vee_{k=1}^n \zeta_k)$ *if $\xi_k \prec \zeta_k$ for $k = 1, \ldots, n$.*

Proof. Since η is strictly concave,
$$\sum_k \eta(\mu(X_k)) \leq n \eta\left(\sum_k \frac{\mu(X_k)}{n} \right) = \log n,$$
moreover, equality holds if and only if all $\mu(X_k)$'s are equal, that is, $\mu(X_k) = 1/n$. Thus (i) is proved.

We have by Proposition 1.1.1(ii)
$$H(\xi \vee \zeta \vee \vartheta) - H(\xi \vee \zeta) = H(\vartheta | \xi \vee \zeta)$$
and
$$H(\zeta \vee \vartheta) - H(\zeta) = H(\vartheta | \zeta).$$
Since $H(\vartheta|\zeta) \geq H(\vartheta | \xi \vee \zeta)$, this proves (ii).

Since $H(\xi \vee \zeta) = H(\xi) + H(\zeta|\xi)$, to prove (iii) we have to show that $H(\zeta|\xi) = H(\zeta)$ if and only if ξ and ζ are independent. Let $\zeta = \{X_1, \ldots, X_n\}$. Since η is strictly concave,
$$H(\zeta|\xi) = \sum_k \int_X \eta(E_\xi(\mathbb{1}_{X_k})) d\mu \leq \sum_k \eta\left(\int_X E_\xi(\mathbb{1}_{X_k}) d\mu \right) = \sum_k \eta(\mu(X_k)),$$

and equality holds if and only if $E_\xi(\mathbb{1}_{X_k})$ is constant a.e. for all k, that is, $E_\xi(\mathbb{1}_{X_k}) = \mu(X_k)$. But the latter condition means exactly that $\mu(\mathbb{1}_{X_k}g) = \mu(X_k)\mu(g)$ for any $g \in L^\infty(X/\xi)$, that is, ξ and ζ are independent.

We have by (1.1)

$$H(\vee_k \zeta_k | \vee_k \xi_k) \leq \sum_k H(\zeta_k | \xi_k).$$

On the other hand, assuming $\xi_k \prec \zeta_k$, we have

$$H(\vee_k \zeta_k | \vee_k \xi_k) = H(\vee_k \zeta_k) - H(\vee_k \xi_k)$$

and $H(\zeta_k | \xi_k) = H(\zeta_k) - H(\xi_k)$. This gives (iv). □

It is useful to note that subadditivity of entropy is equivalent to the following number inequality.

Lemma 1.1.10. *Let $\{\lambda_{i_1\ldots i_n}\}_{i_1 \in I_1, \ldots, i_n \in I_n}$ be a finite set of nonnegative numbers such that $\sum_{i_1,\ldots,i_n} \lambda_{i_1\ldots i_n} = 1$. Set $\lambda_{i_k}^{(k)} = \sum_{i_1,\ldots,i_{k-1},i_{k+1},\ldots,i_n} \lambda_{i_1\ldots i_n}$. Then*

$$\sum_{i_1,\ldots,i_n} \eta(\lambda_{i_1\ldots i_n}) \leq \sum_{k=1}^n \sum_{i_k \in I_k} \eta(\lambda_{i_k}^{(k)})$$

Proof. Apply the subadditivity inequality $H(\vee_k \xi_k) \leq \sum_k H(\xi_k)$ to the partitions ξ_k of $I_1 \times \ldots \times I_n$ consisting of the sets $I_1 \times \ldots \times I_{k-1} \times \{i_k\} \times I_{k+1} \times I_n$, $k = 1, \ldots, n$, and the measure μ defined by $\mu(\{(i_1, \ldots, i_n)\}) = \lambda_{i_1\ldots i_n}$. □

1.2 Topological Entropy

Let T be a homeomorphism of a compact separable space X. We call (X, T) a **topological dynamical system**. Let $\mathcal{U} = \{U_i\}_i$ be an open cover of X. Denote by $N(\mathcal{U})$ the minimal number of elements of $\mathcal{U}$ needed to cover X. If $\mathcal{U} = \{U_i\}_i$ and $\mathcal{V} = \{V_j\}_j$ are two open covers, we denote by $\mathcal{U} \vee \mathcal{V}$ the open cover formed by the sets $U_i \cap V_j$. We obviously have $N(\mathcal{U} \vee \mathcal{V}) \leq N(\mathcal{U})N(\mathcal{V})$. Define

$$h(\mathcal{U}; T) = \lim_{n \to \infty} \frac{1}{n} \log N(\mathcal{U} \vee T\mathcal{U} \vee \ldots \vee T^{n-1}\mathcal{U}).$$

The quantity

$$h_{top}(T) = \sup_{\mathcal{U}} h(\mathcal{U}; T),$$

where the supremum is taken over all open covers of X, is called the **topological entropy** of T.

An equivalent definition is given in terms of separating and spanning sets. Let d be a compatible metric on X. Take $\varepsilon > 0$. A subset F of X is called (n, ε)-**spanning** with respect to T if for every $x \in X$ there exists $y \in F$ such that

$$\max_{0 \le k \le n-1} d(T^k x, T^k y) < \varepsilon,$$

and it is called (n,ε)-**separated** if for any $x, y \in F$, $x \ne y$, we have

$$\max_{0 \le k \le n-1} d(T^k x, T^k y) \ge \varepsilon.$$

Let $sp_n(\varepsilon)$ be the minimal cardinality of an (n,ε)-spanning set, and $sr_n(\varepsilon)$ the maximal cardinality of an (n,ε)-separated set. If $\mathcal{U}$ is an open cover with Lebesgue number ε (that is, any open ball of radius ε is contained in one of the elements of $\mathcal{U}$), and $\mathcal{V}$ an open cover with diameter less than ε (that is, the diameter of every element in $\mathcal{V}$ is less than ε), then one can check that

$$N(\vee_{k=0}^{n-1} T^k \mathcal{U}) \le sp_n(\varepsilon) \le sr_n(\varepsilon) \le N(\vee_{k=0}^{n-1} T^k \mathcal{V}).$$

It follows that

$$h_{top}(T) = \lim_{\varepsilon \downarrow 0} \liminf_{n \to \infty} \frac{1}{n} \log sp_n(\varepsilon) = \lim_{\varepsilon \downarrow 0} \limsup_{n \to \infty} \frac{1}{n} \log sp_n(\varepsilon)$$
$$= \lim_{\varepsilon \downarrow 0} \liminf_{n \to \infty} \frac{1}{n} \log sr_n(\varepsilon) = \lim_{\varepsilon \downarrow 0} \limsup_{n \to \infty} \frac{1}{n} \log sr_n(\varepsilon).$$

This gives an intuitive interpretation of topological entropy as a measure of the asymptotic distortion of the iterates of T along orbits.

The main properties of topological entropy are as follows.

Theorem 1.2.1. *For every topological dynamical system (X,T) we have:*

(i) $h_{top}(T^n) = |n| h_{top}(T)$ *for* $n \in \mathbb{Z}$;

(ii) *if $(X,T) \to (Y,S)$ is a factor map then $h_{top}(T) \ge h_{top}(S)$;*

(iii) *if $Y \subset X$ is a closed T-invariant subset then $h_{top}(T) \ge h_{top}(T|_Y)$, and if in addition Y is open then $h_{top}(T) = \max\{h_{top}(T|_Y), h_{top}(T|_{X \setminus Y})\}$;*

(iv) *if (Y,S) is another topological dynamical system then $h_{top}(T \times S) = h_{top}(T) + h_{top}(S)$;*

(v) *if $\mathcal{U}$ is an open cover such that the diameter of $\vee_{k=-n}^{n} T^k \mathcal{U}$ with respect to some compatible metric tends to zero as $n \to \infty$ then $h_{top}(T) = h(\mathcal{U}; T)$.* □

The relation between measure and topological entropy is given by the **variational principle**.

Theorem 1.2.2. *For every topological dynamical system (X,T) we have*

$$h_{top}(T) = \sup_{\mu} h_\mu(T),$$

where the supremum is taken over all T-invariant probability measures. □

A topological dynamical system (X,T) is called **uniquely ergodic** if there exists a unique T-invariant probability measure μ. Then by the variational principle $h_\mu(T) = h_{top}(T)$.

In Sect. 1.1 we remarked that any dynamical system has a topological model. The following theorem gives a much more precise result for ergodic systems.

Theorem 1.2.3. (Jewett-Krieger) *Every ergodic dynamical system has a uniquely ergodic topological model.* □

Finishing our discussion of connections between measure entropy and topological entropy we shall give an equivalent definition of measure entropy in terms of spanning sets.

Let (X,T) be a topological dynamical system, and μ a T-invariant probability measure, which for simplicity we assume to be ergodic. A subset F of X is called (n, ε, δ)-spanning for (T, μ) if the set of $x \in X$ such that there exists $y \in F$ with

$$\max_{0 \leq k \leq n-1} d(T^k x, T^k y) < \varepsilon$$

has measure greater than δ. Denote by $sp_n(\varepsilon; \delta)$ the minimal cardinality of an (n, ε, δ)-spanning set. Then for any $\delta \in (0,1)$

$$h_\mu(T) = \lim_{\varepsilon \downarrow 0} \liminf_{n \to \infty} \frac{1}{n} \log sp_n(\varepsilon; \delta).$$

The proof can be obtained from the following theorem.

Theorem 1.2.4. (Shannon-McMillan-Breiman) *Let (X, μ, T) be an ergodic dynamical system, ξ a finite measurable partition of X. For $x \in X$ and $n \in \mathbb{N}$ denote by $I_n(x)$ the element of the partition $\vee_{k=0}^{n-1} T^k \xi$ containing x. Then the convergence*

$$-\frac{\log \mu(I_n(x))}{n} \to h(\xi; T)$$

holds in mean and almost everywhere. □

1.3 Entropy via Partitions of Unity

We shall in the present section rewrite the definition of $H(\vee_{k=1}^n \xi_k)$ in a form which is more suitable for generalization to the noncommutative case.

Since the algebra generated by noncommuting finite dimensional algebras can be infinite dimensional, we shall consider $H(\vee_{k=1}^n \xi_k)$ as a function of n partitions.

Consider the case $n = 1$, so we are dealing with one partition $\xi = \{X_k\}_k$. Then $1 = \sum_k \mathbb{1}_{X_k}$ is a partition of unity in $L^\infty(X/\xi)$, and

$$H(\xi) = \sum_k \eta(\mu(\mathbb{1}_{X_k})).$$

This expression makes perfect sense for finite dimensional noncommutative algebras. However, in the noncommutative setting it is difficult to form new partitions of unity consisting of projections from given noncommuting ones. Hence a natural question is how we can express $H(\xi)$ using arbitrary finite decompositions $1 = \sum_i f_i$, $f_i \in L^\infty(X/\xi)$, $f_i \geq 0$. The quantity $\sum_i \eta(\mu(f_i))$ can be arbitrarily large. So we have to compensate for the fact that the f_i's are not necessarily projections. The function $\eta(f_i)$ is then a natural ingredient of a correction term as it vanishes exactly when f_i is a projection.

Lemma 1.3.1. *We have*

$$H(\xi) = \sup \sum_i (\eta(\mu(f_i)) - \mu(\eta(f_i))),$$

where the supremum is taken over all finite families $\{f_i\}_i$ of positive functions in $L^\infty(X/\xi)$ with sum 1.

Proof. We just have to check that the supremum is not larger than $H(\xi)$. Let $\xi = \{X_k\}_{k=1}^n$ and $f_i \in L^\infty(X/\xi)$, $0 \leq f_i \leq 1$, $\sum_i f_i = 1$. Then $f_i = \sum_k \lambda_{ik} \mathbb{1}_{X_k}$ with $\sum_i \lambda_{ik} = 1$. Using that $\eta(x+y) \leq \eta(x) + \eta(y)$ for any positive numbers x and y we compute:

$$\sum_i (\eta(\mu(f_i)) - \mu(\eta(f_i))) = \sum_i \eta(\mu(f_i)) - \sum_{i,k} \mu(\eta(\lambda_{ik} \mathbb{1}_{X_k}))$$

$$\leq \sum_{i,k} (\eta(\mu(\lambda_{ik} \mathbb{1}_{X_k})) - \mu(\eta(\lambda_{ik} \mathbb{1}_{X_k})))$$

$$= \sum_{i,k} (\eta(\lambda_{ik} \mu(X_k)) - \eta(\lambda_{ik}) \mu(X_k))$$

$$= \sum_k \eta(\mu(X_k)),$$

which proves the lemma. □

Given a partition of unity $1 = \sum_i f_i$ in $L^\infty(X)$ we can form a partition of unity in $L^\infty(X/\xi)$ by applying the conditional expectation E_ξ. Thus $H(\xi)$ is the supremum of

$$\sum_i (\eta(\mu(f_i)) - \mu(\eta(E_\xi(f_i))))$$

over all partitions of unity $1 = \sum_i f_i$ in $L^\infty(X)$.

Note now that

$$\eta(\mu(f)) - \mu(\eta(f)) = \int_X f(\log f - \log \mu(f)) d\mu.$$

1.3 Entropy via Partitions of Unity

For arbitrary positive functions f and g in $L^1(X,\mu)$ the quantity

$$S(f,g) = \int_X f(\log f - \log g) d\mu$$

is called the **relative entropy** of the measures $f\,d\mu$ and $g\,d\mu$.

Next we turn to the case of n partitions $\xi_1,\ldots,\xi_n$. Let $\xi_k = \{X_{j_k}^{(k)}\}_{j_k \in J_k}$. Then

$$H(\vee_k \xi_k) = \sum_{j_1,\ldots,j_n} \eta(\mu(\mathbb{1}_{X_{j_1}^{(1)}} \cdots \mathbb{1}_{X_{j_n}^{(n)}})).$$

In other words, we consider the partitions of unity consisting of projections $\mathbb{1}_{X_{j_k}^{(k)}}$ and form a new partition consisting of products of these projections. This is not easy to generalize to the noncommutative case. So instead of trying to form a new partition of unity we consider all possible finite partitions $1 = \sum_{i_1,\ldots,i_n} f_{i_1\ldots i_n}$ and think of them as coming from n partitions $1 = \sum_{i_k} f_{i_k}^{(k)}$, $k = 1,\ldots,n$, where $f_{i_k}^{(k)}$ is the sum of the $f_{i_1\ldots i_n}$'s over $i_1,\ldots,i_{k-1},i_{k+1},\ldots,i_n$. We then take $\sum_{i_1,\ldots,i_n} \eta(\mu(f_{i_1\ldots i_n}))$ as the first approximation of the entropy and define the correction term as the sum of the correction terms for the above n partitions of unity. We thus arrive at the following expression.

Proposition 1.3.2. *Let $\xi_1,\ldots,\xi_n$ be finite measurable partitions of X. Then*

$$H(\vee_{k=1}^n \xi_k) = \sup\left\{ \sum_{i_1,\ldots,i_n} \eta(\mu(f_{i_1\ldots i_n})) - \sum_{k=1}^n \sum_{i_k} \mu(\eta(E_{\xi_k}(f_{i_k}^{(k)}))) \right\},$$

where the supremum is taken over all finite families $\{f_{i_1\ldots i_n}\}_{i_1,\ldots,i_n}$ of positive functions in $L^\infty(X)$ with sum 1.

Proof. Let $\xi = \vee_{k=1}^n \xi_k$. Let $1 = \sum_{i_1,\ldots,i_n} f_{i_1\ldots i_n}$ be a partition of unity. Since $\mu \circ E_\xi = \mu$, by Lemma 1.3.1

$$H(\xi) \geq \sum_{i_1,\ldots,i_n} \eta(\mu(f_{i_1\ldots i_n})) - \sum_{i_1,\ldots,i_n} \mu(\eta(E_\xi(f_{i_1\ldots i_n}))).$$

Apply Lemma 1.1.10 to the numbers $E_\xi(f_{i_1\ldots i_n})(x)$, $x \in X$, and then integrate over X. We then see that the above expression is not smaller than

$$\sum_{i_1,\ldots,i_n} \eta(\mu(f_{i_1\ldots i_n})) - \sum_{k=1}^n \sum_{i_k} \mu(\eta(E_\xi(f_{i_k}^{(k)}))).$$

As was already used in the proof of Proposition 1.1.1(iii), since $\xi_k \prec \xi$ and therefore $E_{\xi_k} = E_{\xi_k} \circ E_\xi$, it follows from Jensen's inequality that $\mu(\eta(E_\xi(f))) \leq \mu(\eta(E_{\xi_k}(f)))$ for any $f \geq 0$. Hence the above is expression is not smaller than

$$\sum_{i_1,\ldots,i_n} \eta(\mu(f_{i_1\ldots i_n})) - \sum_{k=1}^n \sum_{i_k} \mu(\eta(E_{\xi_k}(f_{i_k}^{(k)}))).$$

Thus the left side of the formula in the proposition majorizes the right side. For the opposite inequality assume $\xi_k = \{X_{i_k}^{(k)}\}_{i_k}$ and let

$$f_{i_1\ldots i_n} = 1\!\!1_{X_{i_1}^{(1)} \cap \ldots \cap X_{i_n}^{(n)}}.$$

Then $f_{i_k}^{(k)} = 1\!\!1_{X_{i_k}^{(k)}}$, so that $\eta(E_{\xi_k}(f_{i_k}^{(k)})) = 0$. Thus the right side with these $f_{i_1\ldots i_n}$'s equals $H(\xi)$. □

1.4 Notes

As our basic reference on entropy we refer the reader to the book by Glasner [72], which contains all the theorems we stated without proofs. For more on Lebesgue spaces, introduced by Rohlin in [181], one can consult [1] or [184].

Entropy was introduced by Kolmogorov [111], who was inspired by Shannon's information theory [195] and motivated by the isomorphism problem for Bernoulli shifts. The definition was improved by Sinai [199]. The isomorphism theorem for Bernoulli shifts with the same entropy was proved by Ornstein [150]. He developed a powerful theory of finitely determined processes giving several dynamical characterizations of Bernoulli shifts. A more constructive proof was given by Keane and Smorodinsky [104], see also [53] for an exposition of their proof. Theorem 1.1.7, due to Sinai, was proved earlier [200], and then it stated that *any* Bernoulli shift with entropy $h \leq h(T)$ can be obtained as a factor system.

The notion of topological entropy in terms of open coverings was introduced by Adler, Konheim and McAndrew [2]. The definition in terms of spanning and separated sets was given by Bowen [29]. The variational principle, Theorem 1.2.2, was proved by Goodwyn [82] and Dinaburg [55]. A stronger version of the variational principle has been obtained by Blanchard, Glasner and Host [26], [72, Chapter 17]. Theorem 1.2.3 is due to Jewett [96] and Krieger [113]. The definition of measure entropy in terms of spanning sets was proposed by Katok [99], see also [166] for a proof of equivalence of that definition to the usual one. Theorem 1.2.4 was obtained by Shannon [195], McMillan [121] and Breiman [31] who established convergence in measure, in mean and finally almost everywhere, respectively.

The formula for entropy in the last section, obtained by Connes and Størmer [51], will be the basis for the definition of entropy in the noncommutative setting in Chap. 3.

2
Relative Entropy

Relative entropy of measures has appeared naturally in the expression for entropy of partitions given in Proposition 1.3.2. In the present chapter we shall define relative entropy for states on C*-algebras and show its properties needed in the sequel. We shall mainly consider finite dimensional C*-algebras.

2.1 Relative Entropy for Matrix Algebras

Let A be a finite dimensional C*-algebra. Denote by Tr_A the canonical trace on A, so $\operatorname{Tr}_A(p) = 1$ for every minimal projection $p \in A$. The algebra A is a direct sum of full matrix algebras, and the restriction of Tr_A to each summand is the usual matricial trace. We shall write Tr instead of Tr_A if no confusion can arise.

Each positive linear functional φ on A is of the form $\varphi = \operatorname{Tr}(\cdot Q_\varphi)$ for a unique positive element $Q_\varphi \in A$ called the **density operator**.

Definition 2.1.1. *If φ and ψ are positive linear functionals on A, then the relative entropy of φ and ψ is*

$$S(\varphi, \psi) = \begin{cases} \operatorname{Tr}(Q_\varphi(\log Q_\varphi - \log Q_\psi)), & \text{if } \varphi \leq \lambda \psi \text{ for some } \lambda > 0, \\ +\infty, & \text{otherwise.} \end{cases}$$

Since A is finite dimensional, $\varphi \leq \lambda \psi$ for some $\lambda > 0$ if and only if $\operatorname{supp} \varphi \leq \operatorname{supp} \psi$, that is, $\varphi(a) = 0$ for a positive $a \in A$ as soon as $\psi(a) = 0$.

Two identities are immediate from the definition:

$$S(\lambda_1 \varphi, \lambda_2 \psi) = \lambda_1 S(\varphi, \psi) + \lambda_1 \varphi(1) \log \frac{\lambda_1}{\lambda_2} \quad \text{if } \lambda_1, \lambda_2 > 0, \tag{2.1}$$

$$S(\varphi, \psi) = \frac{d}{dt} \operatorname{Tr}(Q_\varphi^t Q_\psi^{1-t})|_{t=1} \quad \text{if } \varphi \leq \lambda \psi. \tag{2.2}$$

We next list the main properties of relative entropy.

Theorem 2.1.2. *We have:*

(i) $S(\varphi,\psi) \geq 0$ *for states φ and ψ, and equality holds if and only if $\varphi = \psi$;*

(ii) $S(\varphi,\psi)$ *is decreasing in ψ, i.e., $S(\varphi,\psi_1) \geq S(\varphi,\psi_2)$ if $\psi_1 \leq \psi_2$;*

(iii) $S(\varphi,\psi)$ *is a lower semicontinuous function of (φ,ψ); moreover, it is continuous on the sets $\{(\varphi,\psi)\,|\,\varphi \leq \lambda\psi\}$, $\lambda > 0$;*

(iv) $S(\varphi,\psi)$ *is a convex function of (φ,ψ);*

(v) *if $\alpha\colon B \to A$ is a unital Schwarz map, then*

$$S(\varphi \circ \alpha, \psi \circ \alpha) \leq S(\varphi,\psi);$$

(vi) *if φ and ψ are states on A, B is a C^*-subalgebra of A, and $E\colon A \to B$ is a ψ-preserving conditional expectation, then*

$$S(\varphi,\psi) = S(\varphi|_B,\psi|_B) + S(\varphi,\varphi \circ E);$$

(vii) *if $\varphi = \sum_{i=1}^n \varphi_i$, then $\sum_i S(\varphi_i,\psi) = S(\varphi,\psi) + \sum_i S(\varphi_i,\varphi)$.*

Recall that if A and B are C^*-algebras, a linear map $\alpha\colon B \to A$ is said to be a **Schwarz map** if $\alpha(b^*b) \geq \alpha(b)^*\alpha(b)$ for each $b \in B$. Then clearly α is positive. The map α is **completely positive** if

$$\alpha \otimes \mathrm{id}\colon B \otimes \mathrm{Mat}_n(\mathbb{C}) \to A \otimes \mathrm{Mat}_n(\mathbb{C})$$

is positive for all natural numbers n, where $\mathrm{Mat}_n(\mathbb{C})$ denotes the complex $n \times n$ matrices. See App. A, where we have collected some basic facts on such maps. In particular, unital completely positive maps are Schwarz maps by A.2.

Recall also that if $B \subset A$, then a **conditional expectation** of A onto B is a unital positive map $E\colon A \to B$ such that $E(bac) = bE(a)c$ for all $b,c \in B$ and $a \in A$. Then E is completely positive, see A.13. If A is finite dimensional, there always exists a unique Tr_A-preserving conditional expectation E_B of A onto B. However, for an arbitrary state ψ on A a ψ-preserving conditional expectation may not exist.

Parts (iv) and (v) of Theorem 2.1.2 require nontrivial estimates. In both cases the proof will be based on (2.2) and the following operator inequality.

Proposition 2.1.3. *Let x_1 and x_2 be bounded positive operators on a Hilbert space K, and y_i, z_i, $i = 1,2$, bounded positive operators on a Hilbert space H such that $x_1x_2 = x_2x_1$, $y_1y_2 = y_2y_1$, $z_1z_2 = z_2z_1$. Suppose $v\colon K \to H$ is a bounded operator such that*

$$x_i \geq v^*(y_i + z_i)v, \quad i = 1,2.$$

Then

$$x_1^t x_2^{1-t} \geq v^*(y_1^t y_2^{1-t} + z_1^t z_2^{1-t})v$$

for $t \in [0,1]$.

2.1 Relative Entropy for Matrix Algebras

Proof. If we can show the proposition for x_i replaced by $x_i + \varepsilon 1$ for $\varepsilon > 0$ then by continuity the conclusion of the proposition follows when $\varepsilon \to 0$. We may thus assume x_1 and x_2 are invertible.

We first consider the case $t = 1/2$. Let
$$x = x_1^{1/2} x_2^{1/2}, \quad y = y_1^{1/2} y_2^{1/2}, \quad z = z_1^{1/2} z_2^{1/2}.$$

Put $c = v^*(y+z)v$, and let $\zeta, \xi \in K$. Then by using the Cauchy-Schwarz inequality twice together with the assumption of the proposition we have

$$\begin{aligned}
|(c\xi, \zeta)| &= |((y+z)v\xi, v\zeta)| \\
&\leq |(y_1^{1/2} v\xi, y_2^{1/2} v\zeta)| + |(z_1^{1/2} v\xi, z_2^{1/2} v\zeta)| \\
&\leq (y_1 v\xi, v\xi)^{1/2} (y_2 v\zeta, v\zeta)^{1/2} + (z_1 v\xi, v\xi)^{1/2} (z_2 v\zeta, v\zeta)^{1/2} \\
&\leq ((y_1 + z_1) v\xi, v\xi)^{1/2} ((y_2 + z_2) v\zeta, v\zeta)^{1/2} \\
&\leq (x_1 \xi, \xi)^{1/2} (x_2 \zeta, \zeta)^{1/2}.
\end{aligned}$$

Replacing ξ by $x_1^{-1/2} \xi$ and ζ by $x_2^{-1/2} \zeta$ in the above inequality, we get

$$|(x_2^{-1/2} c x_1^{-1/2} \xi, \zeta)| \leq \|\xi\| \|\zeta\|.$$

Hence $\|x_2^{-1/2} c x_1^{-1/2}\| \leq 1$. In particular, the spectrum of $x_2^{-1/2} c x_1^{-1/2}$ is contained in the unit disc. Recall that the nonzero parts of the spectra of the operators ab and ba coincide (indeed, if $ab - 1$ is invertible, then $ba - 1$ is invertible and $(ba - 1)^{-1} = b(ab - 1)^{-1} a - 1$). Since x_1 and x_2 commute, we thus have

$$\begin{aligned}
\mathrm{Sp}(x^{-1/2} c x^{-1/2}) \setminus \{0\} &= \mathrm{Sp}(x_1^{-1/4} x_2^{-1/4} c x_1^{-1/4} x_2^{-1/4}) \setminus \{0\} \\
&= \mathrm{Sp}(x_2^{-1/2} c x_1^{-1/2}) \setminus \{0\}.
\end{aligned}$$

Hence $x^{-1/2} c x^{-1/2} \leq 1$. Thus $c \leq x$, so that

$$v^*(y_1^{1/2} y_2^{1/2} + z_1^{1/2} z_2^{1/2}) v \leq x_1^{1/2} x_2^{1/2}.$$

Assuming that the proposition is true for exponents s and t, apply the above inequality to $x_1^t x_2^{1-t}$, $y_1^t y_2^{1-t}$, $z_1^t z_2^{1-t}$, $x_1^s x_2^{1-s}$, $y_1^s y_2^{1-s}$, $z_1^s z_2^{1-s}$ instead of $x_1, y_1, z_1, x_2, y_2, z_2$. Then we see that the proposition holds for the exponent $(s+t)/2$. Hence it holds for all dyadic numbers and by continuity, for every $t \in [0,1]$. □

Proof of Theorem 2.1.2. Note that for any numbers $x, y \geq 0$ we have

$$x(\log x - \log y) \geq x - y, \tag{2.3}$$

and equality holds only if $x = y$, which is just another way of writing the inequality $\log t \leq t - 1$ for $t = y/x$. Now let φ and ψ be states, and $Q_\varphi = \sum_i \lambda_i p_i$ and $Q_\psi = \sum_j \mu_j q_j$ the spectral decompositions. It follows that

$$\lambda_i(\log \lambda_i - \log \mu_j)\mathrm{Tr}(p_i q_j) \geq (\lambda_i - \mu_j)\mathrm{Tr}(p_i q_j).$$

Summing over i and j and assuming that $\varphi(1) = \psi(1) = 1$ we get $S(\varphi, \psi) \geq 0$. Moreover, if equality holds then

$$(\lambda_i - \mu_j)p_i q_j = 0,$$

so that summing over i and j again, we get $Q_\varphi = Q_\psi$. Thus (i) is proved.

Part (ii) is a consequence of operator monotonicity of log, see B.4.

To prove (iii) assume $\varphi_n \to \varphi$, $\psi_n \to \psi$. If ψ is faithful, then ψ_n is also faithful for sufficiently large n, and the equality

$$S(\varphi, \psi) = -\mathrm{Tr}(\eta(Q_\varphi)) - \varphi(\log Q_\psi),$$

where $\eta(t) = -t \log t$, shows that $S(\varphi_n, \psi_n) \to S(\varphi, \psi)$. In general, by (ii) we have for $\varepsilon > 0$

$$S(\varphi_n, \psi_n) \geq S(\varphi_n, \psi_n + \varepsilon \mathrm{Tr}) \to S(\varphi, \psi + \varepsilon \mathrm{Tr}),$$

and since $S(\varphi, \psi + \varepsilon \mathrm{Tr}) = -\mathrm{Tr}(\eta(Q_\varphi)) - \varphi(\log(Q_\psi + \varepsilon 1)) \nearrow S(\varphi, \psi)$ as $\varepsilon \to 0$, we conclude that

$$\liminf_n S(\varphi_n, \psi_n) \geq S(\varphi, \psi).$$

If in addition $\varphi_n \leq \lambda \psi_n$, then

$$0 \leq S(\varphi_n, \psi_n) - S(\varphi_n, \psi_n + \varepsilon \mathrm{Tr}) = \varphi_n(\log(Q_{\psi_n} + \varepsilon 1) - \log Q_{\psi_n})$$
$$\leq \lambda \psi_n(\log(Q_{\psi_n} + \varepsilon 1) - \log Q_{\psi_n}) \to \lambda \psi(\log(Q_\psi + \varepsilon 1) - \log Q_\psi),$$

and since the latter expression tends to zero as $\varepsilon \to 0$, we see that $S(\varphi_n, \psi_n) \to S(\varphi, \psi)$.

To prove (iv), by (2.2) it suffices to show (for $x = 1$) that for $x \in A$ and $t \in [0, 1]$

$$\mathrm{Tr}(x^* a^t x b^{1-t}) \text{ is a concave function of } (a, b) \in A_+ \times A_+. \tag{2.4}$$

For this consider the Hilbert space $H = L^2(A, \mathrm{Tr})$, that is, the space A with inner product $(\Lambda(y), \Lambda(z)) = \mathrm{Tr}(z^* y)$, where we denote by $\Lambda(y)$ the element $y \in A$ viewed as a vector in H. Assume $a = \lambda a_1 + (1 - \lambda)a_2$ and $b = \lambda b_1 + (1 - \lambda)b_2$. Define operators on H by

$$x_1 \Lambda(y) = \Lambda(ay), \quad y_1 \Lambda(y) = \Lambda(\lambda a_1 y), \quad z_1 \Lambda(y) = \Lambda((1 - \lambda)a_2 y),$$

$$x_2 \Lambda(y) = \Lambda(yb), \quad y_2 \Lambda(y) = \Lambda(\lambda y b_1), \quad z_2 \Lambda(y) = \Lambda((1 - \lambda)y b_2).$$

Then $x_1 x_2 = x_2 x_1$, $y_1 y_2 = y_2 y_2$, $z_1 z_2 = z_2 z_1$. We have $y_i + z_i = x_i$, $i = 1, 2$, so by Proposition 2.1.3 applied to $v = 1$

$$x_1^t x_2^{1-t} \geq y_1^t y_2^{1-t} + z_1^t z_2^{1-t}.$$

Hence for $x \in A$

$$\begin{aligned}
\mathrm{Tr}(x^*a^txb^{1-t}) &= (\Lambda(xb^{1-t}), \Lambda(a^tx)) \\
&= (x_1^t x_2^{1-t}\Lambda(x), \Lambda(x)) \\
&\geq ((y_1^t y_2^{1-t} + z_1^t z_2^{1-t})\Lambda(x), \Lambda(x)) \\
&= \lambda \mathrm{Tr}(x^*a_1^t x b_1^{1-t}) + (1-\lambda)\mathrm{Tr}(x^*a_2^t x b_2^{1-t}),
\end{aligned}$$

and (2.4) is proved.

Similarly, by virtue of (2.2) in order to prove (v) it suffices to show (for $x = 1$) that for $x \in B$ and $t \in [0,1]$

$$\mathrm{Tr}_B(x^*Q_{\varphi\circ\alpha}^t x Q_{\psi\circ\alpha}^{1-t}) \geq \mathrm{Tr}_A(\alpha(x)^*Q_\varphi^t \alpha(x) Q_\psi^{1-t}). \tag{2.5}$$

Define a linear operator $v\colon L^2(B, \mathrm{Tr}_B) \to L^2(A, \mathrm{Tr}_A)$ by

$$v\Lambda(y) = \Lambda(\alpha(y)), \quad y \in B.$$

For $y \in B$, $z \in A$ define

$$x_1\Lambda(y) = \Lambda(Q_{\varphi\circ\alpha}y), \quad y_1\Lambda(z) = \Lambda(Q_\varphi z),$$

$$x_2\Lambda(y) = \Lambda(yQ_{\psi\circ\alpha}), \quad y_2\Lambda(z) = \Lambda(zQ_\psi),$$

We have

$$(x_1\Lambda(y), \Lambda(y)) = \mathrm{Tr}_B(y^*Q_{\varphi\circ\alpha}y) = \varphi(\alpha(yy^*)),$$

$$(v^*y_1v\Lambda(y), \Lambda(y)) = \mathrm{Tr}_A(\alpha(y)^*Q_\varphi\alpha(y)) = \varphi(\alpha(y)\alpha(y)^*).$$

Since α is a Schwarz map, it follows that $x_1 \geq v^*y_1v$. Similarly we have $x_2 \geq v^*y_2v$. Hence by Proposition 2.1.3, with $z_1 = z_2 = 0$,

$$(x_1^t x_2^{1-t}\Lambda(x), \Lambda(x)) \geq (v^*y_1^t y_2^{1-t}v\Lambda(x), \Lambda(x)),$$

which is exactly (2.5).

To prove (vi), first of all note that both sides of the equality are simultaneously either finite or infinite. Indeed, if $\mathrm{supp}\,\varphi \leq \mathrm{supp}\,\psi$, then obviously $\mathrm{supp}\,\varphi|_B \leq \mathrm{supp}\,\psi|_B$. Furthermore, if $0 \leq a \leq \mathrm{supp}\,\psi$ then by A.13 the support $s(E(a))$ of $E(a)$ is not smaller than $s(a)$, so $\varphi(E(a)) > 0$ if $\varphi(a) > 0$, and thus $\mathrm{supp}\,\varphi \leq \mathrm{supp}\,\varphi \circ E$. Conversely, suppose $\mathrm{supp}\,\varphi|_B \leq \mathrm{supp}\,\psi|_B$ and $\mathrm{supp}\,\varphi \leq \mathrm{supp}\,\varphi \circ E$. Then if $\varphi(a) > 0$ for some $a \geq 0$, we have $\varphi(E(a)) > 0$, hence $\psi(a) = \psi(E(a)) > 0$, so that $\mathrm{supp}\,\varphi \leq \mathrm{supp}\,\psi$.

Thus we may assume that $\varphi \leq \lambda\psi$, $\varphi \leq \lambda\varphi \circ E$. Let $F\colon A \to B$ be any faithful conditional expectation, and ω a faithful state on A. Then using (iii) we may replace E by $E_\varepsilon = (1-\varepsilon)E + \varepsilon F$ and ψ by $((1-\varepsilon)\psi + \varepsilon\omega) \circ E_\varepsilon$, and thus assume that ψ and E are faithful. Replacing further φ by $(1-\varepsilon)\varphi + \varepsilon\psi$, we may assume that φ is faithful as well.

We claim that

$$Q^t_{\varphi\circ E}Q^{-t}_\psi = Q^t_{\varphi|B}Q^{-t}_{\psi|B}.$$

Then, applying φ to both sides and taking the derivative at $t = 0$, we get $S(\varphi|_B, \psi|_B)$ on the right side and

$$\varphi(\log Q_{\varphi\circ E} - \log Q_\psi) = S(\varphi, \psi) - S(\varphi, \varphi \circ E)$$

on the left side, proving (vi).

It remains to prove the claim. In other words, we must prove that

$$Q^t_\varphi Q^{-t}_\psi = Q^t_{\varphi|B} Q^{-t}_{\psi|B} \qquad (2.6)$$

if $\varphi = \varphi \circ E$ and $\psi = \psi \circ E$. This is a well-known general property of Connes' Radon-Nikodym cocycle, but an elementary proof of our particular case is possible.

We shall reduce the proof of (2.6) to the case when B is abelian. The identity

$$Q^t_\varphi Q^{-t}_\psi = Q^t_\varphi Q^{-t}_\omega (Q^t_\psi Q^{-t}_\omega)^{-1}$$

for a state ω shows that to prove (2.6) for arbitrary φ and ψ, it suffices to consider some fixed ψ. Thus we can assume that $\psi|_B$ is a trace. Choose a maximal abelian subalgebra C of B containing $Q_{\varphi|B}$. Since $Q_{\psi|B}$ is in the center of B, we also have $Q_{\psi|B} \in C$. The algebra C is generated by minimal projections $p_1, \ldots, p_m$ in B. Then $E_C(b) = \sum_i p_i b p_i$ is the unique conditional expectation of B onto C. It is $\varphi|_B$- and $\psi|_B$-preserving. Thus to prove (2.6) it is enough to prove the analogous statement for $E_C \circ E\colon A \to C$ and $E_C\colon B \to C$. In other words, we may assume that $B = C$ is abelian.

The conditional expectation $E\colon A \to B$ is then of the form $E(a) = \sum_i \varphi_i(a) p_i$, where φ_i is a state with $\operatorname{supp} \varphi_i = p_i$. Let Q_i be the density operator of φ_i. Then $Q_\varphi = \sum_i \varphi(p_i) Q_i$ and $Q_\psi = \sum_i \psi(p_i) Q_i$, so that

$$Q^t_\varphi Q^{-t}_\psi = \sum_i \frac{\varphi(p_i)^t}{\psi(p_i)^t} p_i = Q^t_{\varphi|B} Q^{-t}_{\psi|B},$$

and the proof of (vi) is complete.

Part (vii) is immediate as $Q_\varphi = \sum_i Q_{\varphi_i}$. □

Remark 2.1.4.

(i) Concerning Theorem 2.1.2(i), a more careful analysis shows that

$$S(\varphi, \psi) \geq \frac{1}{2} \|\varphi - \psi\|^2$$

for any states φ and ψ.

(ii) If $\varphi \leq \psi$ are positive linear functionals, then $S(\varphi, \psi) \leq 0$, and equality holds if and only if the support p of φ commutes with Q_ψ and $Q_\varphi = pQ_\psi$. Indeed, $S(\varphi, \psi) \leq S(\varphi, \varphi) = 0$ by Theorem 2.1.2(ii). Assume $S(\varphi, \psi) = 0$. As

$Q_\psi \geq Q_\varphi$, we have $(Q_\psi + t)^{-1} \leq (Q_\varphi + t)^{-1}$ by B.3. So using the integral formula $\log a = \int_0^\infty ((1+t)^{-1} - (a+t)^{-1})\, dt$ as in B.4, we get from $S(\varphi, \psi) = 0$ that
$$p((Q_\varphi + t)^{-1} - (Q_\psi + t)^{-1})p = 0$$
for any $t > 0$. In other words, p commutes with $(Q_\psi + t)^{-1}$ and $p(Q_\psi + t)^{-1} = p(Q_\varphi + t)^{-1}$, so $pQ_\psi = pQ_\varphi = Q_\varphi$. ♦

2.2 Von Neumann Entropy

A natural generalization of the entropy of a partition is the following notion.

Definition 2.2.1. *The entropy, or von Neumann entropy, of a positive linear functional φ on a finite dimensional C^*-algebra A is*
$$S(\varphi) = \mathrm{Tr}(\eta(Q_\varphi)) = -S(\varphi, \mathrm{Tr}),$$
where $\eta(t) = -t \log t$.

Before we proceed to the study of this notion, recall that the **centralizer** of a positive linear functional φ is the set A_φ of all $x \in A$ such that $\varphi(xy) = \varphi(yx)$ for any $y \in A$. This is just the algebra of elements commuting with Q_φ.

By the **rank** of A we mean the dimension of a maximal abelian subalgebra of A. Then $\operatorname{rank} A = \mathrm{Tr}_A(1)$.

Theorem 2.2.2. *We have:*
(i) $0 \leq S(\varphi) \leq \log \operatorname{rank} A$ *for any state φ; moreover, $S(\varphi) = 0$ if and only if φ is pure, and $S(\varphi) = \log \operatorname{rank} A$ if and only if φ is the normalized canonical trace;*
(ii) *the entropy is continuous, concave, and*
$$S(\varphi + \psi) \leq S(\varphi) + S(\psi);$$
in particular, if φ, ψ are states and $0 < \lambda < 1$, then
$$S(\lambda \varphi + (1-\lambda)\psi) \leq \lambda S(\varphi) + (1-\lambda) S(\psi) + \eta(\lambda) + \eta(1-\lambda);$$
(iii) *if $\varphi = \sum_i \lambda_i \varphi_i$ is a convex combination of states, then*
$$S(\varphi) \geq \sum_i \lambda_i S(\varphi_i, \varphi),$$
and equality holds if and only if φ_i is pure for every $\lambda_i \neq 0$;
(iv) *if φ is a state on A and ψ is a state on B, then*
$$S(\varphi \otimes \psi) = S(\varphi) + S(\psi);$$

(v) *if φ is a positive linear functional on $A \otimes B \otimes C$, then*

$$S(\varphi) + S(\varphi|_B) \le S(\varphi|_{A \otimes B}) + S(\varphi|_{B \otimes C});$$

(vi) *if φ is a state on $A \otimes B$, then*

$$|S(\varphi) - S(\varphi|_B)| \le S(\varphi|_A);$$

(vii) *if B is a maximal abelian subalgebra of A, then $S(\varphi|_B) \ge S(\varphi)$, and equality holds if and only if B is in the centralizer A_φ of φ.*

Most of the properties of von Neumann entropy are simple consequences of properties of relative entropy. In order to prove (vi) we shall need the following result.

Lemma 2.2.3. *We have:*
(i) *if φ is a pure state on the algebra $\mathrm{Mat}_n(\mathbb{C}) \otimes \mathrm{Mat}_m(\mathbb{C}) \cong \mathrm{Mat}_{nm}(\mathbb{C})$, then $S(\varphi|_{\mathrm{Mat}_n(\mathbb{C})}) = S(\varphi|_{\mathrm{Mat}_m(\mathbb{C})})$;*
(ii) *every state on $\mathrm{Mat}_n(\mathbb{C}) \otimes 1 \subset \mathrm{Mat}_n(\mathbb{C}) \otimes \mathrm{Mat}_n(\mathbb{C})$ extends to a pure state on $\mathrm{Mat}_n(\mathbb{C}) \otimes \mathrm{Mat}_n(\mathbb{C})$.*

Proof. If φ is a pure state on $\mathrm{Mat}_n(\mathbb{C}) \otimes \mathrm{Mat}_m(\mathbb{C})$, then the density operator Q_φ is a projection of rank one, so it is the projection onto the line spanned by a unit vector $\sum_{i,j} \lambda_{ij} e_i \otimes e_j \in \ell_2^n \otimes \ell_2^m$. Thus

$$Q_\varphi = \sum_{i,j,k,l} \lambda_{ij} \bar{\lambda}_{kl} e_{ik} \otimes e_{jl},$$

where $\{e_{ik}\}_{i,k}$ are the matrix units. The density matrices of $\varphi|_{\mathrm{Mat}_n(\mathbb{C})}$ and $\varphi|_{\mathrm{Mat}_m(\mathbb{C})}$ are $(\mathrm{id} \otimes \mathrm{Tr}_m)(Q_\varphi)$ and $(\mathrm{Tr}_n \otimes \mathrm{id})(Q_\varphi)$, respectively, where Tr_n is the usual trace on $\mathrm{Mat}_n(\mathbb{C})$. Set $T = \sum_{i,j} \lambda_{ij} e_{ij} \in \mathrm{Mat}_{n \times m}(\mathbb{C})$. Then

$$(\mathrm{Tr}_n \otimes \mathrm{id})(Q_\varphi) = (T^*T)',$$

the transpose of the matrix T^*T, and $(\mathrm{id} \otimes \mathrm{Tr}_m)(Q_\varphi) = TT^*$. Since the nonzero eigenvalues of T^*T and TT^* are the same (counting with multiplicities), we get $S(\varphi|_{\mathrm{Mat}_n(\mathbb{C})}) = S(\varphi|_{\mathrm{Mat}_m(\mathbb{C})})$, and (i) is proved.

If φ is a state on $\mathrm{Mat}_n(\mathbb{C})$, we may assume that its density matrix is diagonal, $Q_\varphi = \sum_{i=1}^n \lambda_i e_{ii}$. Then

$$P = \sum_{i,j} \lambda_i^{1/2} \lambda_j^{1/2} e_{ij} \otimes e_{ij}$$

is a rank one projection in $\mathrm{Mat}_n(\mathbb{C}) \otimes \mathrm{Mat}_n(\mathbb{C})$ such that $Q_\varphi = (\mathrm{id} \otimes \mathrm{Tr}_n)(P)$. Thus the pure state with density matrix P has the property that its restriction to $\mathrm{Mat}_n(\mathbb{C}) \otimes 1$ coincides with φ. □

2.2 Von Neumann Entropy

Proof of Theorem 2.2.2. If φ is a state, and $\{\lambda_i\}_i$ are the eigenvalues of Q_φ, then $\sum_i \lambda_i = 1$ and $S(\varphi) = \sum_i \eta(\lambda_i)$. We see that $S(\varphi) = 0$ if and only if $\lambda_i = 1$ for some i and $\lambda_j = 0$ for $j \neq i$, that is, φ is a pure state. The rest of (i) follows from the corresponding properties of entropy of partitions, Proposition 1.1.9(i).

As $S(\varphi) = -S(\varphi, \text{Tr})$, the entropy is concave by Theorem 2.1.2(iv). Since η is continuous, the entropy is also continuous. To prove the first inequality in (ii) we use operator monotonicity of log, B.4, so

$$S(\varphi + \psi) = -\varphi(\log(Q_\varphi + Q_\psi)) - \psi(\log(Q_\varphi + Q_\psi))$$
$$\leq -\varphi(\log Q_\varphi) - \psi(\log Q_\psi) = S(\varphi) + S(\psi).$$

This implies the second inequality in (ii), since

$$S(\lambda\varphi) = \lambda S(\varphi) + \eta(\lambda),$$

if φ is a state.

To prove (iii) note that

$$\sum_i \lambda_i S(\varphi_i, \varphi) = \sum_i \lambda_i \varphi_i(\log Q_{\varphi_i} - \log Q_\varphi) = S(\varphi) - \sum_i \lambda_i S(\varphi_i).$$

Since $S(\varphi_i) \geq 0$, and $S(\varphi_i) = 0$ if and only if φ_i is pure, we get (iii).

Part (iv) is immediate.

To prove (v) note that

$$S(\varphi|_{A\otimes B}) - S(\varphi|_B) = -S(\varphi|_{A\otimes B}, \text{Tr}_A \otimes \varphi|_B)$$

and similarly

$$S(\varphi) - S(\varphi|_{B\otimes C}) = -S(\varphi, \text{Tr}_A \otimes \varphi|_{B\otimes C}).$$

Since $S(\varphi, \text{Tr}_A \otimes \varphi|_{B\otimes C}) \geq S(\varphi|_{A\otimes B}, \text{Tr}_A \otimes \varphi|_B)$ by Theorem 2.1.2(v) (applied to the inclusion mapping $A \otimes B \to A \otimes B \otimes C$), we get (v).

Part (v) implies in particular that if φ is a state on $A \otimes B$, then

$$S(\varphi) \leq S(\varphi|_A) + S(\varphi|_B).$$

So to prove (vi) it is enough to show that

$$S(\varphi|_B) \leq S(\varphi) + S(\varphi|_A). \tag{2.7}$$

Note that if D is a finite dimensional C*-algebra, then there exists an embedding $D \hookrightarrow \text{Mat}_n(\mathbb{C})$ with $n = \text{rank}\, D$ such that $\text{Tr}_D = \text{Tr}_n|_D$, where Tr_n is the usual trace on $\text{Mat}_n(\mathbb{C})$. This allows us to assume without loss of generality that A and B are full matrix algebras. Let $C \cong A \otimes B$. By Lemma 2.2.3(ii) there exists a pure state $\tilde{\varphi}$ on $A \otimes B \otimes C$ extending φ. Then $S(\tilde{\varphi}|_{B \otimes C}) = S(\tilde{\varphi}|_A)$ by Lemma 2.2.3(i). Hence (2.7) follows from (v) applied to the state $\tilde{\varphi}$.

To prove (vii) note that as B is maximal abelian in A, we have $\operatorname{Tr}_{A|B} = \operatorname{Tr}_B$. In particular, if $E_B \colon A \to B$ is the trace preserving conditional expectation, then $Q_{\varphi \circ E_B} = Q_{\varphi|B}$. Hence

$$S(\varphi|B) - S(\varphi) = \varphi(\log Q_\varphi - \log Q_{\varphi|B}) = S(\varphi, \varphi \circ E_B).$$

Of course, this also follows from Theorem 2.1.2(vi). Hence $S(\varphi|B) \geq S(\varphi)$. If $Q_\varphi \in B$, then $Q_{\varphi|B} = Q_\varphi$ as $\operatorname{Tr}_{A|B} = \operatorname{Tr}_B$, so $S(\varphi|B) = S(\varphi)$. Conversely, if $S(\varphi|B) = S(\varphi)$, then $S(\varphi, \varphi \circ E_B) = 0$. By Theorem 2.1.2(i) we get $\varphi = \varphi \circ E_B$, so $Q_\varphi = Q_{\varphi \circ E_B} = Q_{\varphi|B} \in B$, that is, B is in the centralizer of φ. □

Part (vii) of the above theorem implies that if a subalgebra B of A contains a maximal abelian subalgebra of A, then $S(\varphi|B) \geq S(\varphi)$. The following lemma gives a useful estimate of the difference.

Lemma 2.2.4. *Let $e_1, \ldots, e_n$ be projections in A with sum 1, φ a state on A. Let $B = \oplus_{i=1}^n B_i$, where $B_i = e_i A e_i$. Then*

$$S(\varphi|B) - S(\varphi) \leq \sum_i \eta(\varphi(e_i)).$$

Proof. Consider a decomposition $\varphi = \sum_k \lambda_k \varphi_k$ of φ into a convex combination of pure states. Then by monotonicity of relative entropy, Theorem 2.1.2(v), and Theorem 2.2.2(iii) we have

$$S(\varphi) = \sum_k \lambda_k S(\varphi_k, \varphi) \geq \sum_k \lambda_k S(\varphi_k|B, \varphi|B)$$
$$= S(\varphi|B) - \sum_k \lambda_k S(\varphi_k|B) = S(\varphi|B) - \sum_i \sum_k \lambda_k S(\varphi_k|B_i).$$

The restriction of φ_k to $B_i = e_i A e_i$ is a scalar multiple of a pure state, whence $S(\varphi_k|B_i) = \eta(\varphi_k(e_i))$. By concavity of η we then get

$$\sum_k \lambda_k S(\varphi_k|B_i) = \sum_k \lambda_k \eta(\varphi_k(e_i)) \leq \eta\left(\sum_k \lambda_k \varphi_k(e_i)\right) = \eta(\varphi(e_i)),$$

which completes the proof of the lemma. □

In Chap. 9 we shall use the following form of positivity of relative entropy, called the **thermodynamic inequality**.

Proposition 2.2.5. *If φ is a state on a finite dimensional C^*-algebra A and H a self-adjoint element in A, then we have*

$$S(\varphi) + \varphi(H) \leq \log \operatorname{Tr}(e^H),$$

and equality holds if and only if $Q_\varphi = \operatorname{Tr}(e^H)^{-1} e^H$.

Proof. Let ψ be the state with density operator $\operatorname{Tr}(e^H)^{-1}e^H$. Then
$$S(\varphi,\psi) = -S(\varphi) - \varphi(\log Q_\psi) = -S(\varphi) - \varphi(H) + \log \operatorname{Tr}(e^H).$$
So the result follows from Theorem 2.1.2(i). $\square$

Finishing our discussion of von Neumann entropy we prove the following estimate.

Proposition 2.2.6. *If φ and ψ are states on a finite dimensional C^*-algebra A, $d = \operatorname{rank} A$, and $\|\varphi - \psi\| \le \varepsilon \le 1/e$, then*
$$|S(\varphi) - S(\psi)| \le \varepsilon \log d + \eta(\varepsilon).$$

For this we need an elementary inequality
$$|\eta(s) - \eta(t)| \le \eta(s-t) \quad \text{when} \quad s,t \in [0,1],\ 0 \le s - t \le \frac{1}{2}. \tag{2.8}$$

To show this note that as the function $f(x) = \eta(x+(s-t)) - \eta(x)$ is decreasing, we have for x taking the increasing values $0, t, 1-(s-t)$
$$\eta(s-t) \ge \eta(s) - \eta(t) \ge -\eta(1-(s-t)).$$
If $1/e \le s - t \le 1/2$, then $\eta(1-(s-t)) \le \eta(s-t)$ as η is decreasing on $[1/e, 1]$, so (2.8) is proved in this case. On the other hand, if $s - t \le 1/e$, then using that $\eta(x) + x$ is increasing on $[0,1]$, we get $\eta(s) - \eta(t) \ge t - s$. But $t - s \ge -\eta(s-t)$ whenever $0 \le s - t \le 1/e$, so (2.8) follows in this case too.

We also need the following well-known inequality. Recall that the trace norm is defined by $\|x\|_1 = \operatorname{Tr}(|x|)$. Then if a and b are positive trace-class operators, $\lambda_1 \ge \lambda_2 \ge \ldots$ and $\mu_1 \ge \mu_2 \ge \ldots$ the eigenvalues of a and b, respectively, we have
$$\|a-b\|_1 \ge \sum_i |\lambda_i - \mu_i|. \tag{2.9}$$

Indeed, let x and y be positive operators such that $xy = 0$ and $x - y = a - b$. Then $\|a-b\|_1 = \operatorname{Tr}(x) + \operatorname{Tr}(y)$. Let $c = b + x = a + y$, and let $\nu_1 \ge \nu_2 \ge \ldots$ be the eigenvalues of c. Then $\nu_i \ge \lambda_i, \mu_i$, hence $|\lambda_i - \mu_i| \le 2\nu_i - \lambda_i - \mu_i$. Thus
$$\sum_i |\lambda_i - \mu_i| \le \sum_i (2\nu_i - \lambda_i - \mu_i) = \operatorname{Tr}(2c - a - b) = \operatorname{Tr}(x) + \operatorname{Tr}(y) = \|a-b\|_1,$$
and (2.9) is proved.

Proof of Proposition 2.2.6. Let $\lambda_1 \ge \ldots \ge \lambda_d$ and $\mu_1 \ge \ldots \ge \mu_d$ be the eigenvalues of Q_φ and Q_ψ, respectively. Put $\delta_i = |\lambda_i - \mu_i|$, $\delta = \sum_i \delta_i$. By (2.8)
$$|S(\varphi) - S(\psi)| = \left|\sum_i (\eta(\lambda_i) - \eta(\mu_i))\right| \le \sum_i |\eta(\lambda_i) - \eta(\mu_i)| \le \sum_i \eta(\delta_i).$$

Then $\sum_i \eta(\delta_i) = \delta \sum_i \eta(\delta_i/\delta) - \delta \log \delta \le \delta \log d + \eta(\delta)$. By (2.9) we have $\delta \le \|\varphi - \psi\|$, so $\delta \le \varepsilon \le 1/e$ and $\eta(\delta) \le \eta(\varepsilon)$, proving the proposition. $\square$

2.3 Relative Entropy for General C*-algebras

A few times, notably in Chap. 5, we shall need relative entropy to be defined for arbitrary C*-algebras. The shortest way of introducing it is the following variational expression. Let φ and ψ be positive linear functional on a unital C*-algebra A. Then

$$S(\varphi,\psi) = \sup_{n\in\mathbb{N}} \sup_{x} \left\{ \varphi(1)\log n - \int_{1/n}^{\infty} (\varphi(y(t)^*y(t)) + t^{-1}\psi(x(t)x(t)^*))\frac{dt}{t} \right\}, \tag{2.10}$$

where the second supremum is taken over all step functions $x\colon (1/n,\infty) \to A$ with finite range, and where $y(t) = 1 - x(t)$. It suffices to consider functions x with values in a fixed weakly dense subspace of A containing the unit.

It is not at all obvious that in the finite dimensional case the above expression gives the same value of relative entropy as before. Since all our computations will be done by a reduction to the finite dimensional case anyway, instead of developing the whole theory we confine ourselves to listing the main properties of relative entropy, see Notes at the end of the chapter for references.

Theorem 2.3.1. *For positive linear functionals on unital C^*-algebras we have:*

(i) $S(\lambda_1\varphi, \lambda_2\psi) = \lambda_1 S(\varphi,\psi) + \lambda_1\varphi(1)\log\frac{\lambda_1}{\lambda_2}$ *if* $\lambda_1,\lambda_2 > 0$;

(ii) $S(\varphi,\psi) \geq 0$ *for states* φ *and* ψ, *and equality holds if and only if* $\varphi = \psi$;

(iii) $S(\varphi,\psi)$ *is decreasing in* ψ, *i.e.,* $S(\varphi,\psi_1) \geq S(\varphi,\psi_2)$ *if* $\psi_1 \leq \psi_2$;

(iv) $S(\varphi,\psi)$ *is a weakly* lower semicontinuous function of* (φ,ψ);

(v) $S(\varphi,\psi)$ *is a convex function of* (φ,ψ);

(vi) *if* $\alpha\colon B \to A$ *is a unital Schwarz mapping, then*

$$S(\varphi\circ\alpha, \psi\circ\alpha) \leq S(\varphi,\psi);$$

(vii) *if M is a von Neumann algebra, φ and ψ are normal states on M, N a von Neumann subalgebra of M, and $E\colon M \to N$ a ψ-preserving faithful normal conditional expectation, then*

$$S(\varphi,\psi) = S(\varphi|_N, \psi|_N) + S(\varphi, \varphi\circ E);$$

(viii) *if* $\varphi = \sum_{i=1}^n \varphi_i$, *then* $\sum_i S(\varphi_i,\psi) = S(\varphi,\psi) + \sum_i S(\varphi_i,\varphi)$;

(ix) *if ψ is a state on A, and $\bar\psi$ its normal extension to $\pi_\psi(A)''$, where $\pi_\psi\colon A \to B(H_\psi)$ is the GNS-representation, then*

$$S(\varphi,\psi) = \begin{cases} S(\bar\varphi,\bar\psi), & \text{if } \varphi = \bar\varphi\circ\pi_\psi \text{ for a normal linear functional } \bar\varphi, \\ +\infty, & \text{otherwise;} \end{cases}$$

(x) *if τ is a finite trace on A, a and b positive elements in A such that $a \leq \lambda b$ for some $\lambda > 0$, then $S(\tau(\cdot\, a), \tau(\cdot\, b)) = \tau(a(\log a - \log b))$.* □

2.3 Relative Entropy for General C*-algebras

We finish the chapter with several corollaries of Theorem 2.3.1 (or of Theorem 2.1.2 in the finite dimensional case).

Corollary 2.3.2. *If $\{\varphi_i\}_{i\in I}$ and $\{\psi_i\}_{i\in I}$ are finite sets of positive linear functionals then*

$$S\left(\sum_i \varphi_i, \sum_i \psi_i\right) \leq \sum_i S(\varphi_i, \psi_i).$$

Proof. This follows from the scaling (i) and convexity (v) properties in Theorem 2.3.1. □

Corollary 2.3.3. *Let A be a C^*-algebra and $E\colon A \to B$ be a conditional expectation onto a subalgebra B, φ and ψ positive linear functionals on A such that $\varphi \circ E = \varphi$ and $\psi \circ E = \psi$. Then $S(\varphi|_B, \psi|_B) = S(\varphi, \psi)$.*

Proof. This is obtained by applying the monotonicity property of relative entropy, Theorem 2.3.1(vi), to $E\colon A \to B$ and to the inclusion map $B \hookrightarrow A$. □

Corollary 2.3.4. *Let B be a C^*-subalgebra of a C^*-algebra A, ψ a state on A, π_ψ the GNS-representation defined by ψ. Assume $\pi_\psi(B)$ is weakly operator dense in $\pi_\psi(A)$. Then $S(\varphi, \psi) = S(\varphi|_B, \psi|_B)$ for any positive linear functional $\varphi \leq \psi$.*

Proof. We can identify $\pi_\psi|_B$ with the GNS-representation defined by $\psi|_B$. Then the result follows from Theorem 2.3.1(ix). □

Corollary 2.3.5. *Let φ and ψ be positive linear functionals on a C^*-algebra A, $\{\gamma_i\colon A_i \to A\}_i$ and $\{\theta_i\colon A \to A_i\}_i$ two nets of unital completely positive maps such that $\varphi \circ \gamma_i \circ \theta_i \to \varphi$ and $\psi \circ \gamma_i \circ \theta_i \to \psi$ weakly*. Then $S(\varphi \circ \gamma_i, \psi \circ \gamma_i) \to S(\varphi, \psi)$.*

Proof. By lower semicontinuity, Theorem 2.3.1(iv),

$$S(\varphi, \psi) \leq \liminf_i S(\varphi \circ \gamma_i \circ \theta_i, \psi \circ \gamma_i \circ \theta_i).$$

On the other hand, by monotonicity, Theorem 2.3.1(vi),

$$S(\varphi \circ \gamma_i \circ \theta_i, \psi \circ \gamma_i \circ \theta_i) \leq S(\varphi \circ \gamma_i, \psi \circ \gamma_i) \leq S(\varphi, \psi).$$

Hence $S(\varphi \circ \gamma_i, \psi \circ \gamma_i) \to S(\varphi, \psi)$. □

A related but easier result is the following.

Corollary 2.3.6. *If $\{A_i\}_i$ is an increasing net of unital C^*-subalgebras of a C^*-algebra A such that $\cup_i A_i$ is norm-dense, then $S(\varphi|_{A_i}, \psi|_{A_i}) \nearrow S(\varphi, \psi)$ for any φ and ψ.*

Proof. This follows immediately from the variational expression (2.10), since it suffices to consider step functions with values in $\cup_i A_i$. □

Recall that a unital C*-algebra is called **nuclear** if there exist two nets $\{\gamma_i \colon A_i \to A\}_i$ and $\{\theta_i \colon A \to A_i\}_i$ of unital completely positive maps with finite dimensional C*-algebras A_i such that $\|(\gamma_i \circ \theta_i)(a) - a\| \to 0$ for every $a \in A$. Then by Corollary 2.3.5 it follows that $S(\varphi, \psi) = \lim_i S(\varphi \circ \gamma_i, \psi \circ \gamma_i)$ for any φ and ψ. This equality could be used as a definition of relative entropy for nuclear algebras, which would in fact be sufficient for us in most cases.

Corollary 2.3.7. *Let A be a separable C*-algebra, (X, μ) a Lebesgue space, $X \ni x \mapsto \varphi_x$ and $X \ni x \mapsto \psi_x$ measurable maps into the state space of A, the state space being considered with the weak* topology. Consider the states φ and ψ on $A \otimes L^\infty(X, \mu)$ defined by*

$$\varphi(a \otimes f) = \int_X \varphi_x(a) f(x) d\mu(x), \quad \psi(a \otimes f) = \int_X \psi_x(a) f(x) d\mu(x).$$

Then $S(\varphi, \psi) = \int_X S(\varphi_x, \psi_x) d\mu(x)$.

Proof. Note that if A is finite dimensional, then $\text{Tr}_A \otimes \mu$ is a finite trace on $A \otimes L^\infty(X, \mu)$, and the result can be deduced from Theorem 2.3.1(x). Using Corollary 2.3.5 one can extend the result to nuclear C*-algebras. In general we can argue as follows.

Elements of the algebraic tensor product $A \odot L^\infty(X, \mu)$ can be considered as A-valued functions on X. Hence any step function $(1/n, \infty) \to A \odot L^\infty(X, \mu)$ defines a family of step functions $(1/n, \infty) \to A$ indexed by elements of X. By the variational expression (2.10) this implies the inequality $\le$.

On the other hand, if ξ is a finite measurable partition of X then

$$S(\varphi, \psi) \ge S(\varphi|_{A \otimes L^\infty(X/\xi, \mu)}, \psi|_{A \otimes L^\infty(X/\xi, \mu)})$$
$$= \sum_{Z \in \xi} S(\varphi|_{A \otimes \mathbb{C} 1_Z}, \psi|_{A \otimes \mathbb{C} 1_Z})$$
$$= \sum_{Z \in \xi} S\left(\int_Z \varphi_y d\mu(y), \int_Z \psi_y d\mu(y)\right)$$
$$= \sum_{Z \in \xi} \mu(Z) S\left(\mu(Z)^{-1} \int_Z \varphi_y d\mu(y), \mu(Z)^{-1} \int_Z \psi_y d\mu(y)\right)$$
$$= \int_X S\left(\mu(Z(x))^{-1} \int_{Z(x)} \varphi_y d\mu(y), \mu(Z(x))^{-1} \int_{Z(x)} \psi_y d\mu(y)\right) d\mu(x),$$

where $Z(x)$ is the element of ξ containing x. Choose an increasing sequence $\{\xi_n\}_n$ of finite measurable partitions such that $\vee_n \xi_n$ is the partition into points. For $x \in X$ denote by $Z_n(x)$ the element of ξ_n containing x. By the martingale convergence theorem for any $f \in L^1(X, \mu)$ we have

2.3 Relative Entropy for General C*-algebras

$\mu(Z_n(x))^{-1} \int_{Z_n(x)} f(y) d\mu(y) \to f(x)$ for a.e. $x \in X$. It follows that

$$\mu(Z_n(x))^{-1} \int_{Z_n(x)} \varphi_y d\mu(y) \to \varphi_x \text{ and } \mu(Z_n(x))^{-1} \int_{Z_n(x)} \psi_y d\mu(y) \to \psi_x$$

weakly* for a.e. $x \in X$. Hence the above inequality together with weak* lower semicontinuity of relative entropy implies the inequality $\geq$ in the corollary. □

Corollary 2.3.8. *Let $\{\varphi_{ij}\}_{i \in I, j \in J}$ be a finite set of positive linear functionals on a C^*-algebra. Put $\varphi = \sum_{i,j} \varphi_{ij}$, $\varphi_i^I = \sum_j \varphi_{ij}$ and $\varphi_j^J = \sum_i \varphi_{ij}$. Then*

$$\sum_{i,j} S(\varphi_{ij}, \varphi) \geq \sum_i S(\varphi_i^I, \varphi) + \sum_j S(\varphi_j^J, \varphi).$$

Proof. Using Theorem 2.3.1(viii) and Corollary 2.3.2 we compute

$$\sum_{i,j} S(\varphi_{ij}, \varphi) - \sum_j S(\varphi_j^J, \varphi) = \sum_j \left(\sum_i S(\varphi_{ij}, \varphi) - S(\varphi_j^J, \varphi) \right)$$

$$= \sum_{i,j} S(\varphi_{ij}, \varphi_j^J)$$

$$\geq \sum_i S \left(\sum_j \varphi_{ij}, \sum_j \varphi_j^J \right)$$

$$= \sum_i S(\varphi_i^I, \varphi),$$

which proves the corollary. □

Property (x) in Theorem 2.3.1 could also be used as a definition of entropy for C*-algebras with finite trace. To prove the main properties of entropy such as convexity and monotonicity one could then use the same technique as in the finite dimensional case. In particular, the following three corollaries do not require the full strength of Theorem 2.3.1 and can be deduced from Proposition 2.1.3.

The first corollary is the noncommutative extension of Lemma 1.1.10. In Chap. 4 we shall prove a stronger result.

Corollary 2.3.9. *Let $\{x_{i_1 \ldots i_n}\}_{i_1, \ldots, i_n}$ be a finite set of positive elements in a C^*-algebra with a finite trace τ. Assume $\sum_{i_1, \ldots, i_n} x_{i_1 \ldots i_n} = 1$. For each k put $x_{i_k}^{(k)} = \sum_{i_1, \ldots, i_{k-1}, i_{k+1}, \ldots, i_n} x_{i_1 \ldots i_n}$. Then*

$$\sum_{i_1, \ldots, i_n} \tau(\eta(x_{i_1 \ldots i_n})) \leq \sum_k \sum_{i_k} \tau(\eta(x_{i_k}^{(k)})).$$

Proof. For $n = 2$ the result follows by applying Corollary 2.3.8 to $\varphi_{i_1 i_2} = \tau(\cdot x_{i_1 i_2})$ and using Theorem 2.3.1(x). The general case is proved by induction. □

Corollary 2.3.10. *Let τ be a finite trace on a C^*-algebra A, a and b positive elements in A such that $a \leq \lambda b$ for some $\lambda > 0$. Then*

$$\tau(a(\log a - \log b)) \geq \tau(a) - \tau(b).$$

Proof. By monotonicity, Theorem 2.3.1(vi),

$$S(\tau(\cdot a), \tau(\cdot b)) \geq S(\tau(\cdot a)|_{\mathbb{C}1}, \tau(\cdot b)|_{\mathbb{C}1}) = \tau(a)(\log \tau(a) - \log \tau(b)).$$

On the other hand, $\tau(a)(\log \tau(a) - \log \tau(b)) \geq \tau(a) - \tau(b)$ by (2.3). □

In a similar way we get the following **Peierls-Bogoliubov inequality**.

Corollary 2.3.11. *Let τ be a finite trace on a C^*-algebra A, a and b self-adjoint elements in A. Then*

$$\log \tau(e^{a+b}) - \log \tau(e^b) \geq \frac{\tau(ae^b)}{\tau(e^b)}.$$

In particular,

$$|\log \tau(e^{a+b}) - \log \tau(e^b)| \leq \|a\|.$$

Proof. Applying

$$S(\varphi, \psi) \geq S(\varphi|_{\mathbb{C}1}, \psi|_{\mathbb{C}1}) = \varphi(1)(\log \varphi(1) - \log \psi(1))$$

to $\varphi = \tau(\cdot e^b)$ and $\psi = \tau(\cdot e^{a+b})$ we get the first inequality. It follows that

$$\log \tau(e^{a+b}) - \log \tau(e^b) \geq -\|a\|.$$

Applying this to $-a$ and $a+b$ instead of a and b we get the second inequality. □

2.4 Notes

Our main sources during the preparation of this chapter were the books by Simon [198] and Ohya and Petz [147]. Chapter 5 of the latter book contains in particular proofs of all the properties listed in Theorem 2.3.1.

Relative entropy of measures was introduced by Kullback and Leibler [114]. The definition for normal states on finite von Neumann algebras was given by Umegaki [218]. One should be aware that in the literature several different conventions are used. Sometimes one uses the opposite order of the arguments or the opposite sign. The statement (2.4) is the famous Lieb concavity. It was conjectured by Wigner and Yanase [235] on the basis of remarks of Dyson and proved by Lieb [116]. As we saw, it implies one of the main properties of relative entropy, joint convexity. The monotonicity property, Theorem 2.1.2(v) and Theorem 2.3.1(vi), was proved by Uhlmann [217]. Our deduction of both properties from Proposition 2.1.3 essentially follows Uhlmann's paper using simplifications made by Simon [198]. The conditional expectation property, Theorem 2.1.2(vi) and Theorem 2.3.1(vii), is due to Hiai, Ohya and Tsukada [89] in a weaker form and to Petz [158], [159] in general.

Although in our exposition entropy of states appears after entropy of partitions, it was introduced by von Neumann [145] more than twenty years earlier than the latter [195]. Of course, both notions have their origin in works of Boltzmann and Gibbs on thermodynamics. Among the properties of von Neumann entropy the one which attracted most attention was strong subadditivity, Theorem 2.2.2(v). It was first proved by Lieb and Ruskai [117] using the Lieb concavity. The triangle inequality, Theorem 2.2.2(vi), was established by Araki and Lieb [11]. Proposition 2.2.6 is due to Fannes [66].

The notion of relative entropy was extended to normal states on arbitrary von Neumann algebras by Araki [9], [10]. If φ and ψ are normal states on a von Neumann algebra, $\varphi \leq \lambda\psi$, then, [158],

$$S(\varphi, \psi) = i\frac{d}{dt}\varphi((D\varphi : D\psi)_t)|_{t=0},$$

where $(D\varphi : D\psi)_t$ is Connes' Radon-Nikodym cocycle. This is an extension of (2.2). A definition for states on arbitrary *-algebras was given by Uhlmann [217]. Later Kosaki [112], using ideas of Pusz and Woronowicz [176], obtained the variational expression (2.10). The inequality in Corollary 2.3.8, sometimes called the marginal inequality, was first used by Connes [49], and the inequality in Corollary 2.3.9 by Connes and Størmer [51].

Finally note that in view of Theorem 2.2.2(iii) one can define the entropy of a state φ on a C*-algebra as the supremum of $\sum_i \lambda_i S(\varphi_i, \varphi)$ over all convex decompositions $\varphi = \sum_i \lambda_i \varphi_i$. It turns out that if $S(\varphi) < \infty$, then $\pi_\varphi(A)''$ is discrete, that is, it is a direct sum of algebras $B(H)$, see [147, Chapter 6].

3
Dynamical Entropy

As we already mentioned in Chap. 1, one of the main problems in defining noncommutative entropy is that two finite dimensional subalgebras of a noncommutative algebra can generate an infinite dimensional subalgebra. To overcome this difficulty we consider the entropy of the join of n partitions as a function of the partitions and try to find its noncommutative analogue for each $n \in \mathbb{N}$. In the present chapter we shall define such an analogue and study the dynamical invariant arising from it.

3.1 Mutual Entropy

In Chap. 1 we succeeded in writing a formula for $H(\xi_1 \vee \ldots \vee \xi_n)$ which does not use joins of the partitions ξ_i, see Proposition 1.3.2. This formula makes sense for finite von Neumann algebras. Namely, let N be a von Neumann algebra and τ a faithful normal trace on N. Let $N_1, \ldots, N_n$ be finite dimensional subalgebras of N, and let $E_{N_k} \colon N \to N_k$ denote the τ-preserving conditional expectation. Then the **mutual entropy** of $N_1, \ldots, N_n \subset N$ with respect to τ is

$$H_\tau(N_1, \ldots, N_n) = \sup \left\{ \sum_{i_1, \ldots, i_n} \eta(\tau(x_{i_1 \ldots i_n})) - \sum_{k=1}^n \sum_{i_k} \tau\left(\eta\left(E_{N_k}(x_{i_k}^{(k)})\right)\right) \right\},$$

where $\eta(t) = -t \log t$, and the supremum is taken over all finite partitions of unity

$$1 = \sum_{i_1 \in I_1, \ldots, i_n \in I_n} x_{i_1 \ldots i_n}, \quad x_{i_1 \ldots i_n} \geq 0,$$

in N, where $x_{i_k}^{(k)} = \sum_{i_1, \ldots, i_{k-1}, i_{k+1}, \ldots, i_n} x_{i_1 \ldots i_n}$.

In the case of one algebra $B \subset N$, the definition says that $H_\tau(B)$ is the supremum of

$$\sum_i \eta(\tau(x_i)) - \sum_i \tau(\eta(E_B(x_i))) = -\sum_i \tau(x_i)\tau(\eta(E_B(\tau(x_i)^{-1}x_i))). \quad (3.1)$$

We shall see soon that this supremum is just $S(\tau|B)$.

This definition does not make sense if τ is not a trace, since in general a state preserving conditional expectation $N \to N_k$ may not exist. However, by Theorem 2.3.1(x) we know that if $x \in N_k$ then $\tau(\eta(x)) = -S(\tau(\cdot x)|_{N_k}, \tau|_{N_k})$. Recalling that the map $x \mapsto \tau(\cdot x)$ establishes a one-to-one correspondence between positive elements $x \in N$, $x \leq 1$, and positive linear functionals $\varphi \leq \tau$ on N, we can thus rewrite the definition of the mutual entropy as

$$H_\tau(N_1,\ldots,N_n) = \sup\left\{\sum_{i_1,\ldots,i_n}\eta(\varphi_{i_1\ldots i_n}(1)) + \sum_k\sum_{i_k}S(\varphi_{i_k}^{(k)}|_{N_k},\tau|_{N_k})\right\},$$

where the supremum is taken over all finite decompositions of τ into a sum of positive linear functionals, and $\varphi_{i_k}^{(k)} = \sum_{i_1,\ldots,i_{k-1},i_{k+1},\ldots,i_n}\varphi_{i_1\ldots i_k}$. This formula already makes sense for arbitrary states.

We shall now extend the definition to channels. Let A be a unital C^*-algebra. By a **channel** in A we mean a unital completely positive map $\gamma\colon B \to A$, where B is a finite dimensional C^*-algebra. The reason to consider channels is that in general a C^*-algebra may have too few finite dimensional subalgebras.

Definition 3.1.1. *Let A be a unital C^*-algebra, $\gamma_k\colon A_k \to A$, $1 \leq k \leq n$, a collection of channels. The mutual entropy $H_\varphi(\gamma_1,\ldots,\gamma_n)$ of $\gamma_1,\ldots,\gamma_n$ with respect to a state φ on A is the supremum of*

$$H_\varphi(\gamma_1,\ldots,\gamma_n;\{\varphi_{i_1\ldots i_n}\}) = \sum_{i_1,\ldots,i_n}\eta(\varphi_{i_1\ldots i_n}(1)) + \sum_k\sum_{i_k}S(\varphi_{i_k}^{(k)}\circ\gamma_k,\varphi\circ\gamma_k)$$

over all finite decompositions $\varphi = \sum_{i_1,\ldots,i_n}\varphi_{i_1\ldots i_n}$, where $\varphi_{i_k}^{(k)}$ is obtained by taking the sum of the elements $\varphi_{i_1\ldots i_n}$ with the k-th index equal to i_k.

We call $H_\varphi(\gamma_1,\ldots,\gamma_n;\{\varphi_{i_1\ldots i_n}\})$ the mutual entropy of $\gamma_1,\ldots,\gamma_n$ with respect to the decomposition $\varphi = \sum_{i_1,\ldots,i_n}\varphi_{i_1\ldots i_n}$. Note that since

$$S(\lambda\psi,\varphi) = \lambda S(\psi,\varphi) - \psi(1)\eta(\lambda),$$

it can also be written as

$$\sum_{i_1,\ldots,i_n}\eta(\varphi_{i_1\ldots i_n}(1)) - \sum_k\sum_{i_k}\eta(\varphi_{i_k}^{(k)}(1)) + \sum_k\sum_{i_k}\varphi_{i_k}^{(k)}(1)S(\hat\varphi_{i_k}^{(k)}\circ\gamma_k,\varphi\circ\gamma_k), \quad (3.2)$$

where for a positive linear functional ψ we denote by $\hat\psi$ the state $\psi(1)^{-1}\psi$. The last expression can equivalently be written as

$$\sum_{i_1,\ldots,i_n}\eta(\varphi_{i_1\ldots i_n}(1)) - \sum_k\sum_{i_k}\eta(\varphi_{i_k}^{(k)}(1)) + \sum_k\sum_{i_k}H_\varphi(\gamma_k;\{\varphi_{i_k}^{(k)}\}). \quad (3.3)$$

3.1 Mutual Entropy

In the case when the A_i's are subalgebras of A and the γ_i's are the inclusion maps, we write $H_\varphi(A_1,\ldots,A_n)$ instead of $H_\varphi(\gamma_1,\ldots,\gamma_n)$.

To get a better understanding of the definition, consider the case of one channel $\gamma \colon B \to A$. Then $H_\varphi(\gamma)$ is the supremum of the quantities

$$\sum_i \eta(\varphi_i(1)) + \sum_i S(\varphi_i \circ \gamma, \varphi \circ \gamma). \tag{3.4}$$

For a state ψ on B, call a decomposition $\psi = \sum_i \psi_i$ **orthogonal** if the density matrices of the ψ_i's are mutually orthogonal. Equivalently, there exists a partition of unity $1 = \sum_i p_i$ consisting of projections in the centralizer of ψ such that $\psi_i = \psi(\cdot p_i)$. By Remark 2.1.4(ii) this is also the same as requiring $S(\psi_i, \psi) = 0$ for all i. Thus the first summand in (3.4) is a classical entropy term, while the second one is a correction term which compensates for nonorthogonality of the decomposition $\varphi \circ \gamma = \sum_i \varphi_i \circ \gamma$. The quantity

$$-\sum_i S(\varphi_i \circ \gamma, \varphi \circ \gamma) = \sum_i S(\varphi_i \circ \gamma) - S(\varphi \circ \gamma)$$
$$= \sum_i \eta(\varphi_i(1)) + \sum_i \varphi_i(1) S(\hat\varphi_i \circ \gamma) - S(\varphi \circ \gamma) \tag{3.5}$$

is also called the **entropy defect** of the decomposition $\varphi \circ \gamma = \sum_i \varphi_i \circ \gamma$.

In the case of several channels we similarly have the classical term $\sum_{i_1,\ldots,i_n} \eta(\varphi_{i_1\ldots i_n}(1))$ plus the sum of the correction terms corresponding to the decompositions $\varphi \circ \gamma_k = \sum_{i_k} \varphi_{i_k}^{(k)} \circ \gamma_k$, $1 \le k \le n$.

If A is a von Neumann algebra and φ a faithful normal state, then instead of decompositions of φ we can talk about partitions of unity in A. To show this we first need to recall a few facts from modular theory, see e.g. [30, Chapter 2.5] or [205] for details.

We may assume $A \subset B(H)$ and φ is a vector state, $\varphi(a) = (a\xi_\varphi, \xi_\varphi)$, where ξ_φ is a cyclic and separating vector in H. The anti-linear map

$$A\xi_\varphi \ni a\xi_\varphi \mapsto a^*\xi_\varphi$$

is closable. Denote by S_φ its closure. Let $S_\varphi = J_\varphi \Delta_\varphi^{1/2}$ be the polar decomposition, where J_φ is an anti-linear isometry and Δ_φ a positive operator, which is in general unbounded. Then $J_\varphi^2 = 1$, Δ_φ is nonsingular and $J_\varphi \xi_\varphi = \Delta_\varphi \xi_\varphi = \xi_\varphi$. The operators J_φ and Δ_φ are called the **modular conjugation** and the **modular operator** defined by φ. The formula

$$\sigma_t^\varphi(a) = \Delta_\varphi^{it} a \Delta_\varphi^{-it}, \quad a \in A,$$

defines a one-parameter automorphism group of A called the **modular group** of φ. The modular group measures how far the state φ is from a tracial state. Namely, the **centralizer** A_φ of φ, that is, the set of all elements $a \in A$ such that $\varphi(ab) = \varphi(ba)$ for any $b \in A$, coincides with the fixed point algebra of the modular group. If B is a von Neumann subalgebra of A, then a normal φ-preserving conditional expectation $A \to B$ exists if and only if $\sigma_t^\varphi(B) = B$ for any $t \in \mathbb{R}$. The most important property of J_φ is that $J_\varphi A J_\varphi = A'$, the commutant of A in $B(H)$.

It follows that any positive linear functional $\psi \le \varphi$ can be written in the form
$$\psi = (\cdot J_\varphi x J_\varphi \xi_\varphi, \xi_\varphi) = (\cdot J_\varphi x \xi_\varphi, \xi_\varphi)$$
for a unique element $x \in A$, $0 \le x \le 1$. Since
$$J_\varphi x \xi_\varphi = J_\varphi S_\varphi x \xi_\varphi = \Delta_\varphi^{1/2} x \xi_\varphi = \Delta_\varphi^{1/2} x \Delta_\varphi^{-1/2} \xi_\varphi,$$
by slightly abusing notation we can write $\psi = \varphi(\cdot \sigma_{-i/2}^\varphi(x))$. Thus any partition of unity $1 = \sum_j x_j$ in A gives rise to a decomposition $\varphi = \sum_j \varphi(\cdot \sigma_{-i/2}^\varphi(x_j))$, and vice versa. We shall write $H_\varphi(\gamma_1, \ldots, \gamma_n; \{x_{j_1 \ldots j_n}\})$ instead of $H_\varphi(\gamma_1, \ldots, \gamma_n; \{\varphi(\cdot \sigma_{-i/2}^\varphi(x_{j_1 \ldots j_n}))\})$.

Yet another equivalent way of describing decompositions of states is as follows. Let C be a finite dimensional abelian C*-algebra, $C_1, \ldots, C_n \subset C$ subalgebras, μ a state on C, and $P \colon A \to C$ a unital positive map such that $\varphi = \mu \circ P$. Then we say that $(C, \mu, \{C_k\}_{k=1}^n, P)$ is an **abelian model** for $(A, \varphi, \{\gamma_k\}_{k=1}^n)$. An abelian model defines a decomposition $\varphi = \sum \varphi_{i_1 \ldots i_n}$ by
$$\varphi_{i_1 \ldots i_n}(a) = \mu(P(a) p_{i_1}^{(1)} \ldots p_{i_n}^{(n)}), \tag{3.6}$$
where $\{p_{i_k}^{(k)}\}_{i_k}$ is the set of atoms of C_k. We call $H_\varphi(\gamma_1, \ldots, \gamma_n; \{\varphi_{i_1 \ldots i_n}\})$ the **entropy of the abelian model**. By definition
$$H_\varphi(\gamma_1, \ldots, \gamma_n; \{\varphi_{i_1 \ldots i_n}\}) = S(\mu|_{\vee_k C_k}) + \sum_k \sum_{i_k} S(\mu((P \circ \gamma_k)(\cdot) p_{i_k}^{(k)}), \varphi \circ \gamma_k). \tag{3.7}$$
Conversely, starting with a decomposition of φ one can easily construct an abelian model defining this decomposition: take $C = C(I_1 \times \ldots \times I_n)$, $C_k = C(I_k)$, then (3.6) determines μ and P.

Note that if C is a finite dimensional abelian C*-algebra, $C_1, \ldots, C_n$ are subalgebras of C, and μ is a state on C, then $(C, \mu, \{C_k\}_{k=1}^n, \mathrm{id}_C)$ is an abelian model for $(C, \mu, \{C_k\}_{k=1}^n)$. The entropy of this model is $S(\mu|_{\vee_k C_k})$.

The setting of abelian models makes it explicit that in our definition of mutual entropy we first project a noncommutative C*-algebra into a commutative one, and then compute the entropy of the images of channels under this projection. More precisely, we have the following.

Lemma 3.1.2. *The mutual entropy $H_\varphi(\gamma_1, \ldots, \gamma_n)$ is the supremum of the entropies $H_\mu(P \circ \gamma_1, \ldots, P \circ \gamma_n)$ over all unital completely positive maps $P \colon A \to C$ of A into abelian C*-algebras and over all states μ such that $\mu \circ P = \varphi$.*

Proof. The inequality $H_\varphi(\gamma_1,\dots,\gamma_n) \geq H_\mu(P \circ \gamma_1,\dots,P \circ \gamma_n)$ will follow from Proposition 3.1.3(ii) below. On the other hand, if $(C,\mu,\{C_k\}_{k=1}^n,P)$ is an abelian model for $(A,\varphi,\{\gamma_k\}_{k=1}^n)$, then $(C,\mu,\{C_k\}_{k=1}^n,\mathrm{id}_C)$ is an abelian model for $(C,\mu,\{P \circ \gamma_k\}_{k=1}^n)$. By definition, see (3.7), the entropies of these two abelian models coincide, so that the entropy of the abelian model $(C,\mu,\{C_k\}_{k=1}^n,P)$ for $(A,\varphi,\{\gamma_k\}_{k=1}^n)$ is not larger than $H_\mu(P \circ \gamma_1,\dots,P \circ \gamma_n)$. Thus we get the result. □

The main properties of mutual entropy are the following.

Proposition 3.1.3. *We have:*

(i) *if $\theta_k \colon B_k \to A_k$, $1 \leq k \leq n$, are unital completely positive maps, then*

$$H_\varphi(\gamma_1 \circ \theta_1,\dots,\gamma_n \circ \theta_n) \leq H_\varphi(\gamma_1,\dots,\gamma_n);$$

(ii) *if $\theta \colon A \to B$ is a unital completely positive map and ψ a state on B, then*

$$H_\psi(\theta \circ \gamma_1,\dots,\theta \circ \gamma_n) \leq H_{\psi \circ \theta}(\gamma_1,\dots,\gamma_n);$$

in particular, $H_\psi(\theta \circ \gamma_1,\dots,\theta \circ \gamma_n) = H_{\psi \circ \theta}(\gamma_1,\dots,\gamma_n)$ if θ is an isomorphism;

(iii) *if $k < n$, then $H_\varphi(\gamma_1,\dots,\gamma_k) \leq H_\varphi(\gamma_1,\dots,\gamma_n)$, and*

$$H_\varphi(\gamma_1,\dots,\gamma_n) \leq H_\varphi(\gamma_1,\dots,\gamma_k) + H_\varphi(\gamma_{k+1},\dots,\gamma_n);$$

(iv) *$H_\varphi(\gamma_1,\dots,\gamma_n)$ depends only on the set $\{\gamma_1,\dots,\gamma_n\}$; in other words, it is invariant under permutations of $\gamma_1,\dots,\gamma_n$, and if $\gamma_{n-1} = \gamma_n$, then*

$$H_\varphi(\gamma_1,\dots,\gamma_n) = H_\varphi(\gamma_1,\dots,\gamma_{n-1}).$$

Proof. Part (i) follows from monotonicity of relative entropy, Theorem 2.1.2(v).

Part (ii) is immediate as any decomposition of the state ψ defines a decomposition of $\psi \circ \theta$.

The inequality $H_\varphi(\gamma_1,\dots,\gamma_k) \leq H_\varphi(\gamma_1,\dots,\gamma_n)$ follows from the fact that any decomposition $\varphi = \sum \varphi_{i_1 \dots i_k}$ can be considered as a decomposition $\varphi = \sum \varphi_{i_1 \dots i_n}$ with single-point index sets $I_{k+1},\dots,I_n$.

Given a decomposition $\varphi = \sum \varphi_{i_1 \dots i_n}$, we can construct two decompositions $\varphi = \sum \varphi'_{i_1 \dots i_k}$ and $\varphi = \sum \varphi''_{i_{k+1} \dots i_n}$, which are more transparently described in terms of abelian models: given an abelian model $(C,\mu,\{C_j\}_{j=1}^n,P)$ for $(A,\varphi,\{\gamma_j\}_{j=1}^n)$ we get the abelian models $(C,\mu,\{C_j\}_{j=1}^k,P)$ and $(C,\mu,\{C_j\}_{j=k+1}^n,P)$ for $(A,\varphi,\{\gamma_j\}_{j=1}^k)$ and $(A,\varphi,\{\gamma_j\}_{j=k+1}^n)$, respectively. Then the subadditivity inequality in part (iii) is a consequence of the classical subadditivity

$$S(\mu|_{\vee_{j=1}^n C_j}) \leq S(\mu|_{\vee_{j=1}^k C_j}) + S(\mu|_{\vee_{j=k+1}^n C_j}).$$

It remains to prove part (iv). The invariance under permutations is obvious. Assume now that $\gamma_{n-1} = \gamma_n$. By (iii) we already know that

$H_\varphi(\gamma_1,\ldots,\gamma_{n-1}) \le H_\varphi(\gamma_1,\ldots,\gamma_n)$. To make the proof of the opposite inequality more transparent, consider first the case $n = 2$. Then given any decomposition $\varphi = \sum_{i_1,i_2} \varphi_{i_1 i_2}$ we have

$$H_\varphi(\gamma_1; \{\varphi_{i_1 i_2}\}) \ge H_\varphi(\gamma_1, \gamma_1; \{\varphi_{i_1 i_2}\})$$

by Corollary 2.3.8, since the classical terms on both sides of the above inequality coincide.

Consider now the general case. Any decomposition $\varphi = \sum \varphi_{i_1\ldots i_n}$ can be considered as a decomposition $\varphi = \sum \varphi_{i'_1\ldots i'_{n-1}}$ with $I'_k = I_k$ for $k < n-1$ and $I'_{n-1} = I_{n-1} \times I_n$. In other words, any abelian model $(C,\mu,\{C_k\}_{k=1}^n, P)$ for $(A,\varphi,\{\gamma_k\}_{k=1}^n)$ defines an abelian model $(C,\mu,\{C'_k\}_{k=1}^{n-1}, P)$ for $(A,\varphi,\{\gamma_k\}_{k=1}^{n-1})$ with $C'_k = C_k$ for $k < n-1$ and $C'_{n-1} = C_{n-1} \vee C_n$. Comparing the entropies of these abelian models, we see that the classical terms as well as the first $n-2$ correction terms coincide. Thus Corollary 2.3.8 applied to $I = I_{n-1}$ and $J = I_n$ shows that the entropy of $(C,\mu,\{C'_k\}_{k=1}^{n-1}, P)$ is not smaller than the entropy of $(C,\mu,\{C_k\}_{k=1}^n, P)$. It follows that $H_\varphi(\gamma_1,\ldots,\gamma_{n-1}) \ge H_\varphi(\gamma_1,\ldots,\gamma_n)$. □

So far we have not given any estimates for H_φ.

Lemma 3.1.4. *For any channel $\gamma\colon B \to A$, we have $0 \le H_\varphi(\gamma) \le S(\varphi \circ \gamma)$. If, moreover, A is a von Neumann algebra, φ a faithful normal state and the image of γ does not consist only of scalars, then $H_\varphi(\gamma) > 0$.*

Together with Proposition 3.1.3(iii) this implies that

$$0 \le H_\varphi(\gamma_1,\ldots,\gamma_n) \le \sum_k H_\varphi(\gamma_k) \le \sum_k S(\varphi \circ \gamma_k),$$

and if $\mathrm{Im}\,\gamma_k \subset B \subset A$ for $1 \le k \le n$, then by Proposition 3.1.3(i),(iv)

$$H_\varphi(\gamma_1,\ldots,\gamma_n) \le H_\varphi(\underbrace{B,\ldots,B}_{n}) = H_\varphi(B) \le S(\varphi|_B).$$

Proof of Lemma 3.1.4. The entropy $H_\varphi(\gamma)$ is the supremum of the quantities (see (3.2))

$$\sum_i \varphi_i(1) S(\hat\varphi_i \circ \gamma, \varphi \circ \gamma) = S(\varphi \circ \gamma) - \sum_i \varphi_i(1) S(\hat\varphi_i \circ \gamma).$$

The left hand side of the above equality shows that $H_\varphi(\gamma) \ge 0$, while the right hand side implies $H_\varphi(\gamma) \le S(\varphi \circ \gamma)$.

If A is a von Neumann algebra and φ is normal and faithful, we may assume that $A \subset B(H)$ and φ is given by a cyclic and separating vector ξ. Then any partition of unity $1 = \sum_i x_i$ in the commutant A' of A defines a decomposition $\varphi = \sum_i \varphi_i$ with $\varphi_i = (\cdot\, x_i \xi, \xi)$. If $H_\varphi(\gamma; \{\varphi_i\}) = 0$, then $S(\hat\varphi_i \circ \gamma, \varphi \circ \gamma) = 0$ for all i with $\varphi_i(1) \ne 0$. By Theorem 2.1.2(i) it follows that $\varphi \circ \gamma = \hat\varphi_i \circ \gamma$. In other words, if $H_\varphi(\gamma) = 0$ and $a \in \mathrm{Im}\,\gamma$, then

$$\varphi(a) = \frac{(ax\xi, \xi)}{(x\xi, \xi)}$$

for any $x \in A'$, $0 \le x \le 1$, $x \ne 0$. Thus $((a - \varphi(a)1)x\xi, \xi) = 0$. Since ξ is cyclic for A', this shows that $a = \varphi(a)1$. □

As opposed to the commutative case, $H_\varphi(B)$ for $B \subset A$ cannot be determined just by the restriction of φ to B. For example, if the state φ on A is pure, then $H_\varphi(B) = 0$, since there are no nontrivial decompositions of φ into a convex combination of states. On the other hand, as we shall see in a moment, $H_{\varphi|B}(B) = S(\varphi|_B)$, which is positive unless $\varphi|_B$ is pure. We have however the following simple but useful result.

Lemma 3.1.5. *If the images of the channels $\gamma_k \colon A_k \to A$, $1 \le k \le n$, are contained in a C^*-subalgebra $B \subset A$, and there exists a φ-preserving conditional expectation $E \colon A \to B$, then*

$$H_{\varphi|B}(\gamma_1, \ldots, \gamma_n) = H_\varphi(\gamma_1, \ldots, \gamma_n).$$

Proof. Clearly $H_\varphi(\gamma_1, \ldots, \gamma_n; \{\varphi_{i_1 \ldots i_n}\}) = H_{\varphi|B}(\gamma_1, \ldots, \gamma_n; \{\varphi_{i_1 \ldots i_n}|_B\})$. Since any decomposition of $\varphi|_B$ can be extended to a decomposition of φ by composing positive functionals on B with E, we get the result. □

In general it is rather difficult to compute H_φ, or at least to find a way of producing optimal decompositions. The following proposition gives one of the few examples when an explicit computation is possible.

Proposition 3.1.6. *Let $A_1, \ldots, A_n \subset B \subset A$. Suppose there exist mutually commuting abelian subalgebras $C_k \subset A_k$, $1 \le k \le n$, such that $\vee_k C_k$ is maximal abelian in the centralizer of $\varphi|_B$. Suppose also that there exists a φ-preserving conditional expectation $E \colon A \to B$. Then*

$$H_\varphi(A_1, \ldots, A_n) = H_\varphi(B) = S(\varphi|_B).$$

Proof. Let $F \colon B \to C = \vee_k C_k$ be a φ-preserving conditional expectation. Then $(C, \varphi|_C, \{C_k\}_k, F \circ E)$ is an abelian model for $(A, \varphi, \{C_k\}_k)$. Since $F \circ E$ is the identity map on $C_k \subset C$, the entropy of this model is the same as the entropy of the model $(C, \varphi|_C, \{C_k\}_k, \mathrm{id}_C)$ for $(C, \varphi|_C, \{C_k\}_k)$. The latter entropy is $S(\varphi|_C)$. Thus

$$H_\varphi(C_1, \ldots, C_n) \ge S(\varphi|_C) = S(\varphi|_B),$$

where the equality follows from Theorem 2.2.2(vii). On the other hand, we have

$$H_\varphi(C_1, \ldots, C_n) \le H_\varphi(A_1, \ldots, A_n) \le H_\varphi(B) \le S(\varphi|_B).$$

Hence the above inequalities are in fact equalities. □

In particular, in the abelian case we recover the classical quantities $H(\vee_k \xi_k)$, which also follows from Proposition 1.3.2.

Further properties of mutual entropy are given in the following proposition.

Proposition 3.1.7. *We have:*

(i) *if $\pi_\varphi \colon A \to B(H_\varphi)$ is the GNS-representation and $\bar\varphi$ is the normal extension of φ to $\pi_\varphi(A)''$, then*
$$H_\varphi(\gamma_1, \ldots, \gamma_n) = H_{\bar\varphi}(\pi_\varphi \circ \gamma_1, \ldots, \pi_\varphi \circ \gamma_n);$$

(ii) *if ψ is another state on A and $0 \le \lambda \le 1$, then*
$$H_{\lambda\varphi+(1-\lambda)\psi}(\gamma_1, \ldots, \gamma_n)$$
$$\ge \lambda H_\varphi(\gamma_1, \ldots, \gamma_n) + (1-\lambda)H_\psi(\gamma_1, \ldots, \gamma_n) - (n-1)(\eta(\lambda) + \eta(1-\lambda));$$

(iii) *if ψ is a state on a C^*-algebra B, $0 \le \lambda \le 1$ and $\theta_k \colon B_k \to B$, $1 \le k \le n$, are channels, then on $A \oplus B$*
$$H_{\lambda\varphi \oplus (1-\lambda)\psi}(\gamma_1 \oplus \theta_1, \ldots, \gamma_n \oplus \theta_n)$$
$$= \lambda H_\varphi(\gamma_1, \ldots, \gamma_n) + (1-\lambda)H_\psi(\theta_1, \ldots, \theta_n) + \eta(\lambda) + \eta(1-\lambda);$$

(iv) *under the assumptions of part (iii),*
$$H_{\varphi \otimes \psi}(\gamma_1 \otimes \theta_1, \ldots, \gamma_n \otimes \theta_n) \ge H_\varphi(\gamma_1, \ldots, \gamma_n) + H_\psi(\theta_1, \ldots, \theta_n)$$
on $A \otimes B$, and equality holds if B is abelian and the θ_k's are injective homomorphisms;

(v) *for channels $\gamma_k \colon A_k \otimes B_k \to A$, $1 \le k \le n$,*
$$H_\varphi(\gamma_1, \ldots, \gamma_n) \le H_\varphi(\gamma_1|_{A_1}, \ldots, \gamma_n|_{A_n}) + 2\sum_k S(\varphi \circ \gamma_k|_{B_k}).$$

As we shall see in Chaps. 11 and 12, the inequality in (iv) can be strict.

Proof of Proposition 3.1.7. Part (i) follows from the fact that for any $\psi \le \varphi$ there exists a unique normal positive linear functional $\bar\psi$ on $\pi_\varphi(A)''$ such that $\psi = \bar\psi \circ \pi_\varphi$, so that any decomposition of φ comes from a decomposition of $\bar\varphi$.

To prove (ii), given abelian models $(C, \mu, \{C_k\}_k, P)$ for $(A, \varphi, \{\gamma_k\}_k)$ and $(D, \nu, \{D_k\}_k, Q)$ for $(A, \psi, \{\gamma_k\}_k)$, we can form the abelian model
$$(C \oplus D, \lambda\mu \oplus (1-\lambda)\nu, \{C_k \oplus D_k\}_k, P \oplus Q)$$
for $(A, \lambda\varphi + (1-\lambda)\psi, \{\gamma_k\}_k)$. In the simplest case $n = 1$ this means that starting with decompositions $\varphi = \sum_i \varphi_i$ and $\psi = \sum_j \psi_j$ we consider the decomposition
$$\lambda\varphi + (1-\lambda)\psi = \sum_i \lambda\varphi_i + \sum_j (1-\lambda)\psi_j.$$

The classical entropy term is then
$$\sum_i \eta(\lambda\varphi_i(1)) + \sum_j \eta((1-\lambda)\psi_j(1))$$

3.1 Mutual Entropy

$$= \lambda \sum_i \eta(\varphi_i(1)) + (1-\lambda) \sum_j \eta(\psi_j(1)) + \eta(\lambda) + \eta(1-\lambda).$$

On the other hand, the correction term is by the first equality in (3.5)

$$S(\lambda \varphi \circ \gamma + (1-\lambda)\psi \circ \gamma) - \sum_i S(\lambda \varphi_i \circ \gamma) - \sum_j S((1-\lambda)\psi_j \circ \gamma)$$

$$= S(\lambda \varphi \circ \gamma + (1-\lambda)\psi \circ \gamma) - \sum_i \lambda S(\varphi_i \circ \gamma) - \sum_j (1-\lambda) S(\psi_j \circ \gamma) - \eta(\lambda) - \eta(1-\lambda).$$

Recalling that by Theorem 2.2.2(ii) von Neumann entropy is concave, we see that by adding the above expressions for the classical entropy term and the correction term, part (ii) is true for $n = 1$. For an arbitrary n we similarly get one summand $\eta(\lambda) + \eta(1-\lambda)$ from the classical term and n summands $-\eta(\lambda) - \eta(1-\lambda)$ from the correction terms. Thus (ii) is proved.

To prove (iii) we repeat the computation in (ii), but now for the channels $\gamma_k \oplus \theta_k : A_k \oplus B_k \to A \oplus B$, $1 \le k \le n$, and the states $\varphi \oplus 0$ and $0 \oplus \psi$. This time, for $n = 1$, the correction term is

$$S(\lambda \varphi \circ \gamma \oplus (1-\lambda)\psi \circ \theta) - \sum_i S(\lambda \varphi_i \circ \gamma) - \sum_j S((1-\lambda)\psi_j \circ \theta)$$

$$= \lambda S(\varphi \circ \gamma) + (1-\lambda) S(\psi \circ \theta) - \sum_i \lambda S(\varphi_i \circ \gamma) - \sum_j (1-\lambda) S(\psi_j \circ \theta).$$

Thus we do not get n summands $-\eta(\lambda) - \eta(1-\lambda)$, which proves the inequality $\ge$ in (iii). To prove the opposite inequality, starting with a decomposition $\lambda \varphi \oplus (1-\lambda)\psi = \sum \omega_{i_1 \ldots i_n}$, we get decompositions $\varphi = \sum \varphi_{i_1 \ldots i_n}$ and $\psi = \sum \psi_{i_1 \ldots i_n}$ such that $\omega_{i_1 \ldots i_n} = \lambda \varphi_{i_1 \ldots i_n} \oplus (1-\lambda) \psi_{i_1 \ldots i_n}$. The correction terms for the first decomposition are just convex combinations of the correction terms for the other two decompositions. To estimate the classical term we use the inequality

$$S(\lambda \mu + (1-\lambda)\nu) \le \lambda S(\mu) + (1-\lambda) S(\nu) + \eta(\lambda) + \eta(1-\lambda)$$

from Theorem 2.2.2(ii). This proves the inequality $\le$ in (iii).

Turning to (iv), note that any decompositions $\varphi = \sum \varphi_{i_1 \ldots i_n}$ and $\psi = \sum \psi_{j_1 \ldots j_n}$ give rise to a decomposition

$$\varphi \otimes \psi = \sum_{(i_1, j_1), \ldots, (i_n, j_n)} \varphi_{i_1 \ldots i_n} \otimes \psi_{j_1 \ldots j_n}.$$

This proves the inequality in (iv). Assume now that B is abelian, $B_k \subset B$ and $\theta_k : B_k \to B$ is the inclusion map, $1 \le k \le n$. Since there exists a ψ-preserving conditional expectation $B \to \vee_k B_k$, by Lemma 3.1.5 we may without loss of generality assume that $B = \vee_k B_k$. Let $\{p_j\}_{j=1}^r$ be the atoms of B, $\{\chi_j\}_{j=1}^r$ the corresponding characters. The algebra $A \otimes B$ can be identified with $\oplus_{j=1}^r A$. Under this identification $\varphi \otimes \psi = \oplus_j \psi(p_j) \varphi$, and

3 Dynamical Entropy

$$\gamma_k(a) \otimes \theta_k(b) = (\chi_1(b)\gamma_k(a), \ldots, \chi_r(b)\gamma_k(a)).$$

Thus the channel $\gamma_k \otimes \theta_k$ factorizes through the channel

$$\oplus_j \gamma_k \colon \oplus_j A_k \to \oplus_j A.$$

By part (iii) (or rather by its obvious generalization to direct sums of r algebras) and Proposition 3.1.3(i), we get

$$
\begin{aligned}
H_{\varphi \otimes \psi}(\gamma_1 \otimes \theta_1, \ldots, \gamma_n \otimes \theta_n) &\leq H_{\oplus_j \psi(p_j) \varphi}(\oplus_j \gamma_1, \ldots, \oplus_j \gamma_n) \\
&= \sum_j \psi(p_j) H_\varphi(\gamma_1, \ldots, \gamma_n) + \sum_j \eta(\psi(p_j)) \\
&= H_\varphi(\gamma_1, \ldots, \gamma_n) + S(\psi),
\end{aligned}
$$

which completes the proof of (iv), as $S(\psi) = H_\psi(B_1, \ldots, B_n)$ by Proposition 3.1.6.

It remains to prove (v). So assume we have channels $\gamma_k \colon A_k \otimes B_k \to A$, $1 \leq k \leq n$. Any abelian model for $(A, \varphi, \{\gamma_k\}_k)$ can be considered as an abelian model for $(A, \varphi, \{\gamma_k|_{A_k}\}_k)$. The classical terms for the entropies of these models are the same. So we have to compare the correction terms. In view of the second equality in (3.5) we have to prove that if $\omega = \sum_i \omega_i$ is a decomposition of a state on $A_k \otimes B_k$, then

$$S(\omega) - \sum_i \omega_i(1) S(\hat{\omega}_i) \leq S(\omega|_{A_k}) - \sum_i \omega_i(1) S(\hat{\omega}_i|_{A_k}) + 2S(\omega|_{B_k}).$$

It suffices to show that each of the quantities

$$S(\omega) - S(\omega|_{A_k}) \quad \text{and} \quad \sum_i \omega_i(1)(S(\hat{\omega}_i|_{A_k}) - S(\hat{\omega}_i))$$

does not exceed $S(\omega|_{B_k})$. By Theorem 2.2.2(vi) we have $|S(\omega) - S(\omega|_{A_k})| \leq S(\omega|_{B_k})$ and $|S(\hat{\omega}_i) - S(\hat{\omega}_i|_{A_k})| \leq S(\hat{\omega}_i|_{B_k})$. It remains to note that

$$\sum_i \omega_i(1) S(\hat{\omega}_i|_{B_k}) \leq S(\omega|_{B_k})$$

by concavity of von Neumann entropy, Theorem 2.2.2(ii). □

Our next goal is to establish continuity of the entropy $H_\varphi(\gamma_1, \ldots, \gamma_n)$ as a function of $\gamma_1, \ldots, \gamma_n$. For this we need to show that there exist models of a fixed size. The idea is quite simple. Given a decomposition $\varphi = \sum \varphi_{i_1 \ldots i_n}$, we define a coarser decomposition by collecting indices $i, k \in I_j$ such that the states $\hat{\varphi}_i^{(j)}$ and $\hat{\varphi}_k^{(j)}$ are close to each other. We shall formulate this procedure in a more refined form, which will be useful later.

We shall first describe one more way of encoding decompositions of states. If we have a finite decomposition $\varphi = \sum \varphi_{i_1 \ldots i_n}$, we can define a state λ on $A \otimes C(X)$, where $X = I_1 \times \ldots \times I_n$, by

$$\lambda(a \otimes \mathbb{1}_{\{(i_1,\ldots,i_n)\}}) = \varphi_{i_1\ldots i_n}(a),$$

where $\mathbb{1}_Z$ denotes the characteristic function of the set Z. If ξ_k is the partition of X with atoms $I_1 \times \ldots \times \{i_k\} \times \ldots \times I_n$, then the original decomposition can be written as

$$\varphi = \sum_{X_1 \in \xi_1, \ldots, X_n \in \xi_n} \lambda(\cdot \otimes \mathbb{1}_{X_1 \cap \ldots \cap X_n}). \tag{3.8}$$

More generally, let (X, μ) be a probability space.

Definition 3.1.8. *A coupling of (A, φ) with (X, μ) is a state λ on the tensor product $A \otimes L^\infty(X, \mu)$ such that $\lambda|_A = \varphi$ and $\lambda|_{L^\infty(X,\mu)} = \mu$.*

If $X = S(A)$ is the state space of A, and μ is a regular Borel measure on X, then a coupling is called canonical if

$$\lambda(a \otimes f) = \int_X \psi(a) f(\psi) \, d\mu(\psi).$$

Thus a canonical coupling is determined by a probability measure μ on $S(A)$ with barycenter φ, that is, $\varphi = \int \psi \, d\mu(\psi)$.

Note that for any $a \in A$ the functional $\lambda(a \otimes \cdot)$ is normal on $L^\infty(X, \mu)$, since if a is positive then it is dominated by a scalar multiple of μ. It follows that if X is a compact space and μ is a regular Borel measure on X, then every coupling λ of (A, φ) with (X, μ) is determined by its restriction to $A \otimes C(X)$. Moreover, any state λ on $A \otimes C(X)$ such that $\lambda|_A = \varphi$ and $\lambda|_{C(X)} = \mu$ extends uniquely to a coupling of (A, φ) with (X, μ).

Now if $\xi_1, \ldots, \xi_n$ are finite measurable partitions of (X, μ), we can define a decomposition of φ by (3.8). For any channels $\gamma_1, \ldots, \gamma_n$ we denote their mutual entropy with respect to this decomposition by $H_\lambda(\gamma_1, \ldots, \gamma_n; \xi_1, \ldots, \xi_n)$. Thus by definition

$$H_\lambda(\gamma_1, \ldots, \gamma_n; \xi_1, \ldots, \xi_n) = H_\mu(\vee_k \xi_k) + \sum_k \sum_{Z \in \xi_k} S(\lambda(\gamma_k(\cdot) \otimes \mathbb{1}_Z), \varphi \circ \gamma_k). \tag{3.9}$$

By virtue of (3.2), the right hand side can be written as

$$H_\mu(\vee_k \xi_k) - \sum_k H_\mu(\xi_k) + \sum_k \sum_{Z \in \xi_k} \mu(Z) S(\mu(Z)^{-1} \lambda(\gamma_k(\cdot) \otimes \mathbb{1}_Z), \varphi \circ \gamma_k). \tag{3.10}$$

Putting $\varphi_Z = \mu(Z)^{-1} \lambda(\cdot \otimes \mathbb{1}_Z)$, we can also write this expression as

$$H_\mu(\vee_k \xi_k) - \sum_k H_\mu(\xi_k) + \sum_k \left(S(\varphi \circ \gamma_k) - \sum_{Z \in \xi_k} \mu(Z) S(\varphi_Z \circ \gamma_k) \right). \tag{3.11}$$

Note that (3.3) now becomes

$$H_\lambda(\gamma_1,\ldots,\gamma_n;\xi_1,\ldots,\xi_n) = H_\mu(\vee_k \xi_k) - \sum_k H_\mu(\xi_k) + \sum_k H_\lambda(\gamma_k;\xi_k). \quad (3.12)$$

Any decomposition $\varphi = \sum_i \varphi_i$ defines a canonical coupling in an obvious way: take the measure on $X = S(A)$ to be $\sum_i \varphi_i(1)\delta_{\hat{\varphi}_i}$, where $\delta_{\hat{\varphi}_i}$ is the δ-measure concentrated at the point $\hat{\varphi}_i$.

To put it differently, given a coupling λ of (A,φ) with (Y,μ), and a finite measurable partition ξ of Y, we can construct a map $f\colon Y \to X$ which sends an atom Z of ξ to the state $\varphi_Z = \mu(Z)^{-1}\lambda(\cdot \otimes 1_Z)$. Consider the image $\mu' = f_*(\mu)$ of the measure μ, and put $\lambda' = \lambda \circ (\mathrm{id}_A \otimes f^*)$, where $f^*\colon L^\infty(X,\mu') \to L^\infty(Y,\mu)$ is the map induced by f. Then λ' is the canonical coupling defined by μ'. Indeed, for any bounded Borel function g on X we have $f^*(g) = \sum_{Z \in \xi} g(\varphi_Z)1_Z$, whence

$$\lambda'(a \otimes g) = \sum_Z g(\varphi_Z)\lambda(a \otimes 1_Z).$$

On the other hand,

$$\int_X g(\psi)\psi(a)\,d\mu'(\psi) = \sum_Z g(\psi_Z)\psi_Z(a)\mu(Z) = \sum_Z g(\psi_Z)\lambda(a \otimes 1_Z).$$

Thus the coupling λ' is canonical. Note further that if ζ is a finite Borel partition of X, then the decompositions of φ defined by the pairs $(\lambda, f^{-1}(\zeta))$ and (λ', ζ) are the same, so that

$$H_\lambda(\gamma; f^{-1}(\zeta)) = H_{\lambda'}(\gamma; \zeta)$$

for any channel $\gamma\colon B \to A$. In particular, if ζ is such that $f^{-1}(\zeta) = \xi$, then $H_\lambda(\gamma;\xi) = H_{\lambda'}(\gamma;\zeta)$. But in fact any sufficiently fine ζ gives $H_{\lambda'}(\gamma;\zeta)$ close to $H_\lambda(\gamma;\xi)$.

Lemma 3.1.9. *Let notation be as above. For $\varepsilon > 0$ choose $\delta > 0$ such that $|S(\psi_1) - S(\psi_2)| < \varepsilon$ for states ψ_1 and ψ_2 on B as soon as $\|\psi_1 - \psi_2\| < \delta$. Assume ζ is a finite Borel partition of $S(A)$ such that $\|\varphi_1 \circ \gamma - \varphi_2 \circ \gamma\| < \delta$ as soon as φ_1 and φ_2 are in the same atom of ζ. Then $|H_\lambda(\gamma;\xi) - H_{\lambda'}(\gamma;\zeta)| < \varepsilon$.*

Proof. For $W \in \zeta$, set $\varphi'_W = \mu'(W)^{-1}\lambda'(\cdot \otimes 1_W)$. Then

$$\varphi'_W = \mu'(W)^{-1} \sum_{Z \in \xi: Z \subset f^{-1}(W)} \lambda(\cdot \otimes 1_Z) = \sum_{Z \in \xi: Z \subset f^{-1}(W)} \frac{\mu(Z)}{\mu(f^{-1}(W))}\varphi_Z.$$

By assumption, the states φ_Z with $Z \subset f^{-1}(W)$ are δ-close to each other when composed with γ. It follows that $\|(\varphi'_W - \varphi_Z) \circ \gamma\| < \delta$, and thus

$$|S(\varphi'_W \circ \gamma) - S(\varphi_Z \circ \gamma)| < \varepsilon$$

for $Z \in \xi$, $Z \subset f^{-1}(W)$. Since

$$S(\varphi \circ \gamma) - H_{\lambda'}(\gamma; \zeta) = \sum_W \mu(f^{-1}(W)) S(\varphi'_W \circ \gamma)$$

by (3.11), and similarly

$$S(\varphi \circ \gamma) - H_\lambda(\gamma; \xi) = \sum_Z \mu(Z) S(\varphi_Z \circ \gamma) = \sum_W \sum_{Z \subset f^{-1}(W)} \mu(Z) S(\varphi_Z \circ \gamma),$$

we see that $|H_\lambda(\gamma; \xi) - H_{\lambda'}(\gamma; \zeta)| < \varepsilon$. □

Note that a partition ζ satisfying the conditions of the previous lemma can be chosen such that the number $|\zeta|$ of elements of ζ is not bigger than the number of balls of diameter δ which are needed to cover the state space of B. Thus, what we have proved, is that if we want to compute $H_\varphi(\gamma)$ up to ε, it suffices to consider canonical couplings and one Borel partition of $S(A)$ of size depending only on dim B and ε.

More generally, if λ is a coupling of (A, φ) with (Y, μ), and $\xi_1, \ldots, \xi_n$ are finite measurable partitions of Y, we can define a measure μ' on X^n, where $X = S(A)$, and a coupling λ' of (A, φ) with (X^n, μ') as follows. Let $f_k: Y \to X$ be the map defined by λ and ξ_k, so f_k maps an atom Z of ξ_k to the state $\mu(Z)^{-1}\lambda(\cdot \otimes \mathbb{1}_Z) \in X$. Then the map $f = (f_1, \ldots, f_n): Y \to X^n$ is measurable, and we set $\mu' = f_*(\mu)$ and $\lambda' = \lambda \circ (\mathrm{id} \otimes f^*)$. The properties of the coupling λ' are summarized in the following proposition.

Proposition 3.1.10. *Under the above assumptions, fix $\varepsilon > 0$ and choose $\delta > 0$ such that $|S(\psi_1) - S(\psi_2)| < \varepsilon$ for states ψ_1 and ψ_2 on A_k as soon as $\|\psi_1 - \psi_2\| < \delta$, $1 \le k \le n$. For each k let ζ_k be a finite Borel partition of X such that $\|\varphi_1 \circ \gamma_k - \varphi_2 \circ \gamma_k\| < \delta$ as soon as φ_1 and φ_2 are in the same atom of ζ_k. Then*

(i) $f_k^{-1}(\zeta_k) \prec \xi_k$;

(ii) *the coupling $\lambda'_k = \lambda' \circ (\mathrm{id} \otimes \mathrm{pr}_k^*) = \lambda \circ (\mathrm{id} \otimes f_k^*)$ of (A, φ) with (X, μ'_k), where $\mathrm{pr}_k: X^n \to X$ is the projection onto the k-th factor and $\mu'_k = (\mathrm{pr}_k)_*(\mu') = (f_k)_*(\mu)$, is canonical and $|H_\lambda(\gamma_k; \xi_k) - H_{\lambda'_k}(\gamma_k; \zeta_k)| < \varepsilon$;*

(iii) $H_\lambda(\gamma_1, \ldots, \gamma_n; \xi_1, \ldots, \xi_n) < H_{\lambda'}(\gamma_1, \ldots, \gamma_n; \mathrm{pr}_1^{-1}(\zeta_1), \ldots, \mathrm{pr}_n^{-1}(\zeta_n)) + n\varepsilon$.

Proof. Part (i) is obvious as by definition f_k maps the atoms of ξ_k to points. Part (ii) follows from Lemma 3.1.9 and the discussion preceding it.

To prove (iii) note that since $H_{\lambda'}(\gamma_1, \ldots, \gamma_n; \mathrm{pr}_1^{-1}(\zeta_1), \ldots, \mathrm{pr}_n^{-1}(\zeta_n))$ is nothing but $H_\lambda(\gamma_1, \ldots, \gamma_n; f_1^{-1}(\zeta_1), \ldots, f_n^{-1}(\zeta_n))$, and $H_\lambda(\gamma_k; f_k^{-1}(\zeta_k))$ equals $H_{\lambda'_k}(\gamma_k; \zeta_k)$, in view of (ii) and equality (3.12), it suffices to show that

$$H_\mu(\vee_k \xi_k) - \sum_k H_\mu(\xi_k) \le H_\mu(\vee_k f_k^{-1}(\zeta_k)) - \sum_k H_\mu(f_k^{-1}(\zeta_k)).$$

Since $f_k^{-1}(\zeta_k) \prec \xi_k$, this is true by Proposition 1.1.9(iv). □

The previous proposition in its full generality will be used in subsequent chapters. What is important for us at the moment, is that it implies existence of models of a fixed size. Namely, given $\varepsilon > 0$ and $d \in \mathbb{N}$ there exists $r > 0$ such that for any $n \in \mathbb{N}$ and any channels $\gamma_k \colon A_k \to A$, $1 \leq k \leq n$, such that $\dim A_k \leq d$, there exist a coupling λ of (A, φ) with a probability space (Y, μ), and finite measurable partitions $\zeta_1, \ldots, \zeta_n$ such that $|\zeta_k| \leq r$, $1 \leq k \leq n$, and

$$H_\varphi(\gamma_1, \ldots, \gamma_n) < H_\lambda(\gamma_1, \ldots, \gamma_n; \zeta_1, \ldots, \zeta_n) + n\varepsilon.$$

Explicitly, fix $\varepsilon' < \varepsilon$. Then take r to be the number of balls of diameter δ which are needed to cover the state space of a C*-algebra B with $\dim B \leq d$, where δ is such that $|S(\psi_1) - S(\psi_2)| < \varepsilon'$ for states ψ_1 and ψ_2 on B as soon as $\|\psi_1 - \psi_2\| < \delta$. Take a coupling Λ and partitions $\xi_1, \ldots, \xi_n$ such that $H_\Lambda(\gamma_1, \ldots, \gamma_n; \xi_1, \ldots, \xi_n)$ is close to $H_\varphi(\gamma_1, \ldots, \gamma_n)$ up to $\varepsilon - \varepsilon'$. Then applying the procedure described above we get λ and $\zeta_1, \ldots, \zeta_n$.

For channels $\gamma, \gamma' \colon B \to A$, put

$$\|\gamma - \gamma'\|_\varphi = \sup_{x \in B, \|x\| \leq 1} \|\gamma(x) - \gamma'(x)\|_\varphi,$$

where $\|a\|_\varphi = \varphi(a^*a)^{1/2}$.

Proposition 3.1.11. *For every $\varepsilon > 0$ and $d \geq 1$ there exists $\delta > 0$ such that for any C*-algebra A with a state φ, $n \in \mathbb{N}$ and channels $\gamma_k, \gamma_k' \colon A_k \to A$ such that $\dim A_k \leq d$ and $\|\gamma_k - \gamma_k'\|_\varphi < \delta$, $1 \leq k \leq n$, we have*

$$|H_\varphi(\gamma_1, \ldots, \gamma_n) - H_\varphi(\gamma_1', \ldots, \gamma_n')| < n\varepsilon.$$

Proof. By Proposition 3.1.10 and the discussion following it we can find $r \in \mathbb{N}$ depending only on ε and d, and a decomposition $\varphi = \sum_{i_1, \ldots, i_n \in I} \varphi_{i_1 \ldots i_n}$ with $|I| \leq r$ such that

$$H_\varphi(\gamma_1, \ldots, \gamma_n; \{\varphi_{i_1 \ldots i_n}\}) > H_\varphi(\gamma_1, \ldots, \gamma_n) - \frac{n\varepsilon}{2}.$$

To estimate the difference between the mutual entropies of $\gamma_1, \ldots, \gamma_n$ and $\gamma_1', \ldots, \gamma_n'$ with respect to the decomposition $\varphi = \sum_{i_1, \ldots, i_n} \varphi_{i_1 \ldots i_n}$ we have to estimate the differences between the correction terms. Thus, what we need to prove is the following: there exists $\delta > 0$ such that if $\gamma, \gamma' \colon B \to A$ are channels, $\dim B \leq d$, $\|\gamma - \gamma'\|_\varphi < \delta$ and $\varphi = \sum_{i \in I} \varphi_i$ with $|I| \leq r$, then

$$\left| \sum_i S(\varphi_i \circ \gamma, \varphi \circ \gamma) - \sum_i S(\varphi_i \circ \gamma', \varphi \circ \gamma') \right| < \frac{\varepsilon}{2}. \qquad (3.13)$$

We have

$$\sum_i S(\varphi_i \circ \gamma, \varphi \circ \gamma) - \sum_i S(\varphi_i \circ \gamma', \varphi \circ \gamma')$$

$$= S(\varphi \circ \gamma) - S(\varphi \circ \gamma') - \sum_i \varphi_i(1)(S(\hat{\varphi}_i \circ \gamma) - S(\hat{\varphi}_i \circ \gamma')).$$

Choose $\delta_0 > 0$ such that $|S(\psi_1) - S(\psi_2)| < \varepsilon/6$ for states ψ_1 and ψ_2 on B as soon as $\|\psi_1 - \psi_2\| < \delta_0$ and $\dim B \le d$. If $\delta < \delta_0$, then as $|\varphi(x)| \le \|x\|_\varphi$, we have

$$\|\varphi \circ \gamma - \varphi \circ \gamma'\| \le \|\gamma - \gamma'\|_\varphi < \delta_0,$$

so that

$$|S(\varphi \circ \gamma) - S(\varphi \circ \gamma')| < \frac{\varepsilon}{6}.$$

Set $I_0 = \{i \in I \mid \varphi_i(1) > (\delta/\delta_0)^2\}$. Then for $i \in I_0$ we have

$$\|\gamma - \gamma'\|_{\hat{\varphi}_i} = \varphi_i(1)^{-1/2}\|\gamma - \gamma'\|_{\varphi_i} \le \varphi_i(1)^{-1/2}\|\gamma - \gamma'\|_\varphi < \delta_0,$$

so that $\|\hat{\varphi}_i \circ \gamma - \hat{\varphi}_i \circ \gamma'\| < \delta_0$, and therefore

$$\left| \sum_{i \in I_0} \varphi_i(1)(S(\hat{\varphi}_i \circ \gamma) - S(\hat{\varphi}_i \circ \gamma')) \right| < \frac{\varepsilon}{6}.$$

Finally,

$$\left| \sum_{i \in I \setminus I_0} \varphi_i(1)(S(\hat{\varphi}_i \circ \gamma) - S(\hat{\varphi}_i \circ \gamma')) \right| \le \sum_{i \in I \setminus I_0} \varphi_i(1) \log d \le \left(\frac{\delta}{\delta_0}\right)^2 r \log d.$$

Thus we see that (3.13) is true if $\delta < \delta_0$ and $(\delta/\delta_0)^2 r \log d < \varepsilon/6$. □

Another important consequence of existence of models of a fixed size is upper semicontinuity of mutual entropy as a function of the state.

Proposition 3.1.12. *Let A be a C^*-algebra, $\gamma_k \colon A_k \to A$, $1 \le k \le n$, channels into A. Then the function $\varphi \mapsto H_\varphi(\gamma_1, \ldots, \gamma_n)$ on the state space of A is weakly* upper semicontinuous.*

Proof. By Proposition 3.1.10, for any $\varepsilon > 0$ there exists a finite index set I depending only on ε and the dimensions of the A_k's such that for any state φ there is a decomposition $\varphi = \sum_{i_1,\ldots,i_n \in I} \varphi_{i_1 \ldots i_n}$ with

$$H_\varphi(\gamma_1, \ldots, \gamma_n; \{\varphi_{i_1 \ldots i_n}\}) > H_\varphi(\gamma_1, \ldots, \gamma_n) - n\varepsilon.$$

Now if $\{\varphi_j\}_{j \in J}$ is a net converging to φ we can choose decompositions of the φ_j's as above, and passing to a subnet may assume that these decompositions converge to a decomposition of φ. Since the map $(\omega, \psi) \mapsto S(\omega \circ \gamma_k, \psi \circ \gamma_k)$ is continuous on the set $\{\omega \le \psi\}$ by Theorem 2.1.2(iii), it follows that

$$H_\varphi(\gamma_1, \ldots, \gamma_n) \ge \limsup_j H_{\varphi_j}(\gamma_1, \ldots, \gamma_n) - n\varepsilon.$$

Since $\varepsilon > 0$ was arbitrary, we get the result. □

3.2 Entropy of Dynamical Systems

We are now ready to define entropy of dynamical systems.

By a **C*-dynamical system** we mean a triple (A, φ, α), where A is a unital C*-algebra, φ a state on A, α a φ-preserving automorphism of A. If A is a von Neumann algebra and the state is normal, we call (A, φ, α) a **W*-dynamical system**.

Definition 3.2.1. *The entropy of α with respect to a channel $\gamma \colon B \to A$ is*

$$h_\varphi(\gamma; \alpha) = \lim_{n \to \infty} \frac{1}{n} H_\varphi(\gamma, \alpha \circ \gamma, \ldots, \alpha^{n-1} \circ \gamma).$$

The dynamical entropy $h_\varphi(\alpha)$ is the supremum of $h_\varphi(\gamma; \alpha)$ for all possible γ's.

Note that if we denote $H_\varphi(\gamma, \alpha \circ \gamma, \ldots, \alpha^{n-1} \circ \gamma)$ by a_n, then $a_{n+m} \le a_n + a_m$ by properties (ii) and (iii) in Proposition 3.1.3. Hence the limit in the above definition exists by Lemma 1.1.2.

Note also that the definition makes sense for any unital completely positive φ-preserving map α.

Theorem 3.2.2. *Let (A, φ, α) be a C*-dynamical system. Then*

(i) *if $\beta \colon A \to B$ is an isomorphism, then $h_{\varphi \circ \beta^{-1}}(\beta \circ \alpha \circ \beta^{-1}) = h_\varphi(\alpha)$;*

(ii) *if $\pi_\varphi \colon A \to B(H)$ is the GNS-representation and $\bar\varphi$ and $\bar\alpha$ are the normal state and the automorphism of $\pi_\varphi(A)''$ defined by φ and α, respectively, then $h_{\bar\varphi}(\bar\alpha) = h_\varphi(\alpha)$;*

(iii) $h_\varphi(\alpha^{-1}) = h_\varphi(\alpha)$, *and* $h_\varphi(\alpha^n) \le |n| h_\varphi(\alpha)$ *for any* $n \in \mathbb{Z}$;

(iv) *for any C*-dynamical system (B, ψ, β),*

$$h_{\lambda \varphi \oplus (1-\lambda) \psi}(\alpha \oplus \beta) = \lambda h_\varphi(\alpha) + (1-\lambda) h_\psi(\beta) \quad \text{and}$$
$$h_{\varphi \otimes \psi}(\alpha \otimes \beta) \ge h_\varphi(\alpha) + h_\psi(\beta);$$

(v) *if B is an α-invariant subalgebra of A and there exists a φ-preserving conditional expectation $A \to B$, then $h_{\varphi|_B}(\alpha|_B) \le h_\varphi(\alpha)$.*

Proof. Part (i) follows immediately from the definitions.

Proposition 3.1.7(i) shows that $h_{\bar\varphi}(\pi_\varphi \circ \gamma; \bar\alpha) = h_\varphi(\gamma; \alpha)$ for any channel $\gamma \colon B \to A$. By Proposition 3.1.11 we also know that if two channels $\gamma, \gamma' \colon B \to \pi_\varphi(A)''$ are close in the $\|\cdot\|_{\bar\varphi}$-seminorm, then $h_{\bar\varphi}(\gamma; \bar\alpha)$ and $h_{\bar\varphi}(\gamma'; \bar\alpha)$ are close. Thus to prove (ii) we have to show that for any channel $\gamma \colon B \to \pi_\varphi(A)''$ there exists a channel $\tilde\gamma \colon B \to A$ such that $\|\gamma - \pi_\varphi \circ \tilde\gamma\|_{\bar\varphi}$ is arbitrarily small. This is possible by A.10 and A.11, and (ii) is proved.

By Proposition 3.1.3(ii) we have

$$H_\varphi(\gamma, \alpha \circ \gamma, \ldots, \alpha^n \circ \gamma) = H_\varphi(\alpha^{-n} \circ \gamma, \alpha^{-n+1} \circ \gamma, \ldots, \gamma).$$

It follows that $h_\varphi(\alpha^{-1}) = h_\varphi(\alpha)$. If $n \geq 1$, we have

$$H_\varphi(\gamma, \alpha^n \circ \gamma, \ldots, \alpha^{(m-1)n} \circ \gamma) \leq H_\varphi(\gamma, \alpha \circ \gamma, \ldots, \alpha^{mn-1} \circ \gamma),$$

whence $h_\varphi(\gamma; \alpha^n) \leq nh_\varphi(\gamma; \alpha)$. Thus (iii) is proved.

Part (iv) follows from Proposition 3.1.7(iii)-(iv), while (v) is a consequence of Lemma 3.1.5. □

To proceed further we need to restrict ourselves to systems satisfying the following approximation property. We say that a net of channels $\{\gamma_i \colon A_i \to A\}_i$ is φ-**approximating** if there exists a net $\{\theta_i \colon A \to A_i\}_i$ of unital completely positive maps such that $\|(\gamma_i \circ \theta_i)(x) - x\|_\varphi \to 0$ for any $x \in A$. If $\gamma \colon B \to A$ is an arbitrary channel, then $\|\gamma_i \circ \theta_i \circ \gamma - \gamma\|_\varphi \to 0$. Hence by Proposition 3.1.11

$$\lim_i h_\varphi(\gamma_i \circ \theta_i \circ \gamma; \alpha) = h_\varphi(\gamma; \alpha).$$

On the other hand, $h_\varphi(\gamma_i \circ \theta_i \circ \gamma; \alpha) \leq h_\varphi(\gamma_i; \alpha) \leq h_\varphi(\alpha)$ by Proposition 3.1.3(i). We thus get the following weak analogue of the Kolmogorov-Sinai theorem.

Theorem 3.2.3. *If $\{\gamma_i \colon A_i \to A\}_i$ is a φ-approximating net for a C^*-dynamical system (A, φ, α), then $h_\varphi(\alpha) = \lim_i h_\varphi(\gamma_i; \alpha)$.* □

Approximating nets exist for nuclear C^*-algebras and injective von Neumann algebras, and in fact existence of a φ-approximating net is close to $\pi_\varphi(A)''$ being injective. If $\{A_i\}_i$ is an increasing net of finite dimensional C^*-subalgebras of a C^*-algebra A with $\cup_i A_i$ norm-dense in A, then the net $\{A_i \hookrightarrow A\}_i$ is φ-approximating (take arbitrary conditional expectations $A \to A_i$ for θ_i's). If A is a von Neumann algebra, φ is normal and $\cup_i A_i$ is dense in A in the strong operator topology, then the net $\{A_i \hookrightarrow A\}_i$ is again φ-approximating, but this is not so simple, see e.g. [91]. Note that we do not really need this fact, since what is important for Theorem 3.2.3, is that any channel into A can be approximated by channels into $\cup_i A_i$, which is easy by A.10.

We also have the following related approximation result.

Proposition 3.2.4. *Let (A, φ, α) be a C^*-dynamical system, $\{A_i\}_{i \in I}$ an increasing net of α-invariant C^*-subalgebras of A such that $\cup_i \pi_\varphi(A_i)$ is strongly operator dense in $\pi_\varphi(A)$. Then $h_\varphi(\alpha) \leq \liminf_i h_{\varphi|A_i}(\alpha|A_i)$.*

Proof. By A.10 and A.11 any channel into A can be approximated in the $\|\cdot\|_\varphi$-seminorm by channels into $\cup_i A_i$. But if $\gamma \colon B \to A_i$ is a channel, then $h_\varphi(\gamma; \alpha) \leq h_{\varphi|A_i}(\gamma; \alpha|A_i)$ by Theorem 3.1.3(ii) applied to the inclusion map $\theta \colon A_i \to A$. □

We now show how approximation properties translate into properties of dynamical entropy.

Theorem 3.2.5. Let (A, φ, α) be a C^*-dynamical system having a φ-approximating net. Then

(i) $h_\varphi(\alpha^n) = |n| h_\varphi(\alpha)$ for any $n \in \mathbb{Z}$; moreover, if $\{\alpha_t\}_{t \in \mathbb{R}}$ is a φ-preserving one-parameter automorphism group of A such that $\alpha_1 = \alpha$, then $h_\varphi(\alpha_t) = |t| h_\varphi(\alpha)$ for any $t \in \mathbb{R}$;

(ii) if $\varphi = \lambda \omega + (1 - \lambda) \psi$, where ω and ψ are α-invariant states, then

$$h_\varphi(\alpha) \geq \lambda h_\omega(\alpha) + (1 - \lambda) h_\psi(\alpha);$$

(iii) if (B, ψ, β) is an abelian C^*-dynamical system, then

$$h_{\varphi \otimes \psi}(\alpha \otimes \beta) = h_\varphi(\alpha) + h_\psi(\beta);$$

(iv) if ψ is a state on a C^*-algebra B, and there exists a ψ-approximating net, then $h_{\varphi \otimes \psi}(\alpha \otimes \mathrm{id}_B) = h_\varphi(\alpha)$.

In Sect. 3.3 we shall extend (iii) to type I algebras.

Proof of Theorem 3.2.5. We already know from Theorem 3.2.2(iii) that $h_\varphi(\alpha^n) \leq |n| h_\varphi(\alpha)$. To prove the opposite inequality we may by the same theorem assume $n \geq 1$. Let $\{\gamma_i : A_i \to A\}_i$ be a φ-approximating net, and $\theta_i : A \to A_i$, $i \in I$, be maps as in the definition of such a net. Fix a channel $\gamma : B \to A$ and $\delta > 0$, and choose $i \in I$ such that

$$\|\gamma_i \circ \theta_i \circ \alpha^k \circ \gamma - \alpha^k \circ \gamma\|_\varphi < \delta \quad \text{for} \quad 0 \leq k \leq n - 1.$$

Then $\|\alpha^{ln} \circ \gamma_i \circ \theta_i \circ \alpha^k \circ \gamma - \alpha^{ln+k} \circ \gamma\|_\varphi < \delta$ for any $l \geq 0$ and $0 \leq k \leq n-1$. It follows by Propositions 3.1.11 and 3.1.3(i),(iv) that

$$H_\varphi(\gamma, \alpha \circ \gamma, \ldots, \alpha^{nm-1} \circ \gamma)$$
$$\leq H_\varphi(\{\alpha^{ln} \circ \gamma_i \circ \theta_i \circ \alpha^k \circ \gamma\}_{0 \leq l \leq m-1, 0 \leq k \leq n-1}) + nm\varepsilon$$
$$\leq H_\varphi(\gamma_i, \alpha^n \circ \gamma_i, \ldots, \alpha^{n(m-1)} \circ \gamma_i) + nm\varepsilon,$$

where $\varepsilon = \varepsilon(\delta, \dim B) \to 0$ as $\delta \to 0$. Thus $h_\varphi(\gamma; \alpha) \leq n^{-1} h_\varphi(\gamma_i; \alpha^n) + \varepsilon$, so that $h_\varphi(\alpha^n) \geq n h_\varphi(\alpha)$.

Assume now we have a flow $\{\alpha_t\}_t$ with $\alpha_1 = \alpha$. To prove that $t^{-1} h_\varphi(\alpha_t)$ is independent of $t > 0$, it suffices to show that $t_2 h_\varphi(\alpha_{t_1}) \leq t_1 h_\varphi(\alpha_{t_2})$ for any $t_1, t_2 > 0$. Replacing $\{\alpha_t\}_t$ by $\{\alpha_{t_2 t}\}_t$, we just have to show that $h_\varphi(\alpha_t) \leq t h_\varphi(\alpha_1)$ for any $t > 0$. The proof is similar to the argument above. Since the set $\{\alpha_s(a) \mid 0 \leq s \leq 1\}$ is compact for any $a \in A$, given a channel $\gamma : B \to A$ and $\delta > 0$ we can find $i \in I$ such that

$$\|\gamma_i \circ \theta_i \circ \alpha_s \circ \gamma - \alpha_s \circ \gamma\|_\varphi < \delta \quad \text{for} \quad 0 \leq s \leq 1.$$

Then $\|\alpha_1^{[lt]} \circ \gamma_i \circ \theta_i \circ \alpha_{lt-[lt]} \circ \gamma - \alpha_t^l \circ \gamma\|_\varphi < \delta$ for any $l \geq 0$, where $[lt]$ denotes the integer part of lt. Hence

$$H_\varphi(\gamma, \alpha_t \circ \gamma, \ldots, \alpha_t^{n-1} \circ \gamma) \le H_\varphi(\gamma_i, \alpha_1 \circ \gamma_i, \ldots, \alpha_1^{[(n-1)t]} \circ \gamma_i) + n\varepsilon,$$

so that in the limit $h_\varphi(\gamma; \alpha_t) \le th_\varphi(\gamma_i; \alpha_1) + \varepsilon$. Thus (i) is proved.

By Proposition 3.1.7(ii) we have

$$h_\varphi(\alpha) \ge \lambda h_\omega(\alpha) + (1-\lambda) h_\psi(\alpha) - \eta(\lambda) - \eta(1-\lambda).$$

We can apply this inequality to α^n. By virtue of (i), dividing by n and letting $n \to \infty$, we get (ii).

Let (B, ψ, β) be an abelian C*-dynamical system. By Proposition 3.2.2(ii), in proving (iii) we may assume that B is a von Neumann algebra. Then B is in particular an inductive limit of finite dimensional abelian C*-algebras, so there exists a ψ-approximating net $\{\gamma'_k\}_k$ consisting of injective homomorphisms. Then $\{\gamma_i \otimes \gamma'_k\}_{(i,k)}$ is a $(\varphi \otimes \psi)$-approximating net, and (iii) follows from Proposition 3.1.7(iv).

To prove (iv), note that $h_{\varphi \otimes \psi}(\alpha \otimes \mathrm{id}_B) \ge h_\varphi(\alpha)$ by Theorem 3.2.2(v). To prove the opposite inequality consider first the case when B is finite dimensional. If $\{\gamma_i\}_i$ is a φ-approximating net, then the net $\{\gamma_i \otimes \mathrm{id}_B\}_i$ is $(\varphi \otimes \psi)$-approximating. By Proposition 3.1.7(v) we have

$$h_{\varphi \otimes \psi}(\gamma_i \otimes \mathrm{id}_B; \alpha \otimes \mathrm{id}_B) \le h_\varphi(\gamma_i; \alpha) + 2S(\psi),$$

so that $h_{\varphi \otimes \psi}(\alpha \otimes \mathrm{id}_B) \le h_\varphi(\alpha) + 2S(\psi)$. As in the proof of (ii), applying this inequality to α^n we conclude that $h_{\varphi \otimes \psi}(\alpha \otimes \mathrm{id}_B) \le h_\varphi(\alpha)$.

Consider now the general case. Let $\{\gamma'_k: B_k \to B\}_k$ be a ψ-approximating net. Then $\{\gamma_i \otimes \gamma'_k\}_{(i,k)}$ is a $(\varphi \otimes \psi)$-approximating net. By Proposition 3.1.3(ii) and finite dimensionality of the B_k's we have

$$h_{\varphi \otimes \psi}(\gamma_i \otimes \gamma'_k; \alpha \otimes \mathrm{id}_B) \le h_{\varphi \otimes (\psi \circ \gamma'_k)}(\gamma_i \otimes \mathrm{id}_{B_k}; \alpha \otimes \mathrm{id}_{B_k}) \le h_\varphi(\alpha).$$

Hence $h_{\varphi \otimes \psi}(\alpha \otimes \mathrm{id}_B) \le h_\varphi(\alpha)$. □

Similarly to the classical case, the general theory we have developed so far was first applied to compute entropy of noncommutative Bernoulli shifts. We next discuss these systems.

Example 3.2.6.

(i) **Noncommutative Bernoulli shifts.** Let B be a finite dimensional C*-algebra, ψ a state on B, $(M, \varphi) = (B, \psi)^{\otimes \mathbb{Z}}$ the infinite W*-tensor product (that is, if $A = B^{\otimes \mathbb{Z}}$ is the infinite C*-tensor product, then M is the weak operator closure of A in the GNS-representation corresponding to the state $\psi^{\otimes \mathbb{Z}}$), and α the shift to the right on M. Then $h_\varphi(\alpha) = S(\psi)$.

To see this, let $\pi_n: B \to M$ be the canonical homomorphism onto the n-th factor, so that $\alpha(\pi_n(b)) = \pi_{n+1}(b)$. Let $B_{[n,m]}$ be the algebra generated by $\pi_k(B)$ for $n \le k \le m$. If D is a maximal abelian subalgebra in the centralizer of ψ, then the algebras $\pi_k(D)$, $n \le k \le m$, mutually commute and generate a maximal abelian subalgebra in the centralizer of the restriction of φ to $B_{[n,m]}$. By Proposition 3.1.6 it follows that for any $n, m \in \mathbb{N}$ we have

$$H_\varphi(B_{[-n,n]}, \ldots, \alpha^m(B_{[-n,n]})) = S(\varphi|_{B_{[-n,n+m]}}) = (m + 2n + 1)S(\psi).$$

Thus $h_\varphi(B_{[-n,n]}; \alpha) = S(\psi)$, and $h_\varphi(\alpha) = S(\psi)$ by Theorem 3.2.3.

We remark that the same result is true if B is infinite dimensional and we use the definition of entropy of a state mentioned in Notes to Chap. 2.

(ii) **Bernoulli shifts on the hyperfinite II$_1$-factor.** Consider the same system as above, but assume that $B \cong \mathrm{Mat}_n(\mathbb{C})$ and ψ is faithful. Let $N = M_\varphi$ be the centralizer of φ, so $\beta = \alpha|_N$ is an automorphism and $\tau = \varphi|_N$ is a faithful normal β-invariant trace on N. Then $h_\tau(\beta) = S(\psi)$.

To see this we can repeat the above argument with $B_{[n,m]}$ replaced by $N \cap B_{[n,m]}$. Another possibility is to argue as follows. Denote by C the von Neumann subalgebra of M generated by $\pi_n(D)$, $n \in \mathbb{Z}$. There exist φ-preserving conditional expectations $M \to N$ and $N \to C$. Hence $h_{\tau|C}(\beta|_C) \le h_\tau(\beta) \le h_\varphi(\alpha)$. But $(C, \tau|_C, \beta|_C)$ is the classical Bernoulli shift with entropy $S(\psi)$. Since also $h_\varphi(\alpha) = S(\psi)$, we see that $h_\tau(\beta) = S(\psi)$.

Observe next that N is the hyperfinite II$_1$-factor. The hyperfiniteness is clear. We shall give two proofs of the factoriality. The first one is shorter, but the second one can be applied in more general situations.

The group S_∞ of finite permutations of $\mathbb{Z}$ acts naturally on M. Denote this action by $\alpha: S_\infty \to \mathrm{Aut}(M)$. If $x, y \in \cup_n B_{[-n,n]}$, then $\varphi(\alpha_g(x)y) = \varphi(x)\varphi(y)$ for g outside a finite subset of S_∞. It follows that for any $x, y \in M$ we have $\varphi(\alpha_g(x)y) \to \varphi(x)\varphi(y)$ as g goes to infinity, meaning that g eventually leaves every finite subset of S_∞. In particular, the fixed point algebra M^{S_∞} is trivial. On the other hand, since any automorphism of a full matrix algebra is inner, for each $g \in S_\infty$ there exists $u_g \in \cup_n B_{[-n,n]}$ such that $\alpha_g = \mathrm{Ad}\, u_g$. Since φ is α_g-invariant, we have $u_g \in M_\varphi = N$. Hence if z is in the center of N, then $z \in M^{S_\infty}$. Therefore z is a scalar.

For the second proof note that M, being an infinite tensor product of factors with respect to a product-state, is a factor. Thus it suffices to check that the center of N is contained in the center of M. We shall prove that the relative commutant $N' \cap M$ is contained in the center of M. Let $z \in N' \cap M$, $x \in B_{[-m,m]}$, $x \ne 0$. We want to prove that $xz = zx$. Since $\sigma_t^\varphi|_{B_{[-m,m]}}$ has pure point spectrum, without loss of generality we may assume that $\sigma_t^\varphi(x) = \lambda^{it} x$ for $t \in \mathbb{R}$ and some $\lambda > 0$. Then $\sigma_t^\varphi(\alpha^n(x^*)x) = \alpha^n(x^*)x$, so $\alpha^n(x^*)x \in N$. Hence $\alpha^n(x^*)xz = z\alpha^n(x^*)x$, and thus

$$\alpha^n(xx^*)xz = \alpha^n(x)z\alpha^n(x^*)x.$$

We claim that $\alpha^n(xx^*) \to \varphi(xx^*)1$ in the weak operator topology and $[\alpha^n(x), z] \to 0$ in the strong operator topology as $n \to \infty$. Thus letting $n \to \infty$ in the equality above we get $\varphi(xx^*)xz$ on the left hand side and $\varphi(xx^*)zx$ on the right hand side, whence $xz = zx$.

It remains to prove the claims. If $y \in \cup_k B_{[-k,k]}$, then $\varphi(\alpha^n(xx^*)y) = \varphi(xx^*)\varphi(y)$ if n is sufficiently large. Thus any weak operator limit point a of the sequence $\{\alpha^n(xx^*)\}_n$ has the property $\varphi(ay) = \varphi(xx^*)\varphi(y)$. Hence $a = \varphi(xx^*)1$, and the first claim is proved. To prove the second claim, note that $\alpha^n(x)\xi_\varphi = J_\varphi \Delta_\varphi^{1/2} \alpha^n(x^*)\xi_\varphi = \lambda^{-1/2} J_\varphi \alpha^n(x^*)\xi_\varphi$, where we use the same notation as on page 35. Then the equality

$$[\alpha^n(x),y]\xi_\varphi = \alpha^n(x)y\xi_\varphi - y\alpha^n(x)\xi_\varphi = \alpha^n(x)y\xi_\varphi - \lambda^{-1/2} J_\varphi \alpha^n(x^*) J_\varphi y \xi_\varphi$$

shows that to prove that $[\alpha^n(x),y] \to 0$ strongly for any $y \in M$, it suffices to consider y lying in a strongly dense subspace. But if $y \in \cup_k B_{[-k,k]}$ then $[\alpha^n(x),y] = 0$ for all n sufficiently large. This completes our second proof of factoriality of N. ♦

It follows from the discussion prior to Lemma 3.1.5 that the entropy of a system can be smaller than the entropy of a subsystem. Bernoulli shifts provide a simple example. If in Example 3.2.6 we took an arbitrary maximal abelian subalgebra D of B, then we would get an abelian system $(C, \varphi|_C, \alpha|_C)$ with entropy $S(\psi|_D)$, which is strictly larger than $S(\psi)$ unless D is in the centralizer of ψ. Nevertheless if we have a conditional expectation onto a subalgebra commuting with an automorphism, then any state on the subalgebra extends to a state on the algebra such that the entropy of the system becomes at least as large as the entropy of the subsystem. The following result is a simple but useful extension of this fact.

Proposition 3.2.7. *Let α be an automorphism of a unital C^*-algebra A, $B \subset A$ an α-invariant unital C^*-subalgebra, ψ an α-invariant state on B. Assume there exists a ψ-approximating net. Then there exists an α-invariant state φ on A such that $\varphi|_B = \psi$ and $h_\varphi(\alpha) \geq h_\psi(\alpha|_B)$.*

Proof. Let $\{\gamma_i \colon B_i \to B\}_i$ be a ψ-approximating net, and $\theta_i \colon B \to B_i$, $i \in I$, be maps as in the definition of such a net. By Arveson's extension theorem, A.8, we can extend each θ_i to a unital completely positive map $\bar\theta_i \colon A \to B_i$. Consider the GNS-representation $\pi \colon B \to B(H)$ corresponding to ψ, and let $\tilde\psi$ and β be the normal state and the automorphism of $\pi(B)''$ defined by ψ and $\alpha|_B$, respectively. Let $\Phi \colon A \to \pi(B)''$ be any pointwise weak operator limit point of $\{\pi \circ \gamma_i \circ \bar\theta_i\}_i$. Then Φ is a unital completely positive map, and $\Phi(b)\xi_\psi = \pi(b)\xi_\psi$ for any $b \in B$. Replacing further Φ by a pointwise weak operator limit point of $\{n^{-1}\sum_{k=0}^{n-1} \beta^k \circ \Phi \circ \alpha^{-k}\}$, we may also assume that $\beta \circ \Phi = \Phi \circ \alpha$. Then $\varphi = \tilde\psi \circ \Phi$ is an α-invariant state extending ψ. Moreover, if ω is a positive linear functional on B such that $\omega \leq \psi$, then ω extends to a normal positive functional $\bar\omega$ on $\pi(B)''$, and $\bar\omega \circ \Phi$ extends ω. Hence any decomposition of ψ extends to a decomposition of φ, so that $h_\varphi(\alpha) \geq h_\psi(\alpha|_B)$. □

A related result will be proved in Chap. 5.

3.3 Type I Algebras

In this section we consider type I algebras. Not surprisingly the theory of dynamical entropy in this case reduces completely to abelian systems.

Recall that a type I von Neumann algebra is a direct sum of algebras of the form $A\bar\otimes B(H)$ with A abelian. If the algebra is σ-finite, that is, every set of mutually orthogonal projections is at most countable, then one can equivalently say that the algebra is a direct integral of algebras $B(H)$, see App. C.

Theorem 3.3.1. *Let (M,φ,α) be a W^*-dynamical system with M a type I von Neumann algebra. Let $Z = Z(M)$ be the center of M. Then*

$$h_\varphi(\alpha) = h_{\varphi|Z}(\alpha|_Z).$$

More generally, if (B,ψ,β) is a C^*-dynamical system having a ψ-approximating net, then $h_{\psi\otimes\varphi}(\beta\otimes\alpha) = h_\psi(\beta) + h_{\varphi|Z}(\alpha|_Z)$.

Proof. Since the modular group acts trivially on the center, there exists a φ-preserving conditional expectation $M \to Z$, and hence the inequality $\geq$ follows from Theorem 3.2.2(iv),(v). To prove the opposite inequality consider first the case when $M \cong Z \otimes \mathrm{Mat}_n(\mathbb{C})$. The proof in this case is similar to that of Theorem 3.2.5(iv). Let $\{\gamma_i\}_i$ be a ψ-approximating net and $\{\gamma'_k\}_k$ be a $\varphi|_Z$-approximating net. Then the net $\{\gamma_i \otimes \gamma'_k \otimes \mathrm{id}_{\mathrm{Mat}_n(\mathbb{C})}\}_{(i,k)}$ is $(\psi \otimes \varphi)$-approximating. By Proposition 3.1.7(v) we have

$$h_{\psi\otimes\varphi}(\gamma_i \otimes \gamma'_k \otimes \mathrm{id}_{\mathrm{Mat}_n(\mathbb{C})}; \beta \otimes \alpha) \leq h_{\psi\otimes(\varphi|Z)}(\gamma_i \otimes \gamma'_k; \beta \otimes \alpha|_Z) + 2\log n,$$

so that $h_{\psi\otimes\varphi}(\beta \otimes \alpha) \leq h_{\psi\otimes(\varphi|Z)}(\beta \otimes \alpha|_Z) + 2\log n$. Applying this inequality to β^m and α^m we conclude that

$$h_{\psi\otimes\varphi}(\beta \otimes \alpha) \leq h_{\psi\otimes(\varphi|Z)}(\beta \otimes \alpha|_Z) = h_\psi(\beta) + h_{\varphi|Z}(\alpha|_Z),$$

where the equality follows from Theorem 3.2.5(iii).

Consider next the case when M is finite, so M is a (possibly infinite) direct sum of algebras $Z_n \otimes \mathrm{Mat}_n(\mathbb{C})$ with Z_n abelian. The required inequality would follow from the previous case if we could use an infinite analogue of Theorem 3.2.2(iv). In other words, we have to check that if (N,φ,γ) is a W^*-dynamical system and z_n, $n \in \mathbb{N}$, are γ-invariant central projections in N such that $\sum_n z_n = 1$, then

$$h_\varphi(\gamma) = \sum_n \lambda_n h_{\varphi_n}(\gamma|_{Nz_n}),$$

where $\lambda_n = \varphi(z_n)$ and $\varphi_n = \lambda_n^{-1}\varphi|_{Nz_n}$. Set

$$M_n = N(z_1 + \ldots + z_n) + \mathbb{C}(1 - z_1 - \ldots - z_n).$$

3.3 Type I Algebras   55

Then $\cup_n M_n$ is strongly operator dense in N. Since there exist φ-preserving conditional expectations $N \to M_n$, by Theorem 3.2.2(v) and Proposition 3.2.4 we get $h_{\varphi|M_n}(\gamma|M_n) \to h_\varphi(\gamma)$. Since

$$h_{\varphi|M_n}(\gamma|M_n) = \sum_{k=1}^{n} \lambda_k h_{\varphi_k}(\gamma|N z_k)$$

by Theorem 3.2.2(iv), we get the result.

Consider now the general case. Replacing M by pMp, where p is the support of φ, we may assume that φ is faithful, in particular M is σ-finite. We claim that there exists an increasing sequence $\{p_n\}_{n=1}^\infty$ of finite α-invariant projections which belongs to the centralizer M_φ of φ and is such that $\|1 - p_n\|_\varphi \to 0$. Indeed, identify Z with $L^\infty(X, \mu)$, so that $\varphi|_Z$ is defined by μ. Then, see C.3 for the notation, we have a direct integral decomposition

$$M = \int_X^\oplus B(H_x) d\mu(x), \quad \varphi = \int_X^\oplus \varphi_x d\mu(x).$$

Let Q_x be the density operator of the state φ_x, so $\varphi_x = \text{Tr}(\cdot\, Q_x)$. Let $p_n(x)$ be the spectral projection of Q_x corresponding to the interval $[1/n, 1]$. Then $p_n(x)$ is a finite projection in $B(H_x)$ belonging to the centralizer of φ_x, and $\|1 - p_n(x)\|_{\varphi_x} \to 0$. Hence

$$p_n = \int_X^\oplus p_n(x) d\mu(x), \quad n \in \mathbb{N},$$

are the required projections.

Set $M_n = p_n M p_n + \mathbb{C}(1 - p_n)$. Then $a \mapsto p_n a p_n + \varphi(a(1 - p_n))(1 - p_n)$ is a φ-preserving conditional expectation onto M_n. Denote it by E_n. Since $\|1 - p_n\|_\varphi \to 0$, we have $\|a - E_n(a)\|_\varphi \to 0$ for any $a \in M$. It follows that

$$h_{\psi \otimes \varphi}(\beta \otimes \alpha) = \lim_n h_{\psi \otimes (\varphi|M_n)}(\beta \otimes \alpha|M_n).$$

The algebra M_n is finite and of type I. Denote by Z_n its center. By the previous case

$$h_{\psi \otimes (\varphi|M_n)}(\beta \otimes \alpha|M_n) = h_\psi(\beta) + h_{\varphi|Z_n}(\alpha|Z_n).$$

To finish the proof, it thus suffices to check that $h_{\varphi|Z_n}(\alpha|Z_n) \leq h_{\varphi|Z}(\alpha|Z)$ for any n.

If $p_n = 1$, there is nothing to prove. Otherwise let $z_n \in Z$ be the central support of p_n. Then $Z_n = Z p_n \oplus \mathbb{C}(1 - p_n)$ is isomorphic to $Z z_n \oplus \mathbb{C}$. Under this isomorphism the state $\varphi|_{Z_n}$ becomes $\varphi(p_n)\psi_n \oplus \varphi(1 - p_n)$, where $\psi_n = \varphi(p_n)^{-1}\varphi(\cdot\, p_n)|_{Z z_n}$. Hence $h_{\varphi|Z_n}(\alpha|Z_n) = \varphi(p_n) h_{\psi_n}(\alpha|Z z_n)$. On the other hand, consider the state $\varphi_n = \varphi(z_n)^{-1}\varphi|_{Z z_n}$. Since $Z z_n + \mathbb{C}(1 - z_n)$ is an α-invariant subalgebra of Z, we have $h_{\varphi|Z}(\alpha|Z) \geq \varphi(z_n) h_{\varphi_n}(\alpha|Z z_n)$. Hence we just have to prove that

$$h_{\psi_n}(\alpha|Z_{Z_n}) \le \frac{\varphi(z_n)}{\varphi(p_n)} h_{\varphi_n}(\alpha|Z_{Z_n}).$$

Since $\psi_n \le \varphi(p_n)^{-1}\varphi(z_n)\varphi_n$, and the function $\eta(t) = -t\log t$ is monotone for small t, this follows easily from the classical definition of entropy. □

Recall that a C*-algebra is of type I if its second dual is a type I von Neumann algebra. It follows from Theorem 3.3.1 that inner automorphisms of type I C*-algebras have zero entropy. It turns out this is a characteristic property of the type.

Theorem 3.3.2. *Let A be a unital C*-algebra. Then A is of type I if and only if for every unitary $u \in A$ and $(\operatorname{Ad} u)$-invariant state φ on A we have $h_\varphi(\operatorname{Ad} u) = 0$.*

Proof. One implication follows from Theorem 3.3.1. So assuming that A is not of type I we have to show that there exists an inner automorphism with nonzero entropy. By Glimm's theorem, see e.g. [156, Corollary 6.7.4], there exists a unital C*-subalgebra B of A and an ideal $I \subset B$ such that B/I is a UHF-algebra. Assume there exists a unitary $v \in B/I$ such that $h_\varphi(\operatorname{Ad} v) > 0$ for a state φ on B/I. Since the unitary group of an AF-algebra is connected, the unitary v lifts to a unitary u in B by [24, Proposition 3.4.5]. Let ψ be the state on B obtained by composing φ with the quotient map $B \to B/I$. Then $h_\psi(\operatorname{Ad} u|_B) = h_\varphi(\operatorname{Ad} v)$ by Theorem 3.2.2(ii). By Proposition 3.2.7 we can extend the state ψ to a state φ on A such that $h_\varphi(\operatorname{Ad} u) \ge h_\psi(\operatorname{Ad} u|_B) > 0$. Thus it remains to show that UHF-algebras have inner automorphisms with positive entropy. This will be done in the next lemma. □

Lemma 3.3.3. *Let R be the hyperfinite II_1-factor, $u \in R$ a unitary with nonatomic spectral measure, $A \subset R$ a strongly operator dense UHF-subalgebra. Then there exists a unitary $v \in R$ such that $vuv^* \in A$.*

In particular, there exist inner automorphisms of A having positive entropy with respect to the tracial state.

Proof. Let τ be the tracial state on R. We are going to use the fact that if p and q are projections in R such that $\|p - q\|_2 = \|p - q\|_\tau$ is small and $\tau(p) = \tau(q)$, then there exists a partial isometry w such that $w^*w = p$, $ww^* = q$ and $\|w - p\|_2$ is small, see e.g. [214, Lemma XIV.2.1]. Here is a sketch of the proof. Let $qp = w_1|qp|$ be the polar decomposition. Then $\|p - w_1\|_2$ is small. Let w_2 be any partial isometry such that $w_2^*w_2 = p - w_1^*w_1$ and $w_2w_2^* = q - w_1w_1^*$. Put $w = w_1 + w_2$.

Let $\mathcal{F} \subset [0,1]$ be the set of values of τ on projections in A. Since A is UHF, if $s, t \in \mathcal{F}$ and $s+t \le 1$, then $s+t \in \mathcal{F}$. Choose an increasing sequence $\{\xi_n\}_{n=1}^\infty$ of finite partitions of $\mathbb{T}$ into intervals such that if $Y \in \xi_n$ then the length of Y is at most 2^{-n}, and if p_Y is the spectral projection of u corresponding to Y then $\tau(p_Y) \in \mathcal{F}$.

Fix $\varepsilon > 0$. The projections p_Y, $Y \in \xi_1$, can be approximated arbitrarily close in the L^2-norm by mutually orthogonal projections q_Y, $Y \in \xi_1$, in A such that $\tau(p_Y) = \tau(q_Y)$. It follows that there exists a unitary $w_1 \in R$ such that $\|w_1 - 1\|_2 < \varepsilon/2$ and $q_Y = w_1^* p_Y w_1 \in A$ for $Y \in \xi_1$. If $Z \in \xi_2$ and $Z \subset Y$ for some $Y \in \xi_1$, we can approximate $w_1 p_Z w_1^*$ by projections in $q_Y A q_Y \subset A$ and repeat the above argument. By continuing this process we obtain a sequence of unitaries $w_n \in R$, $n \in \mathbb{N}$, such that $\|w_n - 1\|_2 < 2^{-n}\varepsilon$ and $v_n p_Y v_n^* \in A$ for $Y \in \xi_n$, where $v_n = w_n \ldots w_1$, and if $Z \subset Y$ for some $Z \in \xi_{n+1}$ and $Y \in \xi_n$ then $v_{n+1} p_Z v_{n+1}^* \le v_n p_Y v_n^*$. The sequence $\{v_n\}_{n=1}^{\infty}$ converges in the strong operator topology to a unitary $v \in R$. Moreover, $\|1 - v\|_2 < \varepsilon$.

We claim that $vuv^* \in A$. To show this choose a point $t_Y \in Y \subset \mathbb{T}$ for each $Y \in \xi_n$, $n \in \mathbb{N}$, and set $u_n = \sum_{Y \in \xi_n} t_Y p_Y$. Then $\|u - u_n\| \le 2^{-n}$. Since $|t_Z - t_Y| \le 2^{-m}$ and $v_n p_Z v_n^* \le v_m p_Y v_m^*$ for $Z \subset Y$ if $Z \in \xi_n$ and $Y \in \xi_m$, we also have
$$\|v_n u_n v_n^* - v_m u_m v_m^*\| \le 2^{-m}, \quad n \ge m,$$
whence $\|v_n u v_n^* - v_m u_m v_m^*\| \le 2^{-n} + 2^{-m}$, so that
$$\|vuv^* - v_m u_m v_m^*\| \le \liminf_{n \to \infty} \|v_n u_n v_n^* - v_m u_m v_m^*\| \le 2^{-m}.$$

Since $v_m u_m v_m^* \in A$, we see that $vuv^* \in A$.

Since $h_{\tau|A}(\operatorname{Ad} vuv^*|_A) = h_\tau(\operatorname{Ad} vuv^*) = h_\tau(\operatorname{Ad} u)$, to prove the second part of the lemma it suffices to show that there exists a unitary $u \in R$ with nonatomic spectral measure such that $h_\tau(\operatorname{Ad} u) > 0$. Take any ergodic dynamical system (X, μ, T) with positive entropy. Then $L^\infty(X, \mu) \rtimes \mathbb{Z}$ is the hyperfinite II_1-factor, see [214, Chapters XIII and XIV]. For the canonical unitary u in the crossed product we have $h_\tau(\operatorname{Ad} u) \ge h_\tau(\operatorname{Ad} u|_{L^\infty(X,\mu)}) = h_\mu(T) > 0$ (the first inequality is actually equality, as will be shown in Chap. 8). This completes the proof of the lemma and thus also of Theorem 3.3.2. □

It is worth mentioning that there are shorter but less elementary ways of finishing the proof of Theorem 3.3.2. For example, one could use a result of Voiculescu [222] stating that the crossed product of $\operatorname{Mat}_2(\mathbb{C})^{\otimes \mathbb{Z}}$ by the shift automorphism embeds into the UHF-algebra $\otimes_{n \in \mathbb{N}} \operatorname{Mat}_n(\mathbb{C})$.

3.4 Notes

Dynamical entropy for automorphisms preserving a tracial state was introduced by Connes and Størmer [51]. Later Connes [49] emphasized the role of relative entropy and showed how the definition can be extended to arbitrary W*-dynamical system. The formula for $H_\varphi(B)$ was also independently obtained by Narnhofer and Thirring [129]. The definition of dynamical entropy was extended to C*-dynamical systems by Connes, Narnhofer and Thirring [50], who introduced the concept of an abelian model and gave a detailed treatment of the foundations of the theory. The material in

Sects. 3.1 and 3.2 is almost entirely based on their paper. A few properties of entropy observed later include the superadditivity inequality in Proposition 3.1.7(iv) [212], with equality in the case when one of the factors is abelian [141], the apparently new inequality in Proposition 3.1.7(v), the formula for the entropy of a flow [130], Theorem 3.2.5(iii) [141] and Theorem 3.2.5(iv) [79], and Proposition 3.2.7, which is essentially from [34]. The concept of a coupling was introduced by Sauvageot and Thouvenot [188]. It will play a central role in Chap. 5, where we shall give an equivalent definition of dynamical entropy. The notion of an approximating net is borrowed from the Ohya and Petz book [147]. It allows us to simultaneously discuss nuclear C*-algebras and injective von Neumann algebras.

Theorem 3.3.1 is due to the authors [141], and Theorem 3.3.2 to Brown [34].

As we already mentioned, the problem of finding optimal decompositions of a state is quite complicated, see [18], [19] for some results in the finite dimensional case.

Though mutual and dynamical entropies share most properties of their abelian counterparts, there are some properties which are still missing. E.g. if $\{\gamma_i\}_{i \in I}$ is a finite collection of channels, and $I = J \cup K$, it is to the best of our knowledge unknown whether the strong subadditivity inequality

$$H_\varphi(\{\gamma_i\}_{i \in I}) + H_\varphi(\{\gamma_i\}_{i \in J \cap K}) \leq H_\varphi(\{\gamma_i\}_{i \in J}) + H_\varphi(\{\gamma_i\}_{i \in K})$$

holds. It is also unknown whether the entropy $h_\varphi(\alpha)$ is an affine function of the state, though this is stated in [50].

There is one aspect of the theory which may be viewed as a drawback, namely, that it is in a sense too abelian, the entropy being defined by abelian models. We shall see later that the entropy tends to be the smaller the less abelian the system is.

Similarly to abelian systems the notion of entropy can be extended to actions of arbitrary discrete amenable groups. Namely, let $\alpha \colon G \to \mathrm{Aut}(A)$ be an action of a countable amenable group G on A by φ-preserving automorphisms. Then for any channel $\gamma \colon B \to A$ the function m defined on finite subsets of G by

$$m(F) = H_\varphi(\{\alpha_g \circ \gamma\}_{g \in F}) \quad \text{for} \quad F \neq \emptyset, \quad m(\emptyset) = 0,$$

is positive, nondecreasing, left-invariant and subadditive, i.e.,

$$m(F_1 \cup F_2) \leq m(F_1) + m(F_2).$$

By [120, Theorem 6.1] it follows that if $\{F_n\}_n$ is a sequence of finite sets such that

$$\frac{|F_n g \cap F_n|}{|F_n|} \to 1 \quad \text{as} \quad n \to \infty,$$

then the sequence $\{m(F_n)/|F_n|\}_n$ converges, and its limit does not depend on the choice of $\{F_n\}_n$. Denote this limit by $h_\varphi(\gamma; \alpha_G)$. Then we can define

$h_\varphi(\alpha_G)$ by taking supremum over all channels γ into A. Note that if we knew in addition that m is strongly subadditive, by [124, Proposition 3.1.9 and Remark 3.1.7] we would have

$$h_\varphi(\gamma; \alpha_G) = \inf_{F \subset G, F \neq \emptyset} \frac{H_\varphi(\{\alpha_g \circ \gamma\}_{g \in F})}{|F|}.$$

So far the entropy for actions of groups on noncommutative algebras has been considered for very few models, such as shifts on UHF-algebras [92], [102] and Bogoliubov actions [155], [21], [149], [78].

Although the Connes-Størmer-Narnhofer-Thirring definition of entropy is arguably the most successful extension of the classical theory, it is not the only one available. The first definition of entropy for noncommutative systems is due to Emch [59]. It is as follows. Let (M, φ, α) be a W^*-dynamical system with a faithful normal state φ. Then the entropy is defined (compare with (1.3)) as the supremum of the quantities

$$\lim_{n \to \infty} \sum_{i \in I} \varphi(\eta(E_n(p_i))),$$

where the p_i's are projections in the centralizer M_φ of φ such that $\sum_i p_i = 1$, and E_n is the φ-preserving conditional expectation onto the von Neumann algebra generated by $\alpha^{-k}(p_i)$, $i \in I$, $k = 1, \ldots, n$. Very little is known about this entropy. Note also that it depends only on the restriction of the automorphism to the centralizer of the state, which can be quite small. We shall see later examples of systems with trivial centralizer and infinite dynamical entropy.

A different approach was suggested by Lindblad [119] and developed by Alicki and Fannes [4]. Let (A, φ, α) be a C^*-dynamical system, and $A_0 \subset A$ an α-invariant unital $*$-subalgebra. By an operational partition of unity in A_0 one means a finite subset $X = \{x_k\}_{k=1}^n$ of A_0 such that $\sum_k x_k^* x_k = 1$. Such a partition defines a completely positive map $\theta_X \colon \mathrm{Mat}_n(\mathbb{C}) \to A$ by $\theta_X(e_{ij}) = x_i^* x_j$. Denote by $\mathrm{H}[\varphi, X]$ the entropy of the state $\varphi \circ \theta_X$. Note that the density matrix of $\varphi \circ \theta_X$ is $(\varphi(x_j^* x_i))_{i,j}$. If $Y = \{y_l\}_{l=1}^m$ is another operational partition of unity, we can define a new partition $X \circ Y$ consisting of the elements $x_k y_l$. Then put

$$\mathrm{h}[\varphi, \alpha, X] = \limsup_{n \to \infty} \frac{1}{n} \mathrm{H}[\varphi, \alpha^{n-1}(X) \circ \alpha^{n-2}(X) \circ \ldots \circ X].$$

Finally define the entropy $\mathrm{h}[\varphi, \alpha, A_0]$ of the C^*-dynamical system (A, φ, α) with respect to A_0 as the supremum of $\mathrm{h}[\varphi, \alpha, X]$ over all operational partitions of unity in A_0. Computability of $\mathrm{h}[\varphi, \alpha, A_0]$ depends very much on the choice of A_0, and in this respect most choices of A_0 give too large subalgebras. As a result one can hardly use this entropy as an invariant of a dynamical system. On the other hand, in most examples there is a natural choice of A_0.

E.g. for models of quantum statistical mechanics this is the algebra of local observables. We refer the reader to the Alicki and Fannes book [5] for interesting examples.

There is also a completely different class of entropies based on the idea of approximation of finite sets. We shall discuss this topic in Chap. 6.

4
Maximality of Entropy and Commutativity

Since our definition of entropy was based on abelian models, one may expect that the entropy is the larger the more amount of commutativity is present in the system. In this chapter we shall prove several results in this direction. We shall mainly consider tracial states, where proofs are simpler and available results are stronger.

4.1 Maximal Entropy of Subalgebras

If $N_1, \ldots, N_m$ are finite dimensional subalgebras of a C^*-algebra then in general the C^*-algebra $N = \vee_k N_k$ they generate is infinite dimensional. But if N is finite dimensional, a natural question is, under which conditions on the relative positions of the N_k's the entropy of N coincides with the mutual entropy of $N_1, \ldots, N_m$. It follows from Proposition 3.1.6 that if τ is a tracial state and $N_1, \ldots, N_m$ contain pairwise commuting abelian subalgebras $A_1, \ldots, A_n$, respectively, such that $\vee_k A_k$ is maximal abelian in N then $H_\tau(N_1, \ldots, N_m) = H_\tau(N)$. In the present section we show the converse to this result.

Theorem 4.1.1. *Let N be a finite dimensional C^*-algebra with a faithful tracial state τ, $N_1, \ldots, N_m \subset N$. Suppose*

$$H_\tau(N_1, \ldots, N_m) = H_\tau(N).$$

Then there exist mutually commuting abelian subalgebras $A_k \subset N_k$ such that $\vee_k A_k$ is maximal abelian in N.

The key point in the proof of this theorem is the following refinement of the subadditivity inequality from Corollary 2.3.9.

Proposition 4.1.2. *Let $1 = \sum_{i_1,\ldots,i_m} x_{i_1\ldots i_m}$ be a finite partition of unity in a C^*-algebra with a tracial state τ. Then*

$$\sum_k \sum_{i_k} \tau(\eta(x_{i_k}^{(k)})) - \sum_{i_1,\ldots,i_m} \tau(\eta(x_{i_1\ldots i_m})) \geq \frac{1}{2} \max_{k \neq l} \sum_{i_k, i_l} \|[(x_{i_k}^{(k)})^{1/2}, (x_{i_l}^{(l)})^{1/2}]\|_2^2,$$

where $\|x\|_2 = \tau(x^*x)^{1/2}$.

We shall postpone the proof of this inequality. Note that if the elements of the partition lie in a finite dimensional C*-algebra N, then it can be written as

$$\sum_k \left(\sum_{i_k} \eta(\tau(x_{i_k}^{(k)})) - H_\tau(N; \{x_{i_k}^{(k)}\}) \right)$$

$$- \left(\sum_{i_1,\ldots,i_m} \eta(\tau(x_{i_1\ldots i_m})) - H_\tau(N; \{x_{i_1\ldots i_m}\}) \right)$$

$$\geq \frac{1}{2} \max_{k \neq l} \sum_{i_k, i_l} \|[(x_{i_k}^{(k)})^{1/2}, (x_{i_l}^{(l)})^{1/2}]\|_2^2. \qquad (4.1)$$

Proof of Theorem 4.1.1. Let $X = S(N)$ be the state space of N, $\{\xi^{(n)}\}_n$ a sequence of finite Borel partitions of X with diameter tending to zero as $n \to \infty$. By Proposition 3.1.10 we can find a sequence $\{\lambda^{(n)}\}_n$ of states on $N \otimes C(X^m)$ such that $\lambda^{(n)}|_N = \tau$ and

(i) the coupling $\lambda_k^{(n)} = \lambda^{(n)} \circ (\mathrm{id} \otimes \mathrm{pr}_k^*)$ is canonical, where $\mathrm{pr}_k \colon X^m \to X$ is the projection onto the k-th factor, $1 \leq k \leq m$;

(ii) $H_{\lambda^{(n)}}(N_1, \ldots, N_m; \mathrm{pr}_1^{-1}(\xi^{(n)}), \ldots, \mathrm{pr}_m^{-1}(\xi^{(n)})) \to H_\tau(N)$ as $n \to \infty$.

Choosing a subsequence, we may assume that $\{\lambda^{(n)}\}_n$ converges to a state λ. Let $\mu^{(n)}$ (resp. μ) be the measure on X^m defined by $\lambda^{(n)}|_{C(X^m)}$ (resp. by $\lambda|_{C(X^m)}$), $\mu_k^{(n)} = (\mathrm{pr}_k)_*(\mu^{(n)})$, $\mu_k = (\mathrm{pr}_k)_*(\mu)$, and $\lambda_k^{(n)} = \lambda^{(n)} \circ (\mathrm{id} \otimes \mathrm{pr}_k^*)$.

By subadditivity of the function

$$\xi \mapsto H_{\mu^{(n)}}(\xi) - H_{\lambda^{(n)}}(N; \xi) = \sum_{Z \in \xi} S(\lambda^{(n)}(\cdot \otimes 1_Z), \tau),$$

see Corollary 2.3.8, applied to $\xi = \vee_k \mathrm{pr}_k^{-1}(\xi^{(n)})$, we have

$$H_{\mu^{(n)}}(\vee_k \mathrm{pr}_k^{-1}(\xi^{(n)})) - H_{\lambda^{(n)}}(N; \vee_k \mathrm{pr}_k^{-1}(\xi^{(n)}))$$

$$\leq \sum_k (H_{\mu_k^{(n)}}(\xi^{(n)}) - H_{\lambda_k^{(n)}}(N; \xi^{(n)})).$$

Using (3.12), monotonicity of relative entropy, Theorem 2.1.2(v), and the above inequality we then get

$$H_{\lambda^{(n)}}(N_1, \ldots, N_m; \mathrm{pr}_1^{-1}(\xi^{(n)}), \ldots, \mathrm{pr}_m^{-1}(\xi^{(n)}))$$

4.1 Maximal Entropy of Subalgebras

$$= H_{\mu^{(n)}}(\vee_k \mathrm{pr}_k^{-1}(\xi^{(n)})) - \sum_k H_{\mu_k^{(n)}}(\xi^{(n)}) + \sum_k H_{\lambda_k^{(n)}}(N_k; \xi^{(n)})$$

$$\leq H_{\mu^{(n)}}(\vee_k \mathrm{pr}_k^{-1}(\xi^{(n)})) - \sum_k H_{\mu_k^{(n)}}(\xi^{(n)}) + \sum_k H_{\lambda_k^{(n)}}(N; \xi^{(n)})$$

$$\leq H_{\lambda^{(n)}}(N; \vee_k \mathrm{pr}_k^{-1}(\xi^{(n)}))$$

$$\leq H_\tau(N).$$

It follows that the above inequalities become close to equalities as $n \to \infty$. In other words,

$$H_{\lambda_k^{(n)}}(N; \xi^{(n)}) - H_{\lambda_k^{(n)}}(N_k; \xi^{(n)}) \to 0, \quad 1 \leq k \leq m, \tag{4.2}$$

and

$$\sum_k (H_{\mu_k^{(n)}}(\xi^{(n)}) - H_{\lambda_k^{(n)}}(N; \xi^{(n)}))$$

$$- (H_{\mu^{(n)}}(\vee_k \mathrm{pr}_k^{-1}(\xi^{(n)})) - H_{\lambda^{(n)}}(N; \vee_k \mathrm{pr}_k^{-1}(\xi^{(n)}))) \to 0. \tag{4.3}$$

Let $a_\varphi \in N_+^{sa}$ be the density operator of a state φ on N with respect to τ, so that $\varphi = \tau(\cdot a_\varphi)$. Since the coupling $\lambda_k^{(n)}$ is canonical, the decomposition of τ defined by $\lambda_k^{(n)}$ and $\xi^{(n)}$ consists of the functionals $\int_Z \varphi \, d\mu_k^{(n)}(\varphi)$, $Z \in \xi^{(n)}$. In other words, the partition of unity in N defined by $\lambda_k^{(n)}$ and $\xi^{(n)}$ is

$$1 = \sum_{Z \in \xi^{(n)}} \int_Z a_\varphi \, d\mu_k^{(n)}(\varphi).$$

Hence, see (3.1), $H_{\lambda_k^{(n)}}(N_k; \xi^{(n)})$ is equal to

$$- \sum_{Z \in \xi^{(n)}} \mu_k^{(n)}(Z) \tau \left(\eta \left(E_{N_k} \left(\mu_k^{(n)}(Z)^{-1} \int_Z a_\varphi \, d\mu_k^{(n)}(\varphi) \right) \right) \right).$$

Since a_φ depends continuously on φ, and the diameters of the atoms of $\xi^{(n)}$ tend to zero, this expression becomes close to $-\int_X \tau(\eta(E_{N_k}(a_\varphi))) d\mu_k^{(n)}(\varphi)$ as $n \to \infty$, so that

$$H_{\lambda_k^{(n)}}(N_k; \xi^{(n)}) \to -\int_X \tau(\eta(E_{N_k}(a_\varphi))) d\mu_k(\varphi). \tag{4.4}$$

Similarly,

$$H_{\lambda_k^{(n)}}(N; \xi^{(n)}) \to -\int_X \tau(\eta(a_\varphi)) d\mu_k(\varphi).$$

Thus (4.2) implies that

$$\int_X \tau(\eta(E_{N_k}(a_\varphi))) d\mu_k(\varphi) = \int_X \tau(\eta(a_\varphi)) d\mu_k(\varphi).$$

Since $\eta(E_{N_k}(a_\varphi)) \geq E_{N_k}(\eta(a_\varphi))$ by strict operator concavity of η, see B.2 and B.4, we conclude that

$$\eta(E_{N_k}(a_\varphi)) = E_{N_k}(\eta(a_\varphi))$$

for $\varphi \in \operatorname{supp} \mu_k$. By B.5 this implies that $a_\varphi \in N_k$ for $\varphi \in \operatorname{supp} \mu_k$. Denote by B_k the C*-subalgebra of N_k generated by a_φ for $\varphi \in \operatorname{supp} \mu_k$.

Then (4.3) and inequality (4.1) applied to $\{x_{i_k}^{(k)}\} = \{\int_Z a_\varphi \, d\mu_k^{(n)}(\varphi)\}$ imply that for $k \neq l$

$$\sum_{Z,W \in \xi^{(n)}} \left\| \left[\left(\int_Z a_\varphi \, d\mu_k^{(n)}(\varphi) \right)^{1/2}, \left(\int_W a_\psi \, d\mu_l^{(n)}(\psi) \right)^{1/2} \right] \right\|_2^2 \to 0.$$

Since $\mu_k^{(n)}(Z)^{-1} \int_Z a_\varphi \, d\mu_k^{(n)}(\varphi)$ becomes arbitrarily close to a_ψ for $\psi \in Z$, as $n \to \infty$ the above expression gets close to

$$\int_{X \times X} \|[a_\varphi^{1/2}, a_\psi^{1/2}]\|_2^2 \, d\mu_k^{(n)}(\varphi) d\mu_l^{(n)}(\psi),$$

so that in the limit we obtain

$$\int_{X \times X} \|[a_\varphi^{1/2}, a_\psi^{1/2}]\|_2^2 \, d\mu_k(\varphi) d\mu_l(\psi) = 0.$$

Thus $a_\varphi a_\psi = a_\psi a_\varphi$ for $\varphi \in \operatorname{supp} \mu_k$ and $\psi \in \operatorname{supp} \mu_l$. Therefore the algebras $B_1, \ldots, B_m$ mutually commute.

Analogously to (4.4) we have

$$H_{\lambda_k^{(n)}}(B_k; \xi^{(n)}) \to -\int_X \tau(\eta(E_{B_k}(a_\varphi))) d\mu_k(\varphi).$$

But $a_\varphi = E_{N_k}(a_\varphi) = E_{B_k}(a_\varphi)$ for $\varphi \in \operatorname{supp} \mu_k$, so that

$$H_{\lambda_k^{(n)}}(N_k; \xi^{(n)}) - H_{\lambda_k^{(n)}}(B_k; \xi^{(n)}) \to 0.$$

It follows that $H_{\lambda^{(n)}}(B_1, \ldots, B_m; \operatorname{pr}_1^{-1}(\xi^{(n)}), \ldots, \operatorname{pr}_m^{-1}(\xi^{(n)}))$ becomes close to $H_{\lambda^{(n)}}(N_1, \ldots, N_m; \operatorname{pr}_1^{-1}(\xi^{(n)}), \ldots, \operatorname{pr}_m^{-1}(\xi^{(n)}))$ as $n \to \infty$. Since the latter expression converges to $H_\tau(N)$, it follows that

$$H_\tau(B_1, \ldots, B_m) = H_\tau(N).$$

Let A_k be a maximal abelian subalgebra of B_k. Since $B_1, \ldots, B_m$ mutually commute, by Proposition 3.1.6

$$H_\tau(B_1, \ldots, B_m) = S(\tau|_{\vee_k A_k}),$$

hence $H_\tau(N) = S(\tau|_{\vee_k A_k})$. Let A be a maximal abelian subalgebra of N containing $\vee_k A_k$. Then $S(\tau) = S(\tau|_A)$ by Theorem 2.2.2(vii). Thus $S(\tau|_{\vee_k A_k}) =$

$S(\tau|_A)$. Since the atoms of $\vee_k A_k$ are sums of the atoms of A, and $\eta(x+y) < \eta(x)+\eta(y)$ for any numbers $x,y > 0$, this is possible only when $\vee_k A_k = A$. □

In order to prove Proposition 4.1.2 we shall first strengthen the **Golden-Thompson inequality** $\mathrm{Tr}(e^a e^b) \geq \mathrm{Tr}(e^{a+b})$.

Lemma 4.1.3. *For self-adjoint elements a and b in a C^*-algebra with a tracial state τ we have*
$$\tau(e^a e^b) - \tau(e^{a+b}) \geq \frac{1}{2}\|[e^{a/2}, e^{b/2}]\|_2^2.$$

Proof. By the Trotter-Lie formula, see e.g. [177, Theorem VIII.29],
$$(e^{a/2^n} e^{b/2^n})^{2^n} \to e^{a+b} \text{ in norm as } n \to \infty.$$

Note that since $|e^{a/2^n} e^{b/2^n}|^2 = e^{b/2^n} e^{a/2^{n-1}} e^{b/2^n}$, we have
$$|e^{a/2^n} e^{b/2^n}|^{2^n} = (e^{b/2^n} e^{a/2^{n-1}} e^{b/2^n})^{2^{n-1}} = e^{b/2^n} (e^{a/2^{n-1}} e^{b/2^{n-1}})^{2^{n-1}} e^{-b/2^n}.$$

Using the generalized Hölder inequality $\|x_1 \ldots x_p\|_1 \leq \|x_1\|_p \ldots \|x_p\|_p$, see B.6, we then get
$$\tau((e^{a/2^n} e^{b/2^n})^{2^n}) \leq \|(e^{a/2^n} e^{b/2^n})^{2^n}\|_1$$
$$\leq \|e^{a/2^n} e^{b/2^n}\|_{2^n}^{2^n}$$
$$= \tau(|e^{a/2^n} e^{b/2^n}|^{2^n})$$
$$= \tau((e^{a/2^{n-1}} e^{b/2^{n-1}})^{2^{n-1}}).$$

Thus $\tau((e^{a/2^n} e^{b/2^n})^{2^n}) \searrow \tau(e^{a+b})$. In particular, $\tau(e^{a+b}) \leq \tau((e^{a/2} e^{b/2})^2)$. It remains to note that
$$\tau(e^a e^b) - \tau((e^{a/2} e^{b/2})^2) = \frac{1}{2}\|[e^{a/2}, e^{b/2}]\|_2^2.$$

□

Proof of Proposition 4.1.2. Consider first the case $m = 2$. So we have a partition of unity $1 = \sum_{i_1 \in I_1, i_2 \in I_2} x_{i_1 i_2}$. Replacing $x_{i_1 i_2}$ by
$$(1-\varepsilon)x_{i_1 i_2} + \frac{\varepsilon}{|I_1||I_2|} 1,$$
we may assume that the $x_{i_1 i_2}$'s are invertible. Then by Corollary 2.3.10 and Lemma 4.1.3 we have
$$\sum_{i_1} \tau(\eta(x_{i_1}^{(1)})) + \sum_{i_2} \tau(\eta(x_{i_2}^{(2)})) - \sum_{i_1,i_2} \tau(\eta(x_{i_1 i_2}))$$
$$= \sum_{i_1,i_2} \tau(x_{i_1 i_2}(\log x_{i_1 i_2} - \log e^{\log x_{i_1}^{(1)} + \log x_{i_2}^{(2)}}))$$
$$\geq \sum_{i_1,i_2} (\tau(x_{i_1 i_2}) - \tau(e^{\log x_{i_1}^{(1)} + \log x_{i_2}^{(2)}}))$$

$$= 1 - \sum_{i_1,i_2} \tau(e^{\log x_{i_1}^{(1)} + \log x_{i_2}^{(2)}})$$

$$= \sum_{i_1,i_2} (\tau(x_{i_1}^{(1)} x_{i_2}^{(2)}) - \tau(e^{\log x_{i_1}^{(1)} + \log x_{i_2}^{(2)}}))$$

$$\geq \frac{1}{2} \sum_{i_1,i_2} \|[(x_{i_1}^{(1)})^{1/2}, (x_{i_2}^{(2)})^{1/2}]\|_2^2.$$

Consider now the case $m \geq 3$. It is clear that it suffices to prove the inequality for $k = 1$ and $l = 2$. Put

$$y_{i_1 i_2} = \sum_{i_3,\dots,i_m} x_{i_1\dots i_m} \quad \text{and} \quad z_{i_3\dots i_m} = \sum_{i_1,i_2} x_{i_1\dots i_m},$$

so that $y_{i_1}^{(1)} = x_{i_1}^{(1)}$, $y_{i_2}^{(2)} = x_{i_2}^{(2)}$ and $z_{i_n}^{(n)} = x_{i_n}^{(n)}$ for $n = 3, \dots, m$. Then

$$\sum_{i_1,\dots,i_m} \tau(\eta(x_{i_1\dots i_m})) \leq \sum_{i_1,i_2} \tau(\eta(y_{i_1 i_2})) + \sum_{i_3,\dots,i_m} \tau(\eta(z_{i_3\dots i_m}))$$

$$\leq \sum_{i_1,i_2} \tau(\eta(y_{i_1 i_2})) + \sum_{n=3}^{m} \sum_{i_n} \tau(\eta(x_{i_n}^{(n)})),$$

whence

$$\sum_{n=1}^{m} \sum_{i_n} \tau(\eta(x_{i_n}^{(n)})) - \sum_{i_1,\dots,i_m} \tau(\eta(x_{i_1\dots i_m})) \geq \sum_{n=1}^{2} \sum_{i_n} \tau(\eta(y_{i_n}^{(n)})) - \sum_{i_1,i_2} \tau(\eta(y_{i_1 i_2})).$$

Therefore the result follows from the case $m = 2$. □

4.2 Independent Algebras

Let us now consider the problem of maximizing $H_\tau(N_1, \dots, N_m)$ without assuming that $N_1, \dots, N_m$ generate a finite dimensional algebra. By subadditivity we have

$$H_\tau(N_1, \dots, N_m) \leq H_\tau(N_1) + \dots + H_\tau(N_m).$$

In the commutative case, by Proposition 1.1.9(iii), the equality holds exactly when $N_1, \dots, N_m$ are mutually independent. For a C*-algebra A and a faithful tracial state τ on A let us say that C*-subalgebras $A_1, \dots, A_m$ of A are τ-**independent** if they mutually commute and

$$\tau(a_1 \dots a_m) = \tau(a_1) \dots \tau(a_m)$$

for $a_i \in A_i$. Let $N_1, \dots, N_m$ be finite dimensional C*-subalgebras of A. Assume there exist maximal abelian subalgebras $A_i \subset N_i$ such that $A_1, \dots, A_m$ are τ-independent. Since

$$H_\tau(A_1,\ldots,A_m) \le H_\tau(N_1,\ldots,N_m) \le \sum_k H_\tau(N_k) = \sum_k H_\tau(A_k)$$

and

$$H_\tau(A_1,\ldots,A_m) = H_\tau(\vee_k A_k) = \sum_k H_\tau(A_k),$$

we conclude that

$$H_\tau(N_1,\ldots,N_m) = H_\tau(N_1) + \ldots + H_\tau(N_m).$$

One may expect that conversely, if the above equality holds, then there exist τ-independent maximal abelian subalgebras $A_k \subset N_k$. This is indeed the case by Theorem 4.1.1 and Proposition 1.1.9(iii) if we in addition assume that $N_1,\ldots,N_m \subset N$ and $H_\tau(N_1,\ldots,N_m) = H_\tau(N)$. We do not know whether this is true in general. We have, however, the following weaker result.

Theorem 4.2.1. *Let $A_1,\ldots,A_m$ be finite dimensional abelian subalgebras of a von Neumann algebra N with a faithful normal trace τ. Suppose*

$$H_\tau(A_1,\ldots,A_m) = H_\tau(A_1) + \ldots + H_\tau(A_m).$$

Then the algebras $A_1,\ldots,A_m$ are τ-independent.

Note that as soon as the mutual commutativity is established, the independence becomes a purely classical statement, Proposition 1.1.9(iii).

We shall use that for abelian algebras any state has a unique pure state decomposition, and any optimal model arises from this decomposition.

Lemma 4.2.2. *For any $d \in \mathbb{N}$ and $\varepsilon > 0$, there exists $\delta > 0$ such that if A is an abelian C^*-algebra of dimension d with pure states $\varphi_1,\ldots,\varphi_d$, and $\tau = \sum_{i \in I} \lambda_i \psi_i$ is a finite convex combination of states with*

$$\sum_i \lambda_i S(\psi_i,\tau) > S(\tau) - \delta,$$

then there exists a partition $I = I(0) \sqcup \ldots \sqcup I(d)$ of the index set I such that

$$\sum_{i \in I(0)} \lambda_i < \varepsilon, \quad \text{and} \quad \|\psi_i - \varphi_k\| < \varepsilon \quad \text{for} \quad i \in I(k) \quad \text{and} \quad k = 1,\ldots,d.$$

Proof. Since the von Neumann entropy is a continuous function which is zero only on pure states, we can chose δ_0 depending on d and ε such that a state ψ on A is ε-close to a pure state as soon as $S(\psi) < \delta_0$. If

$$\sum_i \lambda_i S(\psi_i) = S(\tau) - \sum_i \lambda_i S(\psi_i,\tau) < \delta,$$

then putting $I(0) = \{i \in I \mid S(\psi_i) \ge \delta_0\}$, we get

$$\sum_{i \in I(0)} \lambda_i < \frac{\delta}{\delta_0},$$

and any state ψ_i for $i \in I \backslash I(0)$ is ε-close to a state φ_k. Thus we can take $\delta = \varepsilon \delta_0$. □

Proof of Theorem 4.2.1. Let $\varphi_1^{(k)}, \ldots, \varphi_{d_k}^{(k)}$ be the pure states of A_k. Suppose we have an almost optimal decomposition $\tau = \sum \omega_{i_1 \ldots i_m}$, so that, see (3.2),

$$\sum_{i_1, \ldots, i_m} \eta(\omega_{i_1 \ldots i_k}(1)) - \sum_k \sum_{i_k \in I_k} \eta(\omega_{i_k}^{(k)}(1)) + \sum_k \sum_{i_k \in I_k} \omega_{i_k}^{(k)}(1) S(\hat{\omega}_{i_k}^{(k)}|_{A_k}, \tau|_{A_k})$$

is close to $H_\tau(A_1) + \ldots + H_\tau(A_m)$. As the first two terms give a nonpositive number by subadditivity, we see that for each k the expression

$$\sum_{i_k \in I_k} \omega_{i_k}^{(k)}(1) S(\hat{\omega}_{i_k}^{(k)}|_{A_k}, \tau|_{A_k}),$$

being not larger than $H_\tau(A_k)$, must be close to $H_\tau(A_k) = S(\tau|_{A_k})$. Thus applying the previous lemma we get partitions $I_k = I_k(0) \cup \ldots \cup I_k(d_k)$, $1 \le k \le m$, such that

$$\sum_{i_k \in I_k(0)} \omega_{i_k}^{(k)}(1) < \varepsilon, \quad \|(\hat{\omega}_{i_k}^{(k)} - \varphi_l^{(k)})|_{A_k}\| < \varepsilon \text{ for } i_k \in I_k(l) \text{ and } l = 1, \ldots, d_k.$$

Consider a coarser decomposition of τ by letting

$$\psi_{j_1 \ldots j_m} = \sum_{i_1 \in I_1(j_1), \ldots, i_m \in I_m(j_m)} \omega_{i_1 \ldots i_m}, \quad 0 \le j_k \le d_k.$$

Then

$$\psi_0^{(k)}(1) < \varepsilon, \text{ and } \|(\hat{\psi}_l^{(k)} - \varphi_l^{(k)})|_{A_k}\| < \varepsilon \text{ for } l = 1, \ldots, d_k,$$

since $\hat{\psi}_l^{(k)}$ is a convex combination of $\hat{\omega}_{i_k}^{(k)}$ for $i_k \in I_k(l)$. Consider a weak* limit point of such decompositions $\tau = \sum_{j_1, \ldots, j_m} \psi_{j_1 \ldots j_m}$ as $\varepsilon \to 0$. We then get a decomposition $\tau = \sum_{j_1, \ldots, j_m} \tau_{j_1 \ldots j_m}$ such that

$$\tau_0^{(k)} = 0, \quad \tau_l^{(k)} \le \varphi_l^{(k)} \text{ on } A_k \text{ for } l = 1, \ldots, d_k.$$

In particular, $\tau_{j_1 \ldots j_m} = 0$ if $j_k = 0$ for some k. We may therefore only consider the indices $1 \le j_k \le d_k$. Let $1 = \sum_{j_1, \ldots, j_m} x_{j_1 \ldots j_m}$ be the corresponding partition of unity in N. Let also $p_l^{(k)}$ be the support of $\varphi_l^{(k)}|_{A_k}$, so that $A_k = \mathbb{C}p_1^{(k)} + \ldots + \mathbb{C}p_{d_k}^{(k)}$. Then $x_l^{(k)} \le p_l^{(k)}$. Hence, if we put $q_{j_1 \ldots j_m} = p_{j_1}^{(1)} \wedge \ldots \wedge p_{j_m}^{(m)}$, we get

$$x_{j_1 \ldots j_m} \le q_{j_1 \ldots j_m}.$$

The projections $q_{j_1 \ldots j_m}$ are mutually orthogonal. Since the $x_{j_1 \ldots j_m}$'s form a partition of unity, we conclude that $x_{j_1 \ldots j_m} = q_{j_1 \ldots j_m}$. By definition $q_{j_1 \ldots j_m}$ is orthogonal to $p_l^{(k)}$ if $j_k \ne l$, and $q_{j_1 \ldots j_m} \le p_l^{(k)}$ if $j_k = l$. Since the projections $q_{j_1 \ldots j_m}$ form a partition of unity, we obtain $p_l^{(k)} = q_l^{(k)}$. Thus the projections $p_l^{(k)}$ lie in the abelian algebra generated by the projections $q_{j_1 \ldots j_m}$. □

4.3 Entropic K-systems

Let (X, μ, T) be an abelian dynamical system. Recall that it is called mixing if for any measurable subsets A and B we have $\mu(A \cap T^n B) \to \mu(A)\mu(B)$ as $n \to \infty$. For any finite measurable partitions ξ and ζ we then have

$$H(\xi \vee T^n \zeta) \to H(\xi) + H(\zeta).$$

This is however not enough to get $h(T) > 0$. The system is called **uniformly mixing** if for any measurable $A \subset X$ and any finite measurable partition ζ we have

$$\lim_{n \to \infty} \sup\{|\mu(A \cap B) - \mu(A)\mu(B)| : B \text{ is } (\vee_{k=n}^\infty \zeta)\text{-measurable}\} = 0.$$

One can then check that $H(\xi | \vee_{k=n}^\infty \zeta) \to H(\xi)$ for any finite measurable partition ξ, which in view of (1.3) implies $h(\xi; T^n) \to H(\xi)$. The latter property is in fact equivalent to uniform mixing. It has a direct generalization to noncommutative systems.

Definition 4.3.1. A W^*-dynamical system (M, φ, α) with faithful state φ and $M \neq \mathbb{C}$ is called an entropic K-system if for any channel $\gamma \colon A \to M$ we have $h_\varphi(\gamma; \alpha^n) \to H_\varphi(\gamma)$ as $n \to \infty$.

Note that since $h_\varphi(\gamma; \alpha^n) \leq n h_\varphi(\gamma; \alpha)$, Lemma 3.1.4 implies that for any γ with $\operatorname{Im} \gamma \neq \mathbb{C}1$ we have $h_\varphi(\gamma; \alpha) > 0$. In particular, $h_\varphi(\alpha) > 0$.

The simplest examples of K-systems in the classical case are Bernoulli shifts. If ξ is the standard generating partition of a Bernoulli shift then the partition $\zeta = \vee_{k=-\infty}^0 T^k \xi$ has the following properties: $\zeta \prec T\zeta$, the algebra $\cup_{n \in \mathbb{Z}} L^\infty(X/T^n \zeta)$ is weakly operator dense in $L^\infty(X)$, and $\cap_{n \in \mathbb{Z}} L^\infty(X/T^n \zeta) = \mathbb{C}1$. It turns out that existence of a partition ζ with such properties is yet another equivalent form of the K-property. This is not the case for noncommutative systems. Nevertheless we have the following result.

Theorem 4.3.2. Let (M, φ, α) be a W^*-dynamical system with faithful φ and $M \neq \mathbb{C}$. Suppose there exists a von Neumann subalgebra M_0 of M such that
(i) $M_0 \subset \alpha(M_0)$;
(ii) $\cup_{n \in \mathbb{N}} (\alpha^{-n}(M_0)' \cap \alpha^n(M_0))$ is weakly operator dense in M;
(iii) $\cap_{n \in \mathbb{Z}} \alpha^n(M_0) = \mathbb{C}1$.
Then (M, φ, α) is an entropic K-system.

For the proof we need two lemmas.

Lemma 4.3.3. Let M be a von Neumann algebra, $\{M_n\}_{n=1}^\infty$ a decreasing sequence of von Neumann subalgebras, φ a normal faithful state on M. Assume $\cap_n M_n = \mathbb{C}1$. Then for any $a \in M$ we have

$$\sup_{b \in M_n \setminus \{0\}} \frac{|\varphi(ab) - \varphi(a)\varphi(b)|}{\|b\|_\varphi + \|b^*\|_\varphi} \to 0 \text{ as } n \to \infty,$$

where $\|x\|_\varphi = \varphi(x^*x)^{1/2}$.

Proof. We may assume that $M \subset B(H)$, and φ is the state defined by a cyclic and separating vector $\xi \in H$, $\varphi(x) = (x\xi, \xi)$. Assume the conclusion of the lemma is false. Then there exist $\varepsilon > 0$, $a \in M$ and a sequence $\{b_n\}_n$ such that $b_n \in M_n$, $\|b_n\|_\varphi + \|b_n^*\|_\varphi = 1$ and

$$|\varphi(ab_n) - \varphi(a)\varphi(b_n)| \geq \varepsilon.$$

Replacing $\{M_n\}_n$ by a subsequence we may assume that the bounded sequences $\{b_n\xi\}_n$ and $\{b_n^*\xi\}_n$ are weakly convergent. Denote by ζ and ϑ their weak limits. Then

$$|(a\zeta, \xi) - (a\xi, \xi)(\zeta, \xi)| \geq \varepsilon. \tag{4.5}$$

On the other hand, if $x \in M'_m$ then for $n \geq m$ we have $(xb_n\xi, \xi) = (x\xi, b_n^*\xi)$, whence

$$(x\zeta, \xi) = (x\xi, \vartheta).$$

Since $\cap_m M_m = \mathbb{C}1$, the algebra $\cup_m M'_m$ is weakly operator dense in $B(H)$. It follows that the above identity holds for all $x \in B(H)$. Then, denoting the projection onto $\mathbb{C}\xi$ by p, for any $x \in B(H)$ we get

$$(p\zeta, x\xi) = (x^*p\zeta, \xi) = (x^*p\xi, \vartheta) = (x^*\xi, \vartheta) = (x^*\zeta, \xi) = (\zeta, x\xi).$$

Hence $p\zeta = \zeta$, so that $\zeta \in \mathbb{C}\xi$. But this contradicts (4.5). □

Lemma 4.3.4. *For any $m \in \mathbb{N}$ and $\varepsilon > 0$ there exists $\delta > 0$ satisfying the following property. If A is a C^*-algebra, $B \subset A$ a C^*-subalgebra, φ a state on A, and $1 = \sum_{i=1}^m x_i$ a partition of unity in $B' \cap A$ such that*

$$|\varphi(x_iy) - \varphi(x_i)\varphi(y)| \leq \delta\|y\| \text{ for any } y \in B, \ i = 1, \ldots, m,$$

then

$$\sum_{i,j} \eta(\varphi(x_iy_j)) > \sum_i \eta(\varphi(x_i)) + \sum_j \eta(\varphi(y_j)) - \varepsilon$$

for any finite partition of unity $1 = \sum_j y_j$ in B.

Proof. Let $\delta > 0$ be such that $|\sum_{i=1}^m \eta(\lambda_i) - \sum_{i=1}^m \eta(\mu_i)| < \varepsilon/2$ for probability distributions $(\lambda_1, \ldots, \lambda_m)$ and $(\mu_1, \ldots, \mu_m)$ as soon as

$$\sum_{i=1}^m |\lambda_i - \mu_i| \leq 2\delta\varepsilon^{-1} m \log m.$$

For every element y of the form $y = \sum_j \pm y_j$ we have $-1 \leq y \leq 1$, and consequently $|\varphi(x_iy) - \varphi(x_i)\varphi(y)| \leq \delta$. Hence

$$\sum_j |\varphi(x_iy_j) - \varphi(x_i)\varphi(y_j)| \leq \delta,$$

and thus

$$\sum_{i,j} |\varphi(x_i y_j) - \varphi(x_i)\varphi(y_j)| \le m\delta.$$

Consider the set J consisting of the indices j such that

$$\sum_i |\varphi(x_i y_j) - \varphi(x_i)\varphi(y_j)| \le 2\varphi(y_j)\varepsilon^{-1} m\delta \log m.$$

Since

$$m\delta \ge \sum_{j \notin J} \sum_i |\varphi(x_i y_j) - \varphi(x_i)\varphi(y_j)| > 2\varepsilon^{-1} m\delta \log m \sum_{j \notin J} \varphi(y_j),$$

we have $\sum_{j \notin J} \varphi(y_j) < \varepsilon/2\log m$. Hence

$$\sum_{j \notin J} \varphi(y_j) \left| \sum_i \eta\left(\frac{\varphi(x_i y_j)}{\varphi(y_j)}\right) - \sum_i \eta(\varphi(x_i)) \right| \le \sum_{j \notin J} \varphi(y_j) \log m < \frac{\varepsilon}{2},$$

since neither sum over i is larger than $\log m$. On the other hand, by our choice of δ, for every $j \in J$ we have

$$\left| \sum_i \eta\left(\frac{\varphi(x_i y_j)}{\varphi(y_j)}\right) - \sum_i \eta(\varphi(x_i)) \right| < \frac{\varepsilon}{2}.$$

It follows that

$$\sum_j \varphi(y_j) \left(\sum_i \eta\left(\frac{\varphi(x_i y_j)}{\varphi(y_j)}\right) - \sum_i \eta(\varphi(x_i)) \right) > -\varepsilon,$$

which is exactly what we need. □

Proof of Theorem 4.3.2. Let $\gamma\colon A \to M$ be a channel. There exists a partition of unity $1 = \sum_{i=1}^m x_i$ in M such that $H_\varphi(\gamma; \{x_i\})$ is arbitrarily close to $H_\varphi(\gamma)$. Since $\cup_N (\alpha^{-N}(M_0)' \cap \alpha^N(M_0))$ is strongly operator dense in M, we may assume that the x_i's belong to $\alpha^{-N}(M_0)' \cap \alpha^N(M_0)$ for some $N \in \mathbb{N}$ (note that one can think of a partition of unity as a unital completely positive map $\mathbb{C}^m \to M$, so the claim follows from A.10). Then $\alpha^n(x_i)$ commutes with x_j for $n \ge 2N$. For $n \ge 2N$ and $k \in \mathbb{N}$ we want to estimate

$$H_\varphi(\gamma, \alpha^n \circ \gamma, \ldots, \alpha^{(k-1)n} \circ \gamma; \{x_{i_1} \alpha^n(x_{i_2}) \ldots \alpha^{(k-1)n}(x_{i_k})\})$$

$$= \sum_{i_1,\ldots,i_k} \eta(\varphi(x_{i_1} \alpha^n(x_{i_2}) \ldots \alpha^{(k-1)n}(x_{i_k}))) - k\sum_i \eta(\varphi(x_i)) + kH_\varphi(\gamma; \{x_i\}),$$

see (3.3). To prove the theorem it suffices to show that for any $\varepsilon > 0$ the above expression is $(k\varepsilon)$-close to $kH_\varphi(\gamma; \{x_i\})$ when n is sufficiently large and k is arbitrary.

72   4 Maximality of Entropy and Commutativity

Choose δ according to Lemma 4.3.4. Since the algebras $\alpha^l(M_0)$ have trivial intersection, by Lemma 4.3.3 there exists $n_0 \leq -N$ such that

$$|\varphi(x_i y) - \varphi(x_i)\varphi(y)| \leq \delta \|y\| \quad \text{for} \quad y \in \alpha^{n_0}(M_0), \ i = 1, \ldots, n.$$

Then by Lemma 4.3.4 for any $l \in \mathbb{Z}$ and any partition of unity $1 = \sum_j y_j$ in $\alpha^{n_0+ln}(M_0)$ we get

$$\sum_{i,j} \eta(\varphi(y_j \alpha^{ln}(x_i))) > \sum_i \eta(\varphi(x_i)) + \sum_j \eta(\varphi(y_j)) - \varepsilon.$$

Since $x_{i_1}\alpha^n(x_{i_2})\ldots\alpha^{(l-1)n}(x_{i_l}) \in \alpha^{N-n+ln}(M_0) \subset \alpha^{n_0+ln}(M_0)$ if $n \geq N-n_0$, applying recursively the above inequality we obtain

$$\sum_{i_1,\ldots,i_k} \eta(\varphi(x_{i_1}\alpha^n(x_{i_2})\ldots\alpha^{(k-1)n}(x_{i_k}))) > k\sum_i \eta(\varphi(x_i)) - (k-1)\varepsilon,$$

which completes the proof of the theorem. □

The simplest examples of entropic K-systems are noncommutative Bernoulli shifts. We shall see more interesting examples in Chaps. 11 and 13.

It is worth noticing that the condition (ii) in the above theorem can not be weakened to the requirement that $\cup_n \alpha^n(M_0)$ is weakly operator dense in M. In fact, as we shall see in Chap. 12, we can then even get a system with zero entropy.

Our next goal is to show that an entropic K-system is necessarily asymptotically abelian, at least when the state is tracial.

Theorem 4.3.5. *Let (N, τ, α) be an entropic K-system, where τ is a faithful normal trace. Then the system is strongly asymptotically abelian and strongly mixing.*

In other words, for any $x, y \in N$ we have $[\alpha^n(x), y] \to 0$ in the strong operator topology and $\tau(\alpha^n(x)y) \to \tau(x)\tau(y)$ as $n \to \infty$.

Proof of Theorem 4.3.5. Let A be a finite dimensional abelian subalgebra of N. Since

$$H_\tau(A, \alpha^n(A), \ldots, \alpha^{(2m-1)n}(A)) \leq mH_\tau(A, \alpha^n(A)),$$

so that $2h_\tau(A; \alpha^n) \leq H_\tau(A, \alpha^n(A))$, it follows from our assumptions that $H_\tau(A, \alpha^n(A))$ converges to $2H_\tau(A)$. We thus need an approximate version of Theorem 4.2.1. A standard trick in such cases is to use ultraproducts.

So let ω be a free ultrafilter on $\mathbb{N}$. Consider the ultraproduct N^ω. By definition this is the quotient of the algebra $\ell_\infty(\mathbb{N}, N)$ of bounded sequences in N by the ideal I_ω consisting of sequences $(x_n)_n$ such that $\|x_n\|_2 \to 0$ as $n \to \omega$. This is a von Neumann algebra with a faithful normal trace τ_ω defined by

$$\tau_\omega((x_n)_n) = \lim_{n \to \omega} \tau(x_n).$$

We identify N with the subalgebra of N^ω consisting of constant sequences. Consider the automorphism θ of N^ω defined by

$$\theta((x_n)_n) = (\alpha^n(x_n))_n.$$

We claim that
$$H_{\tau_\omega}(A, \theta(A)) = 2H_\tau(A).$$

Indeed, by Proposition 3.1.10 and the discussion following it, for any $\varepsilon > 0$ there exists a finite index set I such that for each n there exists a partition of unity $1 = \sum_{i,j \in I} x(n)_{ij}$ in N such that $H_\tau(A, \alpha^n(A); \{x(n)_{ij}\})$ is ε-close to $H_\tau(A, \alpha^n(A))$ (in fact, if we consider only sufficiently large n, then by the proof of Theorem 4.2.1 we may even take I consisting of $d+1$ elements, where d is the dimension of A). Then $y_{ij} = (x(n)_{ij})_n$, $i, j \in I$, form a partition of unity in N^ω such that

$$H_{\tau_\omega}(A, \theta(A); \{y_{ij}\}) = \lim_{n \to \omega} H_\tau(A, \alpha^n(A); \{x(n)_{ij}\}) \geq 2H_\tau(A) - \varepsilon.$$

Thus we can apply Theorem 4.2.1 and conclude that A and $\theta(A)$ are τ_ω-independent. To conclude that N and $\theta(N)$ are τ_ω-independent we need the following lemma.

Lemma 4.3.6. *Let M be a von Neumann algebra with a faithful normal trace τ, $N \subset M$, and θ a τ-preserving automorphism of M. Assume that for any commuting projections p and q in N we have*

$$\theta(p)q = q\theta(p) \quad \text{and} \quad \tau(\theta(p)q) = \tau(p)\tau(q).$$

Then the algebras N and $\theta(N)$ are τ-independent.

Proof. We shall first check that $\tau(x\theta(y)) = \tau(x)\tau(y)$ for $x, y \in N$. Since every self-adjoint operator is approximated by a linear combination of its spectral projections, we have

$$\theta(a)b = b\theta(a) \quad \text{and} \quad \tau(\theta(a)b) = \tau(a)\tau(b)$$

for any commuting self-adjoint elements $a, b \in N$. Let $A \subset N$ be a maximal abelian von Neumann subalgebra of N containing a. Since $\theta(a)$ commutes with any element in N commuting with a, the same is true for $E_N(\theta(a))$. Hence $E_N(\theta(a))$ lies in A. On the other hand,

$$\tau(E_N(\theta(a))b) = \tau(\theta(a)b) = \tau(a)\tau(b).$$

for every $b \in A$. It follows that $E_N(\theta(a))$ is a scalar, so that $\tau(x\theta(a)) = \tau(x)\tau(a)$ for every $x \in N$.

74   4 Maximality of Entropy and Commutativity

To prove that N and $\theta(N)$ commute, it is enough to check that $\theta(p)$ and q commute for any projections p and q in N. Since $[p+q, \theta(p+q)] = 0$, $[p, \theta(p)] = 0$ and $[q, \theta(q)] = 0$, we get

$$[p, \theta(q)] = [\theta(p), q].$$

Using this identity we compute

$$\begin{aligned}
\tau(\theta(p)q\theta(p)q) &= \tau(\theta(p)q\theta(p)) + \tau(\theta(p)q[\theta(p), q]) \\
&= \tau(\theta(p)q) + \tau(\theta(p)q[p, \theta(q)]) \\
&= \tau(\theta(p)q) + \tau(\theta(p)qp\theta(q)) - \tau(\theta(p)q\theta(q)p) \\
&= \tau(\theta(p)q) + \tau(\theta(q)\theta(p)qp) - \tau(\theta(p)\theta(q)qp) \\
&= \tau(p)\tau(q) + \tau(qp)\tau(qp) - \tau(pq)\tau(qp) \\
&= \tau(p)\tau(q),
\end{aligned}$$

whence

$$\|[\theta(p), q]\|_2^2 = \tau(q\theta(p)q) - \tau(\theta(p)q\theta(p)q) - \tau(q\theta(p)q\theta(p)) + \tau(\theta(p)q\theta(p)) = 0.$$

Thus $\theta(p)q = q\theta(p)$. □

Returning to the proof of the theorem, we conclude that N and $\theta(N)$ are τ_ω-independent. That is, for any $x, y \in N$ we get

$$\|[\alpha^n(x), y]\|_2 \to 0 \quad \text{and} \quad \tau(\alpha^n(x)y) \to \tau(x)\tau(y)$$

as $n \to \omega$. Since ω was an arbitrary free ultrafilter, we conclude that these convergences hold as $n \to \infty$. □

4.4 Notes

Theorem 4.1.1 was proved by Haagerup and Størmer [83].

Theorem 4.2.1 is contained in the work of Benatti and Narnhofer [15]. Curiously enough it has apparently never been formulated explicitly.

K-systems (called also Kolmogorov systems, or systems regular in the sense of Kolmogorov) were introduced by Kolmogorov [111], who called them quasi-regular. The main definition is the last one we mentioned. So (X, μ, T) is a K-system if there exists a measurable partition ζ such that $\zeta \prec T\zeta$, $\vee_n T^n \zeta$ is the partition of (X, μ) into points, and the partition $\wedge_n T^n \zeta$ corresponding to $\cap_n L^\infty(X/T^n\zeta)$ is trivial. Equivalent forms of the K-property were obtained by Rohlin and Sinai [183]. For some time the only known examples of K-systems were Bernoulli shifts. It was Ornstein who gave the first example of a K-system which is not a Bernoulli shift [151]. Ornstein and Shields constructed an uncountable family of pairwise nonisomorphic K-systems with the same entropy [152]. A simple example of a non-Bernoullian K-system was given by Kalikow [98]. Katok found examples of smooth non-Bernoullian K-systems [100].

The first appearance of K-systems in the quantum setting is the work of Emch [59], [60], [61]. One says that a W^*-dynamical system (M, φ, α) is an algebraic K-system, the name suggested by Narnhofer and Thirring [132], if there exists a W^*-subalgebra $M_0 \subset M$ such that $M_0 \subset \alpha(M_0)$, $\cup_n \alpha^n(M_0)$ is dense in M, and $\cap_n \alpha^n(M_0) = \mathbb{C}1$. As was already mentioned, in Chap. 12 we shall see examples of algebraic K-systems with zero entropy.

The notion of entropic K-system was proposed by Narnhofer and Thirring [131]. A stronger result than that formulated in Theorem 4.3.2 was stated by Narnhofer [126]. Explicitly Theorem 4.3.2 appeared in the work of Golodets and Neshveyev [76]. Lemma 4.3.3 is due to Powers [170]. Theorem 4.3.5 was proved by Benatti and Narnhofer [15].

An attempt to capture the K-property by imposing certain hyperbolicity conditions was made by Emch, Narnhofer, Sewell and Thirring [62]. See [135], [136] for results in this direction.

5
Dynamical Abelian Models

In Chap. 3 we used abelian models to define mutual entropy of channels into A. It is then natural to consider dynamical abelian models, that is, abelian dynamical systems (X, μ, T) together with equivariant maps $A \to L^\infty(X, \mu)$. In this chapter we shall prove that there always exist enough such models to compute the entropy.

5.1 Entropy via Stationary Couplings

As in Chap. 3 we prefer to work with states on $A \otimes L^\infty(X, \mu)$ rather than with maps $A \to L^\infty(X, \mu)$. Then by a **stationary coupling** of (A, φ, α) with (X, μ, T) we mean an $(\alpha \otimes \beta)$-invariant state λ on $A \otimes L^\infty(X, \mu)$, where $\beta(f) = f \circ T^{-1}$, such that $\lambda|_A = \varphi$ and $\lambda|_{L^\infty(X, \mu)} = \mu$. Given a stationary coupling λ, a channel γ into A and a finite measurable partition ξ of X, we have (see (3.12))

$$H_\lambda(\gamma, \alpha \circ \gamma, \ldots, \alpha^{n-1} \circ \gamma; \xi, T\xi, \ldots, T^{n-1}\xi)$$
$$= H_\mu(\vee_{k=0}^{n-1} T^k \xi) - n H_\mu(\xi) + n H_\lambda(\gamma; \xi). \tag{5.1}$$

This motivates the following definition of entropy.

Definition 5.1.1. *The Sauvageot-Thouvenot entropy of a C^*-dynamical system (A, φ, α) is*

$$h_\varphi^{ST}(\alpha) = \sup\{h_\mu(\xi; T) - H_\mu(\xi) + H_\lambda(A; \xi)\},$$

where the supremum is taken over all stationary couplings λ of (A, φ, α) with abelian dynamical systems (X, μ, T) and over all finite measurable partitions ξ of X.

To make sense of the expression

$$H_\lambda(A;\xi) = H_\mu(\xi) + \sum_{Z\in\xi} S(\lambda(\cdot \otimes 1_Z), \varphi)$$

we need relative entropy to be defined for arbitrary C*-algebras. However, essentially all we have to know in order to deal with, say, nuclear C*-algebras is that if $\{\gamma_i\}_i$ is a φ-approximating net, then $H_\lambda(\gamma_i;\xi) \nearrow H_\lambda(A;\xi)$ by Corollary 2.3.5.

Let us first observe that in some cases just existence of nontrivial couplings, that is, couplings different from product-states, implies positivity of entropy.

Proposition 5.1.2. *If a C*-dynamical system (A, φ, α) has a nontrivial stationary coupling with a classical Bernoulli shift then $h_\varphi(\alpha) > 0$.*

Proof. Let λ be a nontrivial stationary coupling of (A, φ, α) with a Bernoulli shift (X, μ, T). Let ξ be the standard generating partition for (X, μ, T). Since λ is assumed to be nontrivial, there exist a self-adjoint element $a \in A$ and an element $Y \in \vee_{k=-n}^{n} T^k \xi$ such that $\lambda(a \otimes 1_Y) \neq \varphi(a)\mu(Y)$. Since $h_\varphi(\alpha^m) \leq m h_\varphi(\alpha)$ for any $m \in \mathbb{N}$, it suffices to show that $h_\varphi(\alpha^{2n+1}) > 0$. Hence replacing α by α^{2n+1}, ξ by $\vee_{k=-n}^{n} T^k \xi$ and T by T^{2n+1} we may assume $n = 0$.

Since $h_\mu(\xi; T) = H_\mu(\xi)$, by virtue of (5.1) and (3.10) for any channel γ into A we get

$$h_\varphi(\gamma; \alpha) \geq H_\lambda(\gamma; \xi) = \sum_{Z\in\xi} \mu(Z) S(\mu(Z)^{-1} \lambda(\gamma(\cdot) \otimes 1_Z), \varphi \circ \gamma).$$

If $a \in \operatorname{Im}\gamma$, then by assumption $\mu(Y)^{-1}\lambda(\gamma(\cdot) \otimes 1_Y) \neq \varphi \circ \gamma$, and by Theorem 2.1.2(i) we conclude that $h_\varphi(\gamma; \alpha) > 0$. □

In the next section we shall prove a partial converse to the above proposition.

The quantity $H_\mu(\xi) - H_\lambda(A;\xi)$, which is the entropy defect of the decomposition of φ defined by λ and ξ, should be thought of as an analogue of the conditional entropy of $L^\infty(X/\xi)$ with respect to A in $(A \otimes L^\infty(X,\mu), \lambda)$. We shall now show that all the main properties of the classical conditional entropy are satisfied, which allows us to deal efficiently with our new definition of entropy.

For $\gamma \colon B \to A$ denote by $H_{\lambda,\gamma}(\xi)$ the entropy defect of the abelian model defined by λ and ξ, so by (3.9)

$$H_{\lambda,\gamma}(\xi) = H_\mu(\xi) - H_\lambda(\gamma;\xi) = -\sum_{Z\in\xi} S(\lambda(\gamma(\cdot) \otimes 1_Z), \varphi \circ \gamma). \tag{5.2}$$

This is well-defined for an arbitrary unital completely positive mapping $\gamma \colon B \to A$, not only for finite dimensional B's. More generally, for any measurable (not necessarily finite) partition ζ of Y, denote by i_ζ the embedding $L^\infty(Y/\zeta) \hookrightarrow L^\infty(Y)$, and put

$$H_{\lambda,\gamma}(\xi|\zeta) = -\sum_{Z\in\xi} S(\lambda(\gamma(\cdot)\otimes i_\zeta(\cdot)\mathbf{1}_Z), \lambda\circ(\gamma\otimes i_\zeta)).$$

Equivalently, using the diagonal embedding $Y \to Y\times Y$, for any coupling λ of (A,φ) with (Y,μ) we can define a coupling of $(A\otimes L^\infty(Y,\mu),\lambda)$ with (Y,μ), which we again denote by λ, so that $a\otimes f\otimes g \mapsto \lambda(a\otimes fg)$. Then $H_{\lambda,\gamma}(\xi|\zeta) = H_{\lambda,\gamma\otimes i_\zeta}(\xi)$. If ζ is finite then

$$H_{\lambda,\gamma}(\xi|\zeta) = -\sum_{Z\in\xi, W\in\zeta} S(\lambda(\gamma(\cdot)\otimes \mathbf{1}_{Z\cap W}), \lambda(\gamma(\cdot)\otimes \mathbf{1}_W)).$$

Proposition 5.1.3. *For any unital completely positive map $\gamma\colon B \to A$ we have:*
(i) $H_\lambda(\gamma;\xi)$ *is increasing in γ and ξ, that is, $H_\lambda(\gamma\circ\rho;\xi) \leq H_\lambda(\gamma;\vartheta)$ for any unital completely positive mapping ρ into B and any finite partition $\vartheta \succ \xi$;*
(ii) $0 \leq H_\lambda(\gamma;\xi) \leq H_\mu(\xi)$;
(iii) *the function $\lambda \mapsto H_\lambda(\gamma;\xi)$ is convex;*
(iv) $H_{\lambda,\gamma}(\xi|\zeta)$ *is increasing in ξ and decreasing in γ and ζ; moreover, if $\{\zeta_n\}_n$ is an increasing sequence of measurable partitions with $\vee_n \zeta_n = \zeta$, then $H_{\lambda,\gamma}(\xi|\zeta_n) \searrow H_{\lambda,\gamma}(\xi|\zeta)$;*
(v) $0 \leq H_{\lambda,\gamma}(\xi|\zeta) \leq H_\mu(\xi|\zeta)$;
(vi) $H_{\lambda,\gamma}(\xi\vee\vartheta|\zeta) = H_{\lambda,\gamma}(\xi|\zeta) + H_{\lambda,\gamma}(\vartheta|\xi\vee\zeta)$; *in particular, $H_{\lambda,\gamma}(\xi\vee\vartheta) \leq H_{\lambda,\gamma}(\xi) + H_{\lambda,\gamma}(\vartheta)$.*

Proof. Essentially all these statements appeared in one or another form in Chap. 3, but we shall spell out the proofs for convenience.

The fact that $H_\lambda(\gamma;\xi)$ is increasing in γ follows from monotonicity of relative entropy, Theorem 2.3.1(vi). That it is increasing in ξ can be deduced either from convexity of relative entropy or from (v) and (vi) as follows. If $\vartheta \succ \xi$, then

$$\begin{aligned}H_{\lambda,\gamma}(\vartheta) &= H_{\lambda,\gamma}(\xi) + H_{\lambda,\gamma}(\vartheta|\xi)\\ &\leq H_{\lambda,\gamma}(\xi) + H_\mu(\vartheta|\xi)\\ &= H_{\lambda,\gamma}(\xi) + H_\mu(\vartheta) - H_\mu(\xi),\end{aligned}$$

so $H_\lambda(\gamma;\xi) \leq H_\lambda(\gamma;\vartheta)$ by (5.2).

The first inequality in (ii) follows from $S(\psi,\omega) \leq 0$ for $\psi \leq \omega$, Theorem 2.3.1(iii). On the other hand, $S(\psi,\omega) \geq 0$ if ψ and ω are states, Theorem 2.3.1(ii). Hence

$$-S(\lambda(\gamma(\cdot)\otimes\mathbf{1}_Z), \varphi\circ\gamma) = -S(\mu(Z)^{-1}\lambda(\gamma(\cdot)\otimes\mathbf{1}_Z), \varphi\circ\gamma) + \eta(\mu(Z)) \leq \eta(\mu(Z)),$$

which gives the second inequality in (ii).

Part (iii) follows from (3.10) and convexity of relative entropy, Theorem 2.3.1(v).

Since $H_{\lambda,\gamma}(\xi|\zeta) = H_{\lambda,\gamma\otimes i_\zeta}(\xi)$, it follows from (i) that

$$H_{\lambda,\gamma}(\xi|\zeta) = H_\mu(\xi) - H_\lambda(\gamma \otimes i_\zeta;\xi)$$

is decreasing in γ and ζ. Since $H_{\lambda,\gamma}(\vartheta|\xi \vee \zeta) \geq 0$ by (ii), it will follow from (vi) that $H_{\lambda,\gamma}(\xi|\zeta)$ is increasing in ξ. The convergence statement in (iv) follows from Corollaries 2.3.6 and 2.3.4.

We have already mentioned that positivity of $H_{\lambda,\gamma}$ follows from (ii). Since $H_{\lambda,\gamma}(\xi|\zeta)$ is decreasing in γ, to prove the second inequality in (v) we may assume that $B = \mathbb{C}$. Then

$$H_{\lambda,\gamma}(\xi|\zeta) = -\sum_{Z\in\xi} S(\mu(\cdot\mathbb{1}_Z)|_{L^\infty(X/\zeta)}, \mu|_{L^\infty(X/\zeta)}).$$

Since $\mu(\cdot\mathbb{1}_Z) = \mu(\cdot E_\zeta(\mathbb{1}_Z))$ on $L^\infty(X/\zeta)$, where E_ζ is the μ-preserving conditional expectation onto $L^\infty(X/\zeta)$, by Theorem 2.3.1(x) we have

$$S(\mu(\cdot\mathbb{1}_Z)|_{L^\infty(X/\zeta)}, \mu|_{L^\infty(X/\zeta)}) = -\int_X \eta(E_\zeta(\mathbb{1}_Z))d\mu.$$

Hence $H_{\lambda,\gamma}(\xi|\zeta) = H_\mu(\xi|\zeta)$.

To prove (vi) first notice that $H_{\lambda,\gamma}(\vartheta|\xi \vee \zeta) = H_{\lambda,\gamma\otimes i_\xi}(\vartheta|\xi)$, or in other words, $H_{\lambda,\gamma\otimes i_{\xi\vee\zeta}}(\vartheta) = H_{\lambda,\gamma\otimes i_\xi \otimes i_\zeta}(\vartheta)$. This follows from Corollary 2.3.4, since the homomorphism $L^\infty(X/\xi) \otimes L^\infty(X/\zeta) \to L^\infty(X/\xi \vee \zeta)$, $f \otimes g \to fg$, has dense image in the strong operator topology. It follows that in proving (vi) we may replace γ by $\gamma \otimes i_\zeta$ and thus assume that ζ is trivial. By Theorem 2.3.1(viii) we then have

$$\sum_{Z\in\vartheta} S(\lambda(\gamma(\cdot) \otimes \mathbb{1}_{Z\cap W}), \varphi \circ \gamma)$$

$$= S(\lambda(\gamma(\cdot) \otimes \mathbb{1}_W), \varphi \circ \gamma) + \sum_{Z\in\vartheta} S(\lambda(\gamma(\cdot) \otimes \mathbb{1}_{Z\cap W}), \lambda(\gamma(\cdot) \otimes \mathbb{1}_W))$$

for $W \in \xi$. Hence $H_{\lambda,\gamma}(\xi \vee \vartheta) = H_{\lambda,\gamma}(\xi) + H_{\lambda,\gamma}(\vartheta|\xi)$. □

The following formula for $h^{ST}_\varphi(\alpha)$ is reminiscent of (1.3).

Proposition 5.1.4. *For any C^*-dynamical system (A, φ, α) we have*

$$h^{ST}_\varphi(\alpha) = \sup\{H_\mu(\xi|\xi^-) - H_{\lambda,A}(\xi|\xi^-)\},$$

where the supremum is taken over all stationary couplings λ of (A, φ, α) with abelian dynamical systems (Y, μ, T) and over all finite measurable partitions ξ.

Proof. As $H_{\lambda,A}(\xi|\xi^-) \leq H_{\lambda,A}(\xi) = H_\mu(\xi) - H_\lambda(A;\xi)$ and $H_\mu(\xi|\xi^-) = h_\mu(\xi;T)$, the entropy $h^{ST}_\varphi(\alpha)$ is not bigger than the supremum.

5.1 Entropy via Stationary Couplings 81

To prove the opposite inequality, fix a stationary coupling λ. First of all note that using Proposition 5.1.3(iv),(vi) the same argument as in the proof of (1.3) shows that

$$H_{\lambda,A}(\xi|\xi^-) = \lim_{n\to\infty} \frac{1}{n} H_{\lambda,A}(\vee_{k=0}^{n-1} T^k \xi),$$

and hence

$$H_\mu(\xi|\xi^-) - H_{\lambda,A}(\xi|\xi^-) = \lim_{n\to\infty} \frac{1}{n} H_\lambda(A; \vee_{k=0}^{n-1} T^k \xi). \tag{5.3}$$

Fix $\varepsilon > 0$ and choose $n \in \mathbb{N}$ such that

$$\frac{1}{n} H_\lambda(A; \vee_{k=0}^{n-1} T^k \xi) > H_\mu(\xi|\xi^-) - H_{\lambda,A}(\xi|\xi^-) - \varepsilon$$

and

$$\frac{1}{n} H_\mu(\vee_{k=0}^{n-1} T^k \xi) < h_\mu(\xi; T) + \varepsilon.$$

Consider the space $Y' = Y \times (\mathbb{Z}/n\mathbb{Z})$ with the transformation $T'(x, k) = (Tx, k+1)$ and the measure μ' given by the product of μ with the counting measure. By taking the tensor product of λ with the counting measure, we get a stationary coupling λ' of (A, φ, α) with (Y', μ', T'). Let ξ' be the partition of Y' consisting of the elements $Y \times \{1, \ldots, n-1\}$ and $Z \times \{0\}$, where Z runs through the elements of $\vee_{k=0}^{n-1} T^k \xi$. It is easy to check that

$$H_{\mu'}(\xi') = \frac{1}{n} H_\mu(\vee_{k=0}^{n-1} T^k \xi) + \eta\left(\frac{1}{n}\right) + \eta\left(1 - \frac{1}{n}\right) \text{ and } h_{\mu'}(\xi'; T') = h_\mu(\xi; T),$$

so that

$$h_{\mu'}(\xi'; T') - H_{\mu'}(\xi') > -\varepsilon - \eta\left(\frac{1}{n}\right) - \eta\left(1 - \frac{1}{n}\right) > -2\varepsilon$$

if n is sufficiently large. On the other hand, if $\varphi = \sum_i \varphi_i$ is the decomposition defined by λ and $\vee_{k=0}^{n-1} T^k \xi$, then

$$\varphi = \sum_i \frac{1}{n} \varphi_i + \left(1 - \frac{1}{n}\right) \varphi$$

is the decomposition defined by λ' and ξ'. Recalling that we denote by $\hat\varphi_i$ the state $\varphi_i(1)^{-1}\varphi_i$, by virtue of (3.10) we get

$$H_{\lambda'}(A; \xi') = \sum_i \frac{\varphi_i(1)}{n} S(\hat\varphi_i, \varphi) + \left(1 - \frac{1}{n}\right) S(\varphi, \varphi)$$

$$= \frac{1}{n} H_\lambda(A; \vee_{k=0}^{n-1} T^k \xi) > H_\mu(\xi|\xi^-) - H_{\lambda,A}(\xi|\xi^-) - \varepsilon.$$

It follows that

$$h_{\mu'}(\xi';T') - H_{\mu'}(\xi') + H_{\lambda'}(A;\xi') > H_\mu(\xi|\xi^-) - H_{\lambda,A}(\xi|\xi^-) - 3\varepsilon,$$

and the proof is complete. □

We can now formulate the main result of the chapter.

Theorem 5.1.5. *We have $h_\varphi^{ST}(\alpha) \geq h_\varphi(\alpha)$ for any C^*-dynamical system (A, φ, α). If the system has a φ-approximating net, then $h_\varphi^{ST}(\alpha) = h_\varphi(\alpha)$.*

Proof. By (5.1) we have

$$h_\varphi(\gamma;\alpha) \geq h_\mu(\xi;T) - H_\mu(\xi) + H_\lambda(\gamma;\xi).$$

If $\{\gamma_i\}_i$ is a φ-approximating net, then $H_\lambda(\gamma_i;\xi) \nearrow H_\lambda(A;\xi)$ by Corollary 2.3.5. Hence $h_\varphi^{ST}(\alpha) \leq h_\varphi(\alpha)$.

To prove the opposite inequality in general we shall find a channel $\gamma \colon B \to A$, a sequence $\{\lambda_n\}_n$ of nonstationary couplings of (A, φ, α) with (Y, T, μ_n), where Y is a compact metric space, and a partition ξ of Y consisting of clopen sets such that

$$\frac{1}{n}H_{\lambda_n}(\gamma, \alpha \circ \gamma, \ldots, \alpha^{n-1} \circ \gamma; \xi, T\xi, \ldots, T^{n-1}\xi) > h_\varphi(\alpha) - \varepsilon.$$

Then to get a stationary coupling it is natural to take a weak* limit point of the states $n^{-1}\sum_{k=0}^{n-1} \lambda_n \circ (\alpha^k \otimes \beta^k)$, $n \in \mathbb{N}$, on $A \otimes C(Y)$. However, relative entropy is convex, which makes it impossible in general to estimate the entropy of the coupling we get. The key observation allowing to overcome this problem is that although $H_\lambda(\gamma;\xi)$ is a convex function of λ, it is almost affine on canonical couplings, see Definition 3.1.8.

Lemma 5.1.6. *Let $\gamma \colon B \to A$ be a channel. For $\varepsilon > 0$ choose $\delta > 0$ such that $|S(\psi_1) - S(\psi_2)| < \varepsilon$ whenever ψ_1 and ψ_2 are states on B satisfying $\|\psi_1 - \psi_2\| < \delta$. Let ζ be a finite Borel partition of the state space X of A for which $\|\omega_1 \circ \gamma - \omega_2 \circ \gamma\| < \delta$ when ω_1 and ω_2 belong to the same element of ζ. Then if λ_i is a canonical coupling of (A, φ) with (X, μ_i), $i \in I$, and $\lambda = \sum_i \alpha_i \lambda_i$ is a convex combination, we have*

$$\left| H_\lambda(\gamma;\zeta) - \sum_i \alpha_i H_{\lambda_i}(\gamma;\zeta) \right| < \varepsilon.$$

Proof. For any canonical coupling Λ we have by (3.11)

$$H_\Lambda(\gamma;\zeta) = S(\varphi \circ \gamma) - \sum_{Z \in \zeta} \nu(Z) S\left(\nu(Z)^{-1} \int_Z \omega \circ \gamma \, d\nu(\omega)\right),$$

where $\nu = \Lambda|_{C(X)}$. Let $\mu = \sum_i \alpha_i \mu_i$, and set $\varphi_Z = \mu(Z)^{-1} \int_Z \omega \, d\mu(\omega)$. Then by the choice of ζ we have $\|\omega \circ \gamma - \varphi_Z \circ \gamma\| < \delta$ for any $\omega \in Z$. Hence

$$\left| H_\Lambda(\gamma;\zeta) - S(\varphi \circ \gamma) + \sum_{Z \in \zeta} \nu(Z) S(\varphi_Z \circ \gamma) \right| < \varepsilon.$$

In other words, if we put $f(\Lambda) = S(\varphi \circ \gamma) - \sum_{Z \in \zeta} \nu(Z) S(\varphi_Z \circ \gamma)$, then f is an affine function on the space of canonical couplings such that $f(\lambda) = H_\lambda(\gamma, \zeta)$ and $|H_\Lambda(\gamma, \zeta) - f(\Lambda)| < \varepsilon$ for any Λ. This gives the result. □

Returning to the proof of Theorem 5.1.5, fix $h < h_\varphi(\alpha)$ and $\varepsilon > 0$. Choose a channel $\gamma \colon B \to A$ such that $h_\varphi(\gamma; \alpha) > h$. Choose also a finite Borel partition ζ of $X = S(A)$ as in Lemma 5.1.6.

Put $Z = \prod_{-\infty}^{+\infty} X$, and define a homeomorphism T of Z by

$$(T\underline{\omega})_n = \hat{\alpha}(\omega_{n-1}) = \omega_{n-1} \circ \alpha^{-1} \quad \text{for } \underline{\omega} = (\omega_n)_n \in Z.$$

Let β be the corresponding automorphism of $C(Z)$, so $\beta(f) = f \circ T^{-1}$.

Let $\mathrm{pr}_k \colon Z \to X$ be the projection onto the k-th factor. Set $\xi = \mathrm{pr}_0^{-1}(\zeta)$. Then $T^k \xi = \mathrm{pr}_k^{-1}(\hat{\alpha}^k(\zeta))$. By Proposition 3.1.10 applied to the channels $\alpha^k \circ \gamma$ and the partitions $\hat{\alpha}^k(\zeta)$ of X, $0 \le k \le n-1$, there exists a coupling λ_n of (A, φ) with (Z, μ_n) such that

$$\frac{1}{n} H_{\lambda_n}(\gamma, \alpha \circ \gamma, \ldots, \alpha^{n-1} \circ \gamma; \xi, T\xi, \ldots, T^{n-1}\xi) > h - \varepsilon$$

(more precisely, we first get a state on $A \otimes C(\prod_{k=0}^{n-1} X)$, and then extend it arbitrarily to $A \otimes C(Z)$). In other words, by (5.1) and (5.2),

$$\frac{1}{n}\left(H_{\mu_n}(\vee_{k=0}^{n-1} T^k \xi) - \sum_{k=0}^{n-1} H_{\lambda_n, \alpha^k \circ \gamma}(T^k \xi) \right) > h - \varepsilon.$$

By a computation similar to (1.2) this can be written as

$$\frac{1}{n} \sum_{k=0}^{n-1} (H_{\mu_n}(T^k \xi | \vee_{j=0}^{k-1} T^j \xi) - H_{\lambda_n, \alpha^k \circ \gamma}(T^k \xi)) > h - \varepsilon.$$

For any fixed $p \in \mathbb{N}$ the contribution of the summands with $k < p$ becomes negligible as $n \to \infty$. On the other hand, if $k \ge p$ then $\vee_{k-p}^{k-1} T^j \xi \prec \vee_{j=0}^{k-1} T^j \xi$, so that $H_{\mu_n}(T^k \xi | \vee_{j=0}^{k-1} T^j \xi) \le H_{\mu_n}(T^k \xi | \vee_{k-p}^{k-1} T^j \xi)$. Hence

$$\frac{1}{n} \sum_{k=0}^{n-1} (H_{\mu_n}(T^k \xi | \vee_{k-p}^{k-1} T^j \xi) - H_{\lambda_n, \alpha^k \circ \gamma}(T^k \xi)) > h - 2\varepsilon$$

for any fixed p and any n sufficiently large. By Proposition 5.1.3(vi),

$$-H_{\lambda_n, \alpha^k \circ \gamma}(T^k \xi) \le -H_{\lambda_n, \alpha^k \circ \gamma}(\vee_{k-p}^{k} T^j \xi) + H_{\lambda_n, \alpha^k \circ \gamma}(\vee_{k-p}^{k-1} T^j \xi),$$

so that using $H_{\mu_n}(T^k \xi | \vee_{k-p}^{k-1} T^j \xi) = H_{\mu_n}(\vee_{k-p}^{k} T^j \xi) - H_{\mu_n}(\vee_{k-p}^{k-1} T^j \xi)$ we get

84   5 Dynamical Abelian Models

$$\frac{1}{n}\sum_{k=0}^{n-1}(H_{\lambda_n}(\alpha^k \circ \gamma; \vee_{k-p}^k T^j\xi) - H_{\lambda_n}(\alpha^k \circ \gamma; \vee_{k-p}^{k-1} T^j\xi)) > h - 2\varepsilon \qquad (5.4)$$

for any fixed p and any n sufficiently large.

Our goal is to replace $\vee_{k-p}^k T^j\xi$ by $T^k\xi$ in the first summand to be able to use Lemma 5.1.6. Set

$$a_{n,p} = \frac{1}{n}\sum_{k=0}^{n-1} H_{\lambda_n}(\alpha^k \circ \gamma; \vee_{k-p}^k T^j\xi).$$

We have $0 \leq a_{n,p} \leq H_\varphi(\gamma)$. Thus there exists an infinite subset $I \subset \mathbb{N}$ such that the sequence $\{a_{n,p}\}_{n \in I}$ converges as $I \ni n \to \infty$ for any $p \in \mathbb{N}$. Denote its limit by a_p. By Proposition 5.1.3(i) we have $a_{n,p} \geq a_{n,r}$ if $p \geq r$, so $a_p \geq a_r$. Let a_∞ be the limit of $\{a_p\}_p$ as $p \to \infty$. Fix $r \in \mathbb{N}$ such that $a_r > a_\infty - \varepsilon$. As $a_p \leq a_\infty$, for any fixed p and any $n \in I$ sufficiently large we have $a_{n,r} > a_{n,p} - \varepsilon$. Thus (5.4) implies that

$$\frac{1}{n}\sum_{k=0}^{n-1}(H_{\lambda_n}(\alpha^k \circ \gamma; \vee_{k-r}^k T^j\xi) - H_{\lambda_n}(\alpha^k \circ \gamma; \vee_{k-p}^{k-1} T^j\xi)) > h - 3\varepsilon \qquad (5.5)$$

for any fixed p and any $n \in I$ sufficiently large. Now apply Proposition 3.1.10 to the coupling λ_n, the channels $\alpha^k \circ \gamma$, the partitions $\vee_{k-r}^k T^j\xi$ of Z and the partitions $\hat{\alpha}^k(\zeta)$ of X, $0 \leq k \leq n-1$. Thus we get a Borel map $f: Z \to Z$ such that for $0 \leq k \leq n-1$ we have

(i) $f^{-1}(T^k\xi) \prec \vee_{k-r}^k T^j\xi$;
(ii) if we set $\lambda'_n = \lambda_n \circ (\mathrm{id} \otimes f^*)$, then

$$|H_{\lambda_n}(\alpha^k \circ \gamma; \vee_{k-r}^k T^j\xi) - H_{\lambda'_n}(\alpha^k \circ \gamma; T^k\xi)| < \varepsilon;$$

(iii) the coupling $\lambda'_n \circ (\mathrm{id} \otimes \mathrm{pr}_k^*)$ is canonical. (Again, to be more precise, we have maps $f_k: Z \to X$ for $0 \leq k \leq n-1$, but then we take arbitrary Borel maps $Z \to X$ for all other k's.)

By virtue of (i) we have $f^{-1}(\vee_{k-p+r}^{k-1} T^j\xi) \prec \vee_{k-p}^{k-1} T^j\xi$ if $k \geq p - r > 0$ and $k \leq n-1$. By Proposition 5.1.3(i) we thus have

$$H_{\lambda_n}(\alpha^k \circ \gamma; \vee_{k-p}^{k-1} T^j\xi) \geq H_{\lambda_n}(\alpha^k \circ \gamma; f^{-1}(\vee_{k-p+r}^{k-1} T^j\xi))$$
$$= H_{\lambda'_n}(\alpha^k \circ \gamma; \vee_{k-p+r}^{k-1} T^j\xi)$$

if $0 < p - r \leq k \leq n - 1$, so that

$$\frac{1}{n}\sum_{k=0}^{n-1} H_{\lambda_n}(\alpha^k \circ \gamma; \vee_{k-p}^{k-1} T^j\xi) > \frac{1}{n}\sum_{k=0}^{n-1} H_{\lambda'_n}(\alpha^k \circ \gamma; \vee_{k-p+r}^{k-1} T^j\xi) - \varepsilon$$

for $p > r$ and n sufficiently large, since the contribution of the summands with $k < p - r$ becomes negligible as $n \to \infty$. Thus property (ii) of the map f and inequality (5.5) imply

$$\frac{1}{n}\sum_{k=0}^{n-1}(H_{\lambda'_n}(\alpha^k\circ\gamma;T^k\xi)-H_{\lambda'_n}(\alpha^k\circ\gamma;\vee_{k-p+r}^{k-1}T^j\xi))>h-5\varepsilon,$$

or equivalently,

$$\frac{1}{n}\sum_{k=0}^{n-1}(H_{\lambda'_n\circ(\alpha^k\otimes\beta^k)}(\gamma;\xi)-H_{\lambda'_n\circ(\alpha^k\otimes\beta^k)}(\gamma;\vee_{-p+r}^{-1}T^j\xi))>h-5\varepsilon$$

for any fixed $p>r$ and any $n\in I$ sufficiently large. Set

$$\lambda''_n=\frac{1}{n}\sum_{k=0}^{n-1}\lambda'_n\circ(\alpha^k\otimes\beta^k).$$

The couplings $\lambda'_n\circ(\alpha^k\otimes\beta^k)\circ(\mathrm{id}\otimes\mathrm{pr}_0^*)=\lambda'_n\circ(\mathrm{id}\otimes\mathrm{pr}_k^*)\circ(\alpha^{-1}\otimes\hat\alpha^*)^{-k}$, where $\hat\alpha^*(g)=g\circ\hat\alpha$ for $g\in C(X)$, are canonical for $0\le k\le n-1$ by property (iii) of f and the fact that the automorphism $\alpha^{-1}\otimes\hat\alpha^*$ of $A\otimes C(X)$ preserves the set of canonical couplings. Since $\xi=\mathrm{pr}_0^{-1}(\zeta)$, by Lemma 5.1.6 and Proposition 5.1.3(iii) we get

$$H_{\lambda''_n}(\gamma;\xi)-H_{\lambda''_n}(\gamma;\vee_{-p+r}^{-1}T^j\xi)>h-6\varepsilon.$$

Now replace Z by the spectrum Y of a separable β-invariant C*-algebra consisting of bounded Borel functions on Z and containing the characteristic functions of the elements of ξ. Denote the homeomorphism of Y defined by T and the restriction of λ''_n to $A\otimes C(Y)$ by the same symbols. We can think of ξ as a partition of Y consisting of clopen sets. Let λ be a weak* limit point of $\{\lambda''_n\}_{n\in I}$. Since the von Neumann entropy is continuous, we have

$$H_\lambda(\gamma;\xi)-H_\lambda(\gamma;\vee_{-k}^{-1}T^j\xi)\ge h-6\varepsilon$$

for any $k\in\mathbb{N}$. Hence

$$H_\lambda(\gamma;\vee_{-k}^{0}T^j\xi)-H_\lambda(\gamma;\vee_{-k}^{-1}T^j\xi)\ge h-6\varepsilon,$$

equivalently, with $\mu=\lambda|_{C(Y)}$,

$$H_\mu(\vee_{-k}^{0}T^j\xi)-H_\mu(\xi|\vee_{-k}^{-1}T^j\xi)-H_{\lambda,\gamma}(\vee_{-k}^{0}T^j\xi)+H_{\lambda,\gamma}(\vee_{-k}^{-1}T^j\xi)\ge h-6\varepsilon,$$

which by Proposition 5.1.3(vi) can be written as

$$H_\mu(\xi|\vee_{-k}^{-1}T^j\xi)-H_{\lambda,\gamma}(\xi|\vee_{-k}^{-1}T^j\xi)\ge h-6\varepsilon.$$

Letting $k\to\infty$, by Proposition 5.1.3(iv) we get

$$H_\mu(\xi|\xi^-)-H_{\lambda,\gamma}(\xi|\xi^-)\ge h-6\varepsilon.$$

Since $H_{\lambda,\gamma}(\xi|\xi^-)\ge H_{\lambda,A}(\xi|\xi^-)$ by Proposition 5.1.3(iv), using Proposition 5.1.4 we conclude that $h_\varphi^{ST}(\alpha)\ge h-6\varepsilon$, which completes the proof of the theorem. □

In Proposition 3.2.7 we proved that a state on a nuclear subalgebra of a C^*-algebra can be extended to a state on the algebra in an entropy-increasing way. As our first application of existence of abelian models we shall now prove another similar result.

Proposition 5.1.7. *Let α be an automorphism of a nuclear C^*-algebra A, $B \subset A$ an α-invariant C^*-subalgebra, ψ an α-invariant state on B. Then for every $h < h_\psi(\alpha|_B)$ there exists an α-invariant state φ on A such that $\varphi|_B = \psi$ and $h_\varphi(\alpha) > h$.*

Proof. Let λ be a stationary coupling of $(B, \psi, \alpha|_B)$ with (X, μ, T), and ξ a measurable partition of (X, μ) such that $h_\mu(\xi; T) - H_\mu(\xi) + H_\lambda(B; \xi) > h$. Extend λ to an $(\alpha \otimes \beta)$-invariant state Λ on $A \otimes L^\infty(X, \mu)$, where β is the automorphism of $L^\infty(X, \mu)$ defined by T, and put $\varphi = \Lambda|_A$. By Proposition 5.1.3(i) we have $H_\Lambda(A; \xi) \geq H_\lambda(B; \xi)$, hence $h_\varphi(\alpha) = h_\varphi^{ST}(\alpha) > h$. □

If $(X/\xi, \mu, T)$ is a factor system of (X, μ, T) such that the map $X \to X/\xi$ is finite-to-one almost everywhere, then it is not difficult to check that both systems have the same entropy. The following is a noncommutative analogue of this result.

Proposition 5.1.8. *Let (M, φ, α) be a W^*-dynamical system having a φ-approximating net, $N \subset M$ an α-invariant von Neumann subalgebra, $E\colon M \to N$ a φ-preserving faithful normal conditional expectation commuting with α. Assume there exists a constant $c > 0$ such that $E(a) \geq ca$ for any positive $a \in M$. Then $h_\varphi(\alpha) = h_{\varphi|N}(\alpha|_N)$.*

Proof. Using the conditional expectation E, any stationary coupling of the system $(N, \varphi|_N, \alpha|_N)$ with an abelian system can be extended to a coupling of (M, φ, α) with the same system. This implies $h_\varphi(\alpha) \geq h_{\varphi|N}(\alpha|_N)$.

Conversely, let λ be a stationary coupling of (M, φ, α) with (X, μ, T), and ξ a measurable partition of (X, μ). For any normal state ψ on M we have $\psi \circ E \geq c\psi$. By Theorem 2.3.1(vii) we then get

$$S(\psi, \varphi) - S(\psi|_N, \varphi|_N) = S(\psi, \psi \circ E) \leq S(\psi, c\psi) = -\log c.$$

It follows that
$$H_\lambda(M; \xi) - H_\lambda(N; \xi) \leq -\log c,$$
and hence $h_\varphi(\alpha) \leq h_{\varphi|N}(\alpha|_N) - \log c$. Applying this to α^n, dividing by n and letting $n \to \infty$ we get $h_\varphi(\alpha) \leq h_{\varphi|N}(\alpha|_N)$. □

5.2 Zero Entropy Systems

Let (A, φ, α) be a C^*-dynamical system. Consider the space of the GNS-representation defined by φ. On this space the automorphism α is implemented by the unitary U defined by $Ua\xi_\varphi = \alpha(a)\xi_\varphi$ for $a \in A$. We say that the system has **singular spectrum** if the spectral measure of U is singular with respect to the Lebesgue measure on $\mathbb{T}$. If the space is nonseparable, we mean by this that the spectral measure of the restriction of U to any separable invariant subspace is singular. Equivalently, for any vector ξ the measure ν_ξ on $\mathbb{T}$ such that $(U^n \xi, \xi) = \int_{\mathbb{T}} z^n d\nu_\xi(z)$ for $n \in \mathbb{Z}$ is singular.

Theorem 5.2.1. *Let (A, φ, α) be a C^*-dynamical system with singular spectrum. Then $h_\varphi(\alpha) = h_\varphi^{ST}(\alpha) = 0$.*

Proof. We shall first reduce the proof to the case when A separable. Choose an increasing net $\{A_i\}_i$ of separable α-invariant C^*-subalgebras with dense union. Let λ be a stationary coupling of (A, φ, α) with an abelian system (X, μ, T), and ξ a finite measurable partition of (X, μ). By Corollary 2.3.6 it follows that

$$H_\lambda(A; \xi) = \lim_i H_\lambda(A_i; \xi).$$

Hence $h_\varphi^{ST}(\alpha) \leq \liminf_i h_{\varphi|A_i}^{ST}(\alpha|_{A_i})$. So it is enough to show that the entropy of $\alpha|_{A_i}$ is zero for every i. Since the systems $(A_i, \varphi|_{A_i}, \alpha|_{A_i})$ still have singular spectrum, we see that is suffices to prove the theorem for separable C^*-algebras.

By Proposition 5.1.4 we have to show that

$$H_\mu(\xi|\xi^-) = H_{\lambda, A}(\xi|\xi^-) \tag{5.6}$$

for every stationary coupling λ with (X, μ, T) and every finite measurable partition ξ. We shall do this in three steps, first assuming that (X, μ, T) is a Bernoulli shift, then that it is ergodic, and finally in general.

So assume that (X, μ, T) is a Bernoulli shift. We claim that then λ is trivial, $\lambda = \varphi \otimes \mu$. Let β be the automorphism of $L^\infty(X, \mu)$ defined by T, $\beta(f) = f \circ T^{-1}$ for $f \in L^\infty(X, \mu)$. Denote by V the canonical unitary on $L^2(X, \mu)$ implementing β, $Vf = f \circ T^{-1}$ for $f \in L^2(X, d\mu)$. Let ζ be the standard generating partition for (X, μ, T). Then, as we already remarked in Sect. 4.3, $\zeta^- \prec T\zeta^-$, the algebra $\cup_{n \in \mathbb{Z}} L^\infty(X/T^n \zeta^-)$ is weakly operator dense in $L^\infty(X, \mu)$, and $\cap_{n \in \mathbb{Z}} L^\infty(X/T^n \zeta^-) = \mathbb{C}1$. It follows that if we set $L = L^2(X/T\zeta^-, d\mu) \ominus L^2(X/\zeta^-, d\mu)$ then

$$L^2(X, d\mu) \ominus \mathbb{C} = \bigoplus_{n \in \mathbb{Z}} V^n L.$$

Thus on the orthogonal complement of the constants the unitary V has homogeneous Lebesgue spectrum, see C.7. Now let H be the space of the GNS-representation defined by λ, and $v \in H$ the corresponding cyclic vector. Let W be the unitary on H implementing $\alpha \otimes \beta$. According to C.7 the space H decomposes into a direct sum $H_s \oplus H_a$ such that the restriction of W to H_s has singular spectral measure, and the restriction of W to H_a has absolutely continuous spectral measure. The restriction of W to $\overline{Av}$ can be identified with U, while the restriction of W to $\overline{L^\infty(X, \mu)v}$ can be identified with V. It follows that $av \in H_s$ for any $a \in A$, and $fv \in H_a$ for any $f \in L^\infty(X, \mu)$ with $\mu(f) = 0$. Hence av and fv are orthogonal, that is, $\lambda(a \otimes \bar{f}) = 0$. For any $f \in L^\infty(X,\mu)$ we then have

$$\lambda(a \otimes f) = \varphi(a)\mu(f) + \lambda(a \otimes (f - \mu(f)1)) = \varphi(a)\mu(f),$$

and our claim is proved.

Assume now that (X, μ, T) is ergodic. Replacing (X, μ, T) by the factor system $(X/\xi^\pm, \mu, T)$, where $\xi^\pm = \vee_{n \in \mathbb{Z}} T^n \xi$, we may assume that ξ is a generating partition. By Sinai's Theorem 1.1.7 there exists a Bernoullian factor system with the same entropy. In other words, there exists a finite measurable partition ζ such that the partitions $T^n \zeta$, $n \in \mathbb{Z}$, are mutually independent, and $h_\mu(T) = H_\mu(\zeta)$. Since ξ is generating and, as we already observed in the proof of Proposition 5.1.4,

$$H_{\lambda,A}(\vartheta|\vartheta^-) = \lim_{n\to\infty} \frac{1}{n} H_{\lambda,A}(\vee_{k=0}^{n-1} T^k \vartheta)$$

for any finite ϑ, the same proof as that of Lemma 1.1.3 shows that

$$H_{\lambda,A}(\zeta|\zeta^-) \leq H_{\lambda,A}(\xi|\xi^-) + H_{\lambda,A}(\zeta|\xi^\pm) = H_{\lambda,A}(\xi|\xi^-).$$

On the other hand, by the case of Bernoullian systems considered above, we have

$$H_{\lambda,A}(\zeta|\zeta^-) = H_\mu(\zeta|\zeta^-) = h_\mu(T) = H_\mu(\xi|\xi^-).$$

Since we always have $H_{\lambda,A}(\xi|\xi^-) \leq H_\mu(\xi|\xi^-)$ by Proposition 5.1.3(iv), we conclude that (5.6) holds.

Consider the general case. As before, let β be the automorphism of $L^\infty(X, \mu)$ defined by T, $\beta(f) = f \circ T^{-1}$ for $f \in L^\infty(X, \mu)$. Let ϑ be the measurable partition such that $L^\infty(X/\vartheta) = L^\infty(X)^\beta$, the fixed point algebra of β. Then, see C.1-C.4, the space H of the GNS-representation defined by λ decomposes into a direct integral $\int_{X/\vartheta}^\oplus H_y d\mu(y)$, elements of $A \otimes L^\infty(X, \mu)$ are decomposable, and a normal extension Λ of λ to the commutant of $L^\infty(X/\vartheta)$ decomposes into a direct integral $\Lambda = \int_{X/\vartheta}^\oplus \Lambda_y d\mu(y)$. Without loss of generality we may assume that X and X/ϑ are compact metric spaces, $X \mapsto X/\vartheta$ is continuous, and T is a homeomorphism. Then by C.5 the representation of $A \otimes C(X)$ decomposes into a direct integral $\int_{X/\vartheta}^\oplus \pi_y d\mu(y)$ of representations. Set $\lambda_y = \Lambda_y \circ \pi_y$, $\varphi_y = \lambda_y|_A$ and $\mu_y = \lambda_y|_{C(X)}$. The state λ_y is $(\alpha \otimes \beta)$-invariant for a.e. $y \in X/\vartheta$, so that λ_y is a stationary coupling of (A, φ_y, α) with (X, μ_y, T).

If $a \in A$ and ν (resp. ν_y) is the measure on $\mathbb{T}$ such that $\varphi(a^* \alpha^n(a)) = \int_\mathbb{T} z^n d\nu(z)$ (resp. $\varphi_y(a^* \alpha^n(a)) = \int_\mathbb{T} z^n d\nu_y(z)$), then $\nu = \int_{X/\vartheta} \nu_y d\mu(y)$. Since ν is singular by assumption, it follows that ν_y is singular for a.e. y. Thus the system (A, φ_y, α) has singular spectrum for a.e. y.

Next note that (X, μ_y, T) is ergodic for a.e. y. We shall give a sketch of the argument, see e.g. [72, Theorem 3.42] for details. By Birkhoff's ergodic theorem, for any $f \in C(X)$, the sequence $\{n^{-1}\sum_{k=0}^{n-1} f(T^{-k}x)\}_{n=1}^{\infty}$ converges to $E(f)(x)$ for a.e. $x \in X$, where $E\colon L^\infty(X) \to L^\infty(X/\vartheta)$ is the μ-preserving conditional expectation. Since the operator $E(f)$ is diagonal, it follows that for $f \in C(X)$ and a.e. $y \in X/\vartheta$ the sequence $\{n^{-1}\sum_{k=0}^{n-1} \pi_y(\beta^k(f))\}_n$ converges to a scalar operator in the strong operator topology. But if for a fixed y this convergence holds for a countable dense subset of functions $f \in C(X)$, then (X, μ_y, T) is ergodic.

We can thus apply the previous case to the couplings λ_y and conclude that

$$H_{\mu_y}(\xi|\xi^-) = H_{\lambda_y,A}(\xi|\xi^-)$$

for any finite Borel partition ξ of X and a.e. $y \in X/\vartheta$. To prove that (5.6) holds, it is then enough to show that

$$H_\mu(\xi|\xi^-) = \int_{X/\vartheta} H_{\mu_y}(\xi|\xi^-)d\mu(y) \text{ and } H_{\lambda,A}(\xi|\xi^-) = \int_{X/\vartheta} H_{\lambda_y,A}(\xi|\xi^-)d\mu(y).$$

We claim that

$$\int_{X/\vartheta} H_{\lambda_y,A}(\xi|\xi^-)d\mu(y) = H_{\lambda,A}(\xi|\xi^- \vee \vartheta). \tag{5.7}$$

Since $H_{\lambda_y,A}(\xi| \vee_{k=1}^n T^{-k}\xi) \searrow H_{\lambda_y,A}(\xi|\xi^-)$, and similarly $H_{\lambda,A}(\xi|\xi^- \vee \vartheta)$ is the limit of $H_{\lambda,A}(\xi|\vartheta \vee \vee_{k=1}^n T^{-k}\xi)$, it suffices to check that

$$\int_{X/\vartheta} H_{\lambda_y,A}(\xi|\zeta)d\mu(y) = H_{\lambda,A}(\xi|\zeta \vee \vartheta)$$

for any finite Borel partition ζ. By Proposition 5.1.3(vi) it then suffices to check that

$$\int_{X/\vartheta} H_{\lambda_y,A}(\zeta)d\mu(y) = H_{\lambda,A}(\zeta|\vartheta)$$

for any ζ. By (5.2) for this, in turn, it is enough to check that for any Borel subset $Z \subset X$ we have

$$\int_{X/\vartheta} S(\lambda_y(\cdot \otimes \cdot \mathbb{1}_Z), \lambda_y)d\mu(y) = S(\lambda(\cdot \otimes \cdot \mathbb{1}_Z), \lambda). \tag{5.8}$$

Consider the states $\tilde\lambda$ and $\tilde\lambda_Z$ on $A \otimes C(X) \otimes C(X/\vartheta)$ defined by

$$\tilde\lambda(a \otimes f \otimes g) = \int_{X/\vartheta} \lambda_y(a \otimes f)g(y)d\mu(y) \text{ and } \tilde\lambda_Z(a \otimes f \otimes g) = \tilde\lambda(a \otimes f\mathbb{1}_Z \otimes g).$$

Then by Corollary 2.3.7 the left hand side of (5.8) coincides with $S(\tilde\lambda_Z, \tilde\lambda)$, while the right hand side is $S(\tilde\lambda_Z|_{A\otimes C(X)}, \tilde\lambda|_{A\otimes C(X)})$. Since the homomorphism $A \otimes C(X) \otimes C(X/\vartheta) \to A \otimes C(X)$, $a \otimes f \otimes g \mapsto a \otimes fg$, which is also a conditional expectation, preserves both states, both sides indeed coincide by Theorem 2.3.1(ix) or Corollary 2.3.3. Thus (5.7) is established.

Any finite measurable partition $\zeta \prec \vartheta$ of (X, μ) is T-invariant. By Proposition 5.1.3(vi) it follows that

$$H_{\lambda,A}(\xi \vee \zeta | \xi^- \vee \zeta^-) = H_{\lambda,A}(\xi \vee \zeta | \xi^- \vee \zeta) = H_{\lambda,A}(\xi | \xi^- \vee \zeta).$$

On the other hand,

$$H_{\lambda,A}(\xi | \xi^-) = \lim_{n \to \infty} \frac{1}{n} H_{\lambda,A}(\vee_{k=0}^{n-1} T^k \xi)$$
$$\leq \lim_{n \to \infty} \frac{1}{n} H_{\lambda,A}(\vee_{k=0}^{n-1} T^k (\xi \vee \zeta)) = H_{\lambda,A}(\xi \vee \zeta | \xi^- \vee \zeta^-).$$

Consequently

$$H_{\lambda,A}(\xi | \xi^-) = H_{\lambda,A}(\xi | \xi^- \vee \zeta).$$

Choosing an increasing sequence of finite measurable partitions ζ_n, $n \in \mathbb{N}$, such that $\vee_n \zeta_n = \vartheta$, we conclude that

$$H_{\lambda,A}(\xi | \xi^-) = H_{\lambda,A}(\xi | \xi^- \vee \vartheta).$$

Together with (5.7) this shows that $H_{\lambda,A}(\xi | \xi^-) = \int_{X/\vartheta} H_{\lambda_y,A}(\xi | \xi^-) d\mu(y)$. A similar argument gives $H_\mu(\xi | \xi^-) = \int_{X/\vartheta} H_{\mu_y}(\xi | \xi^-) d\mu(y)$, which completes the proof of the theorem. □

Proposition 5.1.2 implies that if a system has a nontrivial stationary coupling with a classical Bernoulli shift, then $h_\varphi^{ST}(\alpha) > 0$. The proof of the above theorem shows that the converse would be true if we could show that if the system does not have nontrivial couplings with Bernoulli shifts, and $\varphi = \int_X \varphi_x d\mu(x)$, then the same is true for (A, φ_x, α) for almost every x. This is obviously true if we do not have nontrivial integral decompositions of φ into α-invariant states. Thus we get the following.

Theorem 5.2.2. *Let (A, φ, α) be a C^*-dynamical system with separable A. Assume φ is ergodic, that is, φ is an extremal α-invariant state. Then $h_\varphi^{ST}(\alpha) > 0$ if and only if there exists a nontrivial stationary coupling with a classical Bernoulli shift.* □

We shall now introduce a class of systems which do not have nontrivial couplings with abelian systems.

Definition 5.2.3. *A system (A, α) consisting of a unital C^*-algebra A and an automorphism α is called asymptotically highly anticommutative if there exists a set $\{x_i\}_{i \in I}$ of elements of A which together with the unit spans a dense subspace of A and satisfies the following property. For any $i \in I$, $\varepsilon > 0$ and $n \in \mathbb{N}$ there exist $m_1, \ldots, m_n \in \mathbb{N}$ such that $\|\alpha^{m_k}(x_i^*)\alpha^{m_l}(x_i) + \alpha^{m_l}(x_i)\alpha^{m_k}(x_i^*)\| < \varepsilon$ for all $k, l \leq n$ with $k \neq l$.*

Proposition 5.2.4. *If (A, α) is asymptotically highly anticommutative then there exists a unique α-invariant state on A.*

Proof. Let φ be an α-invariant state, and $\{x_i\}_{i \in I}$ a set as in the definition of an asymptotically highly anticommutative system. Fix $i \in I$. Let $\varepsilon > 0$ and $m_1, \ldots, m_n \in \mathbb{N}$ be such that $\|\alpha^{m_k}(x_i^*)\alpha^{m_l}(x_i) + \alpha^{m_l}(x_i)\alpha^{m_k}(x_i^*)\| < \varepsilon$ for all $k, l \leq n$ with $k \neq l$. Consider the element $x = n^{-1} \sum_{k=1}^{n} \alpha^{m_k}(x_i)$. Then $x^*x + xx^*$ equals

$$\frac{1}{n^2} \sum_{k=1}^{n} \alpha^{m_k}(x_i^* x_i + x_i x_i^*) + \frac{1}{n^2} \sum_{k \neq l} (\alpha^{m_k}(x_i^*)\alpha^{m_l}(x_i) + \alpha^{m_l}(x_i)\alpha^{m_k}(x_i^*)),$$

whence

$$\|x^*x + xx^*\| \leq \frac{2\|x_i\|^2}{n} + \frac{(n-1)\varepsilon}{n}.$$

Since

$$|\varphi(x_i)| = |\varphi(x)| \leq \frac{1}{2}(\varphi(x^*x)^{1/2} + \varphi(xx^*)^{1/2}) \leq \frac{1}{\sqrt{2}}\varphi(x^*x + xx^*)^{1/2},$$

it follows that $\varphi(x_i) = 0$. Since $\{1\} \cup \{x_i\}_{i \in I}$ spans a dense subspace of A, the condition $\varphi(x_i) = 0$ completely determines the state. □

The above proof implies the following.

Corollary 5.2.5. *If (A, φ, α) is an asymptotically highly anticommutative C^*-dynamical system, then any stationary coupling with an abelian dynamical system is trivial. In particular, $h_\varphi(\alpha) = h_\varphi^{ST}(\alpha) = 0$.*

Proof. Let λ be a stationary coupling of (A, φ, α) with (X, μ, T). Then the same estimate as in the proof of the previous proposition applied to $x_i \otimes f$ instead of x_i shows that $\lambda(x_i \otimes f) = 0 = \varphi(x_i)\mu(f)$. Together with the equality $\lambda(1 \otimes f) = \mu(f)$ this implies $\lambda = \varphi \otimes \mu$. □

We shall see examples of asymptotically highly anticommutative systems in Chaps. 11 and 12. In Chap. 14 we shall consider a system which is not asymptotically highly anticommutative, but which still has only trivial stationary couplings with abelian systems.

5.3 Notes

The notion of a stationary coupling was introduced by Sauvageot and Thouvenot [188], who obtained all the main results of the present chapter. In the classical case stationary couplings are called joinings. They were introduced by Furstenberg [68]. See [72], [184] for an exposition of ergodic theory from the theory of joinings point of view.

Propositions 5.1.7 and 5.1.8 are due to the authors [140], [141].

Theorem 5.2.1, proved by Sauvageot and Thouvenot [188], is an extension to the noncommutative case of a result of Rohlin and Sinai [183]. In the classical case the result states a bit more: if a system has positive entropy then the spectrum has a countably multiple Lebesgue component.

Asymptotically highly anticommutative systems were introduced by Narnhofer, Størmer and Thirring [128], who proved Proposition 5.2.4 and Corollary 5.2.5. The name was coined later by Narnhofer and Thirring [134].

6
Topological Entropy

In this chapter we shall study the extension of topological entropy to C*-algebras. The main problem is to find the right analogue of the cardinality $N(\mathcal{U})$ of a minimal subcover for an open cover $\mathcal{U}$. If X is a totally disconnected compact space and $\mathcal{U}$ a clopen partition of X, then $N(\mathcal{U})$ is the rank of the abelian algebra spanned by the characteristic functions of the elements of $\mathcal{U}$. In the general case instead of the algebra generated by the characteristic functions of the elements of $\mathcal{U}$ we shall consider finite collections of functions which are almost constant on the elements of $\mathcal{U}$. It turns out that instead of $N(\mathcal{U})$ one can then take the ranks of completely positive approximations of these collections.

6.1 Rank of a Completely Positive Approximation

Let A be a unital C*-algebra, Ω a finite subset of A.

Definition 6.1.1. *For $\delta > 0$ denote by $\mathrm{rcp}_A(\Omega, \delta)$ the minimal number $n \in \mathbb{N}$ for which there exist a unital injective homomorphism $\pi\colon A \to B$ for some C*-algebra B and unital completely positive maps $\theta\colon A \to \mathrm{Mat}_n(\mathbb{C})$ and $\gamma\colon \mathrm{Mat}_n(\mathbb{C}) \to B$ such that $\|(\gamma \circ \theta)(a) - \pi(a)\| < \delta$ for any $a \in \Omega$. Put $\mathrm{rcp}_A(\Omega, \delta) = +\infty$ if no such approximation exists.*

We shall write $\mathrm{rcp}(\Omega, \delta)$ instead of $\mathrm{rcp}_A(\Omega, \delta)$ when no confusion can arise.

If D is a finite dimensional C*-algebra, then D can be imbedded into $\mathrm{Mat}_n(\mathbb{C})$, where n is the rank of D. A unital completely positive map $D \to B$ can be extended to a unital completely positive map $\mathrm{Mat}_n(\mathbb{C}) \to B$ by composing it with the trace preserving conditional expectation $\mathrm{Mat}_n(\mathbb{C}) \to D$. So we could define $\mathrm{rcp}_A(\Omega, \delta)$ as the minimal number $n \in \mathbb{N}$ for which there exist an injective homomorphism $\pi\colon A \to B$ for some C*-algebra B and unital completely positive maps $\theta\colon A \to D$ and $\gamma\colon D \to B$ such that $n = \mathrm{rank}\, D$ and $\|(\gamma \circ \theta)(a) - \pi(a)\| < \delta$ for any $a \in \Omega$.

Observe next that there is no need to consider all possible embeddings of A.

Lemma 6.1.2. *In computing* $\mathrm{rcp}_A(\Omega, \delta)$ *it is enough to consider one faithful representation* $\pi \colon A \to B(H)$. *It also suffices to take a nuclear embedding* $\pi \colon A \hookrightarrow B$, *if one exists.*

Recall that a homomorphism $\pi \colon A \to B$ is called **nuclear** if there exist two nets $\{\theta_i \colon A \to \mathrm{Mat}_{n_i}(\mathbb{C})\}_{i \in I}$ and $\{\gamma_i \colon \mathrm{Mat}_{n_i}(\mathbb{C}) \to B\}_{i \in I}$ of unital completely positive maps such that $\|(\gamma_i \circ \theta_i)(a) - \pi(a)\| \to 0$ for any $a \in A$. The algebras admitting nuclear embeddings are said to be nuclearly embeddable, or **exact**. If $\mathrm{rcp}(\Omega, \delta)$ is finite for any $\Omega \subset A$ and any $\delta > 0$, then Lemma 6.1.2 implies that any faithful representation $\pi \colon A \to B(H)$ is nuclear, so that A is exact.

If A is nuclear, then the identity map $A \to A$ is nuclear, so we can take $B = A$. At least one reason to consider the class of exact C*-algebras is that subalgebras of nuclear algebras are not necessarily nuclear, but always exact. The class of exact algebras is quite large. It is obviously closed under taking subalgebras and minimal tensor products. In Chap. 8 we shall show that it is closed under taking crossed products by actions of amenable locally compact groups, while in Chap. 14 we shall prove that reduced free products of exact algebras are exact. Quotients of exact algebras are also exact. This is nontrivial, and will not be used in the sequel.

Proof of Lemma 6.1.2. Denote by $\mathrm{rcp}(\pi, \Omega, \delta)$ the quantity defined exactly as $\mathrm{rcp}(\Omega, \delta)$ but using a fixed embedding $\pi \colon A \hookrightarrow B$.

Let $\pi \colon A \to B(H)$ be a faithful representation. Given an arbitrary embedding $\pi_0 \colon A \hookrightarrow B$, by Arveson's extension theorem A.8 there exists a unital completely positive map $\Phi \colon B \to B(H)$ such that $\Phi \circ \pi_0 = \pi$. Given a unital completely positive map $\gamma \colon \mathrm{Mat}_n(\mathbb{C}) \to B$ we can now consider $\Phi \circ \gamma \colon \mathrm{Mat}_n(\mathbb{C}) \to B(H)$. Since $\|(\Phi \circ \gamma \circ \theta)(a) - \pi(a)\| = \|\Phi((\gamma \circ \theta)(a) - \pi_0(a))\|$, we get $\mathrm{rcp}(\pi, \Omega, \delta) \leq \mathrm{rcp}(\pi_0, \Omega, \delta)$.

Similarly, assume $\pi \colon A \hookrightarrow B$ is a nuclear embedding. Let $\{\theta_i \colon A \to \mathrm{Mat}_{n_i}(\mathbb{C})\}_{i \in I}$ and $\{\gamma_i \colon \mathrm{Mat}_{n_i}(\mathbb{C}) \to B\}_{i \in I}$ be nets of unital completely positive maps such that $\|(\gamma_i \circ \theta_i)(a) - \pi(a)\| \to 0$ for any $a \in A$. Let $\pi_0 \colon A \hookrightarrow B_0$ be another embedding, and let $\theta_0 \colon A \to \mathrm{Mat}_n(\mathbb{C})$ and $\gamma_0 \colon \mathrm{Mat}_n(\mathbb{C}) \to B_0$ be unital completely positive maps such that $\|(\gamma_0 \circ \theta_0)(a) - \pi_0(a)\| < \delta$ for $a \in \Omega$. Again by Arveson's extension theorem there exist unital completely positive maps $\bar{\theta}_i \colon B_0 \to \mathrm{Mat}_{n_i}(\mathbb{C})$, $i \in I$, such that $\bar{\theta}_i \circ \pi_0 = \theta_i$. Consider the maps $\gamma_i \circ \bar{\theta}_i \circ \gamma_0 \colon \mathrm{Mat}_n(\mathbb{C}) \to B$. We have

$$\|(\gamma_i \circ \bar{\theta}_i \circ \gamma_0 \circ \theta_0)(a) - \pi(a)\|$$
$$\leq \|(\gamma_i \circ \bar{\theta}_i \circ \gamma_0 \circ \theta_0)(a) - (\gamma_i \circ \bar{\theta}_i \circ \pi_0)(a)\| + \|(\gamma_i \circ \bar{\theta}_i \circ \pi_0)(a) - \pi(a)\|.$$
$$\leq \|(\gamma_0 \circ \theta_0)(a) - \pi_0(a)\| + \|(\gamma_i \circ \theta_i)(a) - \pi(a)\|.$$

Since $\|(\gamma_i \circ \theta_i)(a) - \pi(a)\| \to 0$, we see that for sufficiently large i we get $\|(\gamma_i \circ \bar{\theta}_i \circ \gamma_0 \circ \theta_0)(a) - \pi(a)\| < \delta$ for any $a \in \Omega$. Thus $\mathrm{rcp}(\pi, \Omega, \delta) \leq \mathrm{rcp}(\pi_0, \Omega, \delta)$. □

6.1 Rank of a Completely Positive Approximation

In order to obtain a better understanding of the definition consider the abelian case.

Lemma 6.1.3. *Let X be a compact space, $\Omega \subset C(X)$ a finite subset, and $\delta > 0$. Let $\mathcal{U}$ be a finite open cover of X such that $|f(x) - f(y)| < \delta$ for every $f \in \Omega$ and $x, y \in X$ lying in the same element of $\mathcal{U}$. Then $\mathrm{rcp}(\Omega, \delta) \leq N(\mathcal{U})$.*

Proof. Let $\mathcal{V}$ be a subcover of $\mathcal{U}$ with minimal number $N = N(\mathcal{U})$ of elements. Let $\mathcal{V} = \{V_1, \ldots, V_N\}$, and for each i choose $x_i \in V_i$. Denote the subset of X consisting of the x_i's by Y. Let further $\{\chi_i\}_{i=1}^{N}$ be a partition of unity subordinate to $\mathcal{V}$, that is, $\mathrm{supp}\,\chi_i \subset V_i$ and $\sum_i \chi_i = 1$. We set $B = C(Y)$, and define
$$\theta \colon C(X) \to B, \quad \theta(f) = f|_Y,$$
and
$$\gamma \colon B \to C(X), \quad \gamma(g) = \sum_{i=1}^{N} g(x_i)\chi_i.$$
If $f \in \Omega$ then for any $x \in X$
$$|f(x) - ((\gamma \circ \theta)(f))(x)| \leq \sum_{i=1}^{N} \chi_i(x)|f(x) - f(x_i)|$$
$$= \sum_{i:\, x \in V_i} \chi_i(x)|f(x) - f(x_i)| < \delta.$$
Hence $\mathrm{rcp}(\Omega, \delta) \leq \mathrm{rank}\, B = N(\mathcal{U})$. $\square$

We next list some properties of rcp.

Lemma 6.1.4. *For every unital C^*-algebra A we have:*
(i) $\mathrm{rcp}(\Omega, \delta) = \mathrm{rcp}(\Omega \cup \Omega^* \cup \{1\}, \delta)$;
(ii) *if $\Omega \subset \Omega_0$ then $\mathrm{rcp}(\Omega, \delta) \leq \mathrm{rcp}(\Omega_0, \delta)$;*
(iii) *if $\Omega \subset_{\delta_0} \Omega_0$, meaning that for any $a \in \Omega$ there is $a_0 \in \Omega_0$ such that $\|a - a_0\| < \delta_0$, then $\mathrm{rcp}(\Omega, \delta + 2\delta_0) \leq \mathrm{rcp}(\Omega_0, \delta)$;*
(iv) *if each $a \in \Omega$ is a linear combination of at most k elements of Ω_0 with coefficients of modulus not greater than l, then $\mathrm{rcp}(\Omega, kl\delta) \leq \mathrm{rcp}(\Omega_0, \delta)$;*
(v) *if Ω consists of unitary operators and contains the unit, then $\mathrm{rcp}(\Omega^2, 4\delta) \leq \mathrm{rcp}(\Omega, \delta) \leq \mathrm{rcp}(\Omega^2, \delta)$, where $\Omega^2 = \{uv \mid u, v \in \Omega\}$;*
(vi) *if α is an automorphism of A, then $\mathrm{rcp}(\alpha(\Omega), \delta) = \mathrm{rcp}(\Omega, \delta)$;*
(vii) *if $\beta \colon A \to C$ is an injective unital homomorphism, then $\mathrm{rcp}_C(\beta(\Omega), \delta) = \mathrm{rcp}_A(\Omega, \delta)$.*

Proof. Parts (i)-(iv) are straightforward. For example, part (iii) follows from the inequality
$$\|(\gamma \circ \theta)(a) - \pi(a)\| \leq \|(\gamma \circ \theta)(a_0) - \pi(a_0)\| + \|(\gamma \circ \theta)(a - a_0)\| + \|\pi(a - a_0)\|.$$

To prove (v) note that the second inequality is trivial as $\Omega \subset \Omega^2$. To prove the other inequality suppose $A \subset B$, and let $\theta\colon A \to \mathrm{Mat}_n(\mathbb{C})$ and $\gamma\colon \mathrm{Mat}_n(\mathbb{C}) \to B$ be unital completely positive maps such that $\|(\gamma \circ \theta)(a) - a\| < \delta$ for $a \in \Omega$. Put $\Phi = \gamma \circ \theta$. Then for $u \in \Omega$

$$\|\Phi(u)\Phi(u)^* - 1\| \le \|(\Phi(u) - u)\Phi(u)^*\| + \|u(\Phi(u)^* - u^*)\| < 2\delta,$$

and similarly $\|\Phi(u)\Phi(v) - uv\| < 2\delta$ and $\|\Phi(v)^*\Phi(v) - 1\| < 2\delta$ when furthermore $v \in \Omega$. Thus by A.5 we have

$$\begin{aligned}
\|\Phi(uv) - uv\| &< \|\Phi(uv) - \Phi(u)\Phi(v)\| + 2\delta \\
&\le \|\Phi(uu^*) - \Phi(u)\Phi(u)^*\|^{1/2} \|\Phi(v^*v) - \Phi(v)^*\Phi(v)\|^{1/2} + 2\delta \\
&= \|1 - \Phi(u)\Phi(u)^*\|^{1/2} \|1 - \Phi(v)^*\Phi(v)\|^{1/2} + 2\delta \\
&< (2\delta)^{1/2}(2\delta)^{1/2} + 2\delta = 4\delta.
\end{aligned}$$

It follows that $\mathrm{rcp}(\Omega^2, 4\delta) \le \mathrm{rcp}(\Omega, \delta)$.

Part (vi) is a particular case of (vii).

To prove (vii), by Lemma 6.1.2 it suffices to take a faithful representation $\pi\colon C \to B(H)$ and check that $\mathrm{rcp}(\pi \circ \beta, \Omega, \delta) = \mathrm{rcp}(\pi, \beta(\Omega), \delta)$. This follows again from Arveson's extension theorem, which implies that every unital completely positive map $A \to \mathrm{Mat}_n(\mathbb{C})$ extends to C. □

6.2 Topological Entropy

We are now in position to define topological entropy.

Definition 6.2.1. *Let A be a unital C^*-algebra, and α an automorphism of A. Let*

$$\begin{aligned}
ht(\Omega, \delta; \alpha) &= \limsup_{n\to\infty} \frac{1}{n} \log \mathrm{rcp}_A(\Omega \cup \alpha(\Omega) \cup \ldots \cup \alpha^{n-1}(\Omega), \delta), \\
ht(\Omega; \alpha) &= \sup_{\delta>0} ht(\Omega, \delta; \alpha), \\
ht(\alpha) &= \sup_{\Omega} ht(\Omega; \alpha),
\end{aligned}$$

where the last supremum is taken over all finite subsets Ω of A. The quantity $ht(\alpha)$ is called the topological entropy of α.

Note that if $\delta_1 < \delta_2$, then $\mathrm{rcp}(\Omega, \delta_1) \ge \mathrm{rcp}(\Omega, \delta_2)$. So the supremum in the definition of $ht(\Omega; \alpha)$ can be replaced by the limit as $\delta \to 0$.

We know that $\mathrm{rcp}(\Omega, \delta)$ is finite for any $\delta > 0$ and any finite $\Omega \subset A$ if and only if A is exact. Thus, even though our definition of topological entropy makes sense for arbitrary C*-algebras, it is infinity for any automorphism of a nonexact algebra.

Theorem 6.2.2. *Let α be an automorphism of an exact C^*-algebra A. Then*
(i) *if $\beta\colon A \to C$ is an isomorphism then $ht(\alpha) = ht(\beta \circ \alpha \circ \beta^{-1})$;*
(ii) *$ht(\alpha^n) = |n|ht(\alpha)$ for every $n \in \mathbb{Z}$; if $\{\alpha_t\}_{t\in\mathbb{R}}$ is a one-parameter automorphism group of A such that $\alpha_1 = \alpha$, then $ht(\alpha_t) = |t|ht(\alpha)$ for $t \in \mathbb{R}$;*
(iii) *if B is an α-invariant unital subalgebra of A, then $ht(\alpha) \geq ht(\alpha|_B)$;*
(iv) *if β is an automorphism of a unital exact C^*-algebra B, then*
$$ht(\alpha \otimes \beta) \leq ht(\alpha) + ht(\beta);$$
in particular, $ht(\alpha \otimes \mathrm{id}_B) = ht(\alpha)$;
(v) *under the assumptions of (iv), $ht(\alpha \oplus \beta) = \max\{ht(\alpha), ht(\beta)\}$.*

Proof. Parts (i) and (iii) are obvious.

By Lemma 6.1.4(vi) we have
$$\mathrm{rcp}(\cup_{k=0}^{m-1}\alpha^k(\Omega), \delta) = \mathrm{rcp}(\cup_{k=0}^{m-1}\alpha^{-k}(\Omega), \delta),$$
which shows $ht(\alpha) = ht(\alpha^{-1})$. Thus to prove the first part of (ii) we may assume $n > 0$. We have $\cup_{k=0}^{m-1}\alpha^{nk}(\Omega) \subset \cup_{k=0}^{n(m-1)}\alpha^k(\Omega)$, so that
$$\mathrm{rcp}(\cup_{k=0}^{m-1}\alpha^{nk}(\Omega), \delta) \leq \mathrm{rcp}(\cup_{k=0}^{n(m-1)}\alpha^k(\Omega), \delta),$$
which implies $ht(\Omega, \delta; \alpha^n) \leq n\, ht(\Omega, \delta; \alpha)$, and thus $ht(\alpha^n) \leq n\, ht(\alpha)$. For the reverse inequality note that if $\Omega \subset A$, then $\Omega_0 = \cup_{k=0}^{n-1}(\Omega)$ satisfies
$$\bigcup_{k=0}^{m-1} \alpha^k(\Omega) \subset \bigcup_{k=0}^{[m/n]} \alpha^{nk}(\Omega_0).$$

Thus we have $n\, ht(\Omega, \delta; \alpha) \leq ht(\Omega_0, \delta; \alpha^n)$. Since Ω is an arbitrary finite subset of A, it follows that $n\, ht(\alpha) \leq ht(\alpha^n)$.

Let now $\{\alpha_t\}_{t\in\mathbb{R}}$ be a one-parameter automorphism group of A such that $\alpha_1 = \alpha$. As in the proof of Theorem 3.2.5(i), to prove that $t^{-1}ht(\alpha_t)$ is independent of $t > 0$, it suffices to check that $ht(\alpha_t) \leq t\, ht(\alpha)$. Given a finite subset $\Omega \subset A$ and $\delta > 0$ find a finite $\Omega_0 \subset A$ such that $\alpha_s(\Omega) \subset_\delta \Omega_0$ for $0 \leq s \leq 1$. Then by Lemma 6.1.4(iii)
$$\mathrm{rcp}(\cup_{k=0}^{n-1}\alpha_t^k(\Omega), 3\delta) \leq \mathrm{rcp}(\cup_{k=0}^{[(n-1)t]}\alpha^k(\Omega_0), \delta),$$
so that $ht(\Omega, 3\delta; \alpha_t) \leq t\, ht(\Omega_0, \delta; \alpha) \leq t\, ht(\alpha)$. Hence $ht(\alpha_t) \leq t\, ht(\alpha)$.

To prove the inequality in (iv) note that by Lemma 6.1.4(iii),(iv) in order to compute the entropy it suffices to consider finite subsets of $A \otimes B$ of the form $\Omega = \Omega_1 \otimes \Omega_2$, where Ω_1 and Ω_2 are finite subsets of the unit balls of A and B, respectively. Since $\mathrm{Mat}_k(\mathbb{C}) \otimes \mathrm{Mat}_m(\mathbb{C}) \cong \mathrm{Mat}_{km}(\mathbb{C})$, it is then clear that

$$\mathrm{rcp}(\cup_{j=0}^{n-1}(\alpha\otimes\beta)^j(\Omega_1\otimes\Omega_2),2\delta)\leq\mathrm{rcp}(\cup_{j=0}^{n-1}\alpha^j(\Omega_1),\delta)\,\mathrm{rcp}(\cup_{j=0}^{n-1}\beta^j(\Omega_2),\delta).$$

This in turn gives $ht(\Omega_1\otimes\Omega_2,2\delta;\alpha\otimes\beta)\leq ht(\Omega_1,\delta;\alpha)+ht(\Omega_2,\delta;\beta)$, and hence we get the desired conclusion.

In particular, $ht(\alpha\otimes\mathrm{id}_B)\leq ht(\alpha)$ if B is exact, since $ht(\mathrm{id}_B)=0$ in this case. Since $A\subset A\otimes B$, we also have the opposite inequality.

It remains to prove (v). By considering direct sums of unital completely positive maps it is clear that if $\Omega_1\subset A$ and $\Omega_2\subset B$, then

$$\mathrm{rcp}(\Omega_1\oplus\Omega_2,\delta)\leq\mathrm{rcp}_A(\Omega_1,\delta)+\mathrm{rcp}_B(\Omega_2,\delta).$$

Since $\log(s+t)\leq\log 2+\max\{\log s,\log t\}$ for any positive numbers s,t, we conclude that $ht(\Omega_1\oplus\Omega_2,\delta;\alpha\oplus\beta)\leq\max\{ht(\Omega_1,\delta;\alpha),ht(\Omega_2,\delta;\beta)\}$.

If $A\subset B(H)$ and $B\subset B(K)$, then $A\oplus B\subset B(H\oplus K)$. Now given $\theta\colon A\oplus B\to\mathrm{Mat}_n(\mathbb{C})$ and $\gamma\colon\mathrm{Mat}_n(\mathbb{C})\to B(H\oplus K)$, fix a state φ on A and consider $\tilde\theta\colon A\to\mathrm{Mat}_n(\mathbb{C})$ defined by $\tilde\theta(a)=\theta(a\oplus\varphi(a)1)$, and the composition $\tilde\gamma\colon\mathrm{Mat}_n(\mathbb{C})\to B(H)$ of γ with the compression map $B(H\oplus K)\to B(H)$. It follows that $\mathrm{rcp}_A(\Omega,\delta)\leq\mathrm{rcp}_{A\oplus B}(\tilde\Omega,\delta)$, where $\tilde\Omega$ consists of the elements $a\oplus\varphi(a)1$, $a\in\Omega$. Thus $ht(\alpha)\leq ht(\alpha\oplus\beta)$, and analogously $ht(\beta)\leq ht(\alpha\oplus\beta)$. □

Recall that by Theorem 3.2.2(iv) the inequality for the dynamical entropy of a tensor product automorphism goes the opposite way.

We next show two Kolmogorov-Sinai type results. We need the following simple lemma.

Lemma 6.2.3. *Let $\Omega\subset A$ be finite, $k\in\mathbb{N}$, $\alpha\in\mathrm{Aut}(A)$, and $\Omega_k=\cup_{j=-k}^{k}\alpha^j(\Omega)$. Then $ht(\Omega_k,\delta;\alpha)=ht(\Omega,\delta;\alpha)$.*

Proof. Since $\cup_{j=0}^{n-1}\alpha^j(\Omega_k)=\alpha^{-k}(\cup_{j=0}^{n+2k-1}\alpha^j(\Omega))$, by Lemma 6.1.4(vi) we have

$$\mathrm{rcp}(\cup_{j=0}^{n-1}\alpha^j(\Omega_k),\delta)=\mathrm{rcp}(\cup_{j=0}^{n+2k-1}\alpha^j(\Omega),\delta).$$

Taking the logarithm, dividing by n and letting $n\to\infty$, we get the result. □

Theorem 6.2.4. *Let A be a C^*-algebra and $\alpha\in\mathrm{Aut}(A)$. Suppose $\{\Omega_\lambda\}_{\lambda\in\Lambda}$ is an increasing net of finite subsets of A such that the linear span of $\cup_{\lambda\in\Lambda,n\in\mathbb{Z}}\alpha^n(\Omega_\lambda)$ is dense in A. Then*

$$ht(\alpha)=\lim_\lambda ht(\Omega_\lambda;\alpha).$$

Proof. By Lemma 6.1.4(iii),(iv) in order to compute the entropy it suffices to consider finite subsets of $\cup_{n\in\mathbb{Z}}\cup_{\lambda\in\Lambda}\alpha^n(\Omega_\lambda)$. Given such a subset Ω we can find λ_0 and $n\in\mathbb{N}$ such that $\Omega\subset\cup_{k=-n}^{n}\alpha^k(\Omega_{\lambda_0})$. Then by the previous lemma and Lemma 6.1.4(ii)

$$ht(\Omega,\delta;\alpha)\leq ht(\cup_{k=-n}^{n}\alpha^k(\Omega_{\lambda_0}),\delta;\alpha)=ht(\Omega_{\lambda_0},\delta;\alpha),$$

so that $ht(\Omega;\alpha) \le ht(\Omega_\lambda;\alpha)$ for any $\lambda \succ \lambda_0$. □

If we consider sets consisting of unitary elements, we get a much more satisfactory result.

Lemma 6.2.5. *Assume Ω consists of unitary operators and contains the unit. Put $\Omega^n = \{u_1 \ldots u_n \,|\, u_i \in \Omega\}$. Then $ht(\Omega;\alpha) = ht(\Omega^n;\alpha)$.*

Proof. By Lemma 6.1.4(v) we have $ht(\Omega;\alpha) = ht(\Omega^2;\alpha)$. Thus by iteration $ht(\Omega;\alpha) = ht(\Omega^{2^k};\alpha)$ for all $k \in \mathbb{N}$. Since $\Omega \subset \Omega^n \subset \Omega^{2^k}$ for sufficiently large k the lemma follows from Lemma 6.1.4(ii). □

Theorem 6.2.6. *Let A be a unital C^*-algebra and $\alpha \in \mathrm{Aut}(A)$. Suppose $\{\Omega_\lambda\}_{\lambda \in \Lambda}$ is an increasing net of finite sets of unitary operators in A such that the $*$-algebra generated by $\cup_{\lambda \in \Lambda, n \in \mathbb{Z}} \alpha^n(\Omega_\lambda)$ is dense in A. Then*

$$ht(\alpha) = \lim_\lambda ht(\Omega_\lambda;\alpha).$$

Proof. By Lemma 6.1.4(i) we may assume $1 \in \Omega_\lambda$ and $\Omega_\lambda^* = \Omega_\lambda$ for any $\lambda \in \Lambda$. Set $\Omega_{\lambda,n} = (\cup_{k=-n}^{n} \alpha^k(\Omega_\lambda))^n$. Define an order on the set $\{(\lambda,n)\,|\,\lambda \in \Lambda, n \in \mathbb{N}\}$ by letting $(\lambda_1,n_1) \succ (\lambda_2,n_2)$ if $\lambda_1 \succ \lambda_2$ and $n_1 \ge n_2$. Then $\{\Omega_{\lambda,n}\}_{\lambda,n}$ is an increasing net, and $\cup_{\lambda,n} \Omega_{\lambda,n}$ spans a dense subspace of A. Hence

$$ht(\alpha) = \lim_{(\lambda,n)} ht(\Omega_{\lambda,n};\alpha).$$

Since $ht(\Omega_{\lambda,n};\alpha) = ht(\Omega_\lambda;\alpha)$ by Lemmas 6.2.3 and 6.2.5, the proof of the theorem is complete. □

In the abelian case the entropy of a homeomorphism with respect to an invariant probability measure is majorized by the topological entropy of the homeomorphism. We now show a noncommutative analogue of this result.

Proposition 6.2.7. *Let (A,φ,α) be a C^*-dynamical system with A exact. Then $h_\varphi(\alpha) \le ht(\alpha)$.*

Proof. Assume $A \subset B(H)$. Let N be a finite dimensional C^*-algebra and $\gamma \colon N \to A$ be a channel. Fix $\delta > 0$ and choose a finite subset ω of N such that its convex hull contains the unit ball. Put $\Omega = \gamma(\omega)$.

Let $n \in \mathbb{N}$, and $\theta \colon A \to \mathrm{Mat}_m(\mathbb{C})$ and $\rho \colon \mathrm{Mat}_m(\mathbb{C}) \to B(H)$ be unital completely positive maps such that

$$\|(\rho \circ \theta)(a) - a\| < \delta \quad \text{for} \quad a \in \alpha^k(\Omega),\ 0 \le k \le n-1.$$

Then $\|\rho \circ \theta \circ \alpha^k \circ \gamma - \alpha^k \circ \gamma\| < \delta$ for $0 \le k \le n-1$.

Let $\varphi = \sum_{i_1,\ldots,i_n} \varphi_{i_1\ldots i_n}$ be a finite decomposition. Extend each $\varphi_{i_1\ldots i_n}$ to a positive linear functional $\psi_{i_1\ldots i_n}$ on $B(H)$, and set $\psi = \sum_{i_1,\ldots,i_n} \psi_{i_1\ldots i_n}$. We obviously have

$$H_\varphi(\gamma,\ldots,\alpha^{n-1}\circ\gamma;\{\varphi_{i_1\ldots i_n}\}) = H_\psi(\gamma,\ldots,\alpha^{n-1}\circ\gamma;\{\psi_{i_1\ldots i_n}\}).$$

On the other hand, by Proposition 3.1.11

$$|H_\psi(\gamma,\ldots,\alpha^{n-1}\circ\gamma) - H_\psi(\rho\circ\theta\circ\gamma,\ldots,\rho\circ\theta\circ\alpha^{n-1}\circ\gamma)| < n\varepsilon$$

for some ε depending only on δ and $\dim N$. Since

$$H_\psi(\rho\circ\theta\circ\gamma,\ldots,\rho\circ\theta\circ\alpha^{n-1}\circ\gamma) \le S(\psi\circ\rho) \le \log m$$

by Proposition 3.1.3(i),(iv) and Lemma 3.1.4, we thus get

$$H_\varphi(\gamma,\ldots,\alpha^{n-1}\circ\gamma;\{\varphi_{i_1\ldots i_n}\}) < \log m + n\varepsilon.$$

It follows that $h_\varphi(\gamma;\alpha) \le ht(\Omega,\delta;\alpha) + \varepsilon \le ht(\alpha) + \varepsilon$. Since $\varepsilon = \varepsilon(\delta,\dim N)$ can be made arbitrarily small, we obtain the desired inequality. □

Remark 6.2.8. For the entropy h^{ST} introduced in Chap. 5, which always majorizes the dynamical entropy and coincides with it in the nuclear case by Theorem 5.1.5, we also have $h_\varphi^{ST}(\alpha) \le ht(\alpha)$. Indeed, assume $A \subset B(H)$. We may also assume that α extends to an automorphism $\bar\alpha$ of $B(H)$ (represent $A \rtimes_\alpha \mathbb{Z}$ faithfully and take $\bar\alpha = \mathrm{Ad}\,u$, where u is the canonical unitary in the crossed product). Let λ be a stationary coupling of (A,φ,α) with (X,μ,T). Extend λ to an $(\bar\alpha\otimes\beta)$-invariant state $\bar\lambda$ on $B(H)\otimes C(X)$, where $\beta(f) = f\circ T^{-1}$ for $f \in C(X)$. Set $\bar\varphi = \bar\lambda|_{B(H)}$.

By definition of h^{ST} we have to prove that

$$h_\mu(\xi;T) + \sum_{Y\in\xi} S(\lambda(\cdot\otimes \mathbf{1}_Y),\varphi) \le ht(\alpha)$$

for any finite measurable partition ξ of X. Fix $\varepsilon > 0$. Since relative entropy is weakly* lower semicontinuous by Theorem 2.3.1(iv), there exist $\delta > 0$ and a finite subset $\Omega \subset A$ such that for $Y \in \xi$

$$S(\lambda(\cdot\otimes \mathbf{1}_Y),\varphi) < S(\omega,\psi) + \frac{\varepsilon}{|\xi|}$$

as soon as $|\omega(a) - \lambda(a\otimes \mathbf{1}_Y)| < \delta$ and $|\psi(a) - \varphi(a)| < \delta$ for $a \in \Omega$. Let $n \in \mathbb{N}$, and $\theta\colon A \to \mathrm{Mat}_m(\mathbb{C})$ and $\gamma\colon \mathrm{Mat}_m(\mathbb{C}) \to B(H)$ be unital completely positive maps such that

$$\|(\gamma\circ\theta)(a) - a\| < \delta \quad \text{for} \quad a \in \alpha^k(\Omega),\ 0 \le k \le n-1.$$

Then using monotonicity of relative entropy, Theorem 2.3.1(vi), we get (see (3.9))

$$H_{\bar\lambda}(\gamma,\ldots,\gamma;\xi,T\xi,\ldots,T^{n-1}\xi) - H_\mu(\vee_{k=0}^{n-1}T^k\xi)$$

$$= \sum_{k=0}^{n-1} \sum_{Y \in \xi} S(\bar\lambda(\gamma(\cdot) \otimes 1_{T^k Y}), \bar\varphi \circ \gamma)$$

$$= \sum_{k=0}^{n-1} \sum_{Y \in \xi} S(\bar\lambda((\bar\alpha^{-k} \circ \gamma)(\cdot) \otimes 1_Y), \bar\varphi \circ \gamma)$$

$$\geq \sum_{k=0}^{n-1} \sum_{Y \in \xi} S(\bar\lambda((\bar\alpha^{-k} \circ \gamma \circ \theta \circ \alpha^k)(\cdot) \otimes 1_Y), \bar\varphi \circ \gamma \circ \theta \circ \alpha^k)$$

$$> n \sum_{Y \in \xi} S(\lambda(\cdot \otimes 1_Y), \varphi) - n\varepsilon.$$

On the other hand,

$$H_{\bar\lambda}(\gamma, \dots, \gamma; \xi, \dots, T^{n-1}\xi) \leq H_{\bar\varphi}(\gamma) \leq S(\bar\varphi \circ \gamma) \leq \log m.$$

It follows that

$$\log \operatorname{rcp}(\cup_{k=0}^{n-1} \alpha^k(\Omega), \delta) > H_\mu(\vee_{k=0}^{n-1} T^k \xi) + n \sum_{Y \in \xi} S(\lambda(\cdot \otimes 1_Y), \varphi) - n\varepsilon.$$

Dividing by n and letting $n \to \infty$ we get the desired conclusion, as ε can be made arbitrarily small. ♦

We can now prove that our definition of topological entropy reduces to the classical one in the abelian case.

Proposition 6.2.9. *Let* $T\colon X \to X$ *be a homeomorphism of a compact metric space* X, *and let* α *be the automorphism of* $C(X)$ *induced by* T *via* $\alpha(f) = f \circ T^{-1}$. *Then* $ht(\alpha) = h_{top}(T)$.

Proof. Let Ω be a finite subset of $C(X)$, $\delta > 0$, and let $\mathcal{U}$ be an open cover of X such that $|f(x) - f(y)| < \delta$ for any $f \in \Omega$ and any $x, y \in X$ lying in the same element of $\mathcal{U}$. Let $n \in \mathbb{N}$. Then if $f \in \cup_{j=0}^{n-1} \alpha^j(\Omega)$ and x, y belong to the same element of $\mathcal{U}_n = \vee_{j=0}^{n-1} T^j \mathcal{U}$ we have $|f(x) - f(y)| < \delta$. By Lemma 6.1.3 we get

$$\operatorname{rcp}(\cup_{j=0}^{n-1} \alpha^j(\Omega), \delta) \leq N(\mathcal{U}_n),$$

whence $ht(\Omega, \delta; \alpha) \leq h(\mathcal{U}; T)$. It follows that $ht(\alpha) \leq h_{top}(T)$.

The opposite inequality is more involved. By Proposition 6.2.7, we have $ht(\alpha) \geq h_\varphi(\alpha)$ for every α-invariant state φ on $C(X)$. In other words, $ht(\alpha) \geq h_\mu(T)$ for every T-invariant probability measure μ. Thus by the variational principle, Theorem 1.2.2, we have

$$h_{top}(T) = \sup_\mu h_\mu(T) \leq ht(\alpha),$$

and the proof is complete. □

We consider next two simple examples of computations of topological entropy.

Example 6.2.10.

(i) Let α be an automorphism of $A = K(H) + \mathbb{C}1 \subset B(H)$, where $K(H)$ is the algebra of compact operators on a Hilbert space H. Then $ht(\alpha) = 0$.

It suffices to prove that $ht(\Omega; \alpha) = 0$ for any finite $\Omega \subset pB(H)p$, where p is a finite rank projection. Let $p_n = \vee_{k=0}^{n-1} \alpha^k(p)$. Then p_n is a finite rank projection such that $\cup_{k=0}^{n-1} \alpha^k(\Omega) \subset p_n B(H) p_n$. So if we put $B_n = p_n B(H) p_n + \mathbb{C}(1 - p_n)$, and take $\theta \colon A \to B_n$ to be a conditional expectation (e.g. $\theta(a) = p_n a p_n + \varphi(a)(1-p_n)$, where φ is a state with support $\leq 1-p_n$) and $\gamma \colon B_n \to A$ to be the embedding map, then for any $\delta > 0$ we get

$$\mathrm{rcp}(\cup_{k=0}^{n-1} \alpha^k(\Omega), \delta) \leq \mathrm{rank}\, B_n = \mathrm{Tr}(p_n) + 1 \leq n\mathrm{Tr}(p) + 1.$$

Hence $ht(\Omega, \delta; \alpha) = 0$.

(ii) Let B be a finite dimensional C*-algebra, $A = B^{\otimes \mathbb{Z}}$ the infinite C*-tensor product, α the shift automorphism of A. Then $ht(\alpha) = \log \mathrm{rank}\, B$.

The inequality $\leq$ follows easily from $\mathrm{rank}\, B^{\otimes n} = (\mathrm{rank}\, B)^n$. On the other hand, if τ_0 is the normalized canonical trace on B and $\tau = \tau_0^{\otimes \mathbb{Z}}$, then by Example 3.2.6(i) we have $h_\tau(\alpha) = S(\tau_0) = \log \mathrm{rank}\, B$. Since $h_\tau(\alpha) \leq ht(\alpha)$, this gives the opposite inequality. ♦

We finish the chapter with some technical remarks.

Remark 6.2.11.

(i) Even when dealing with unital algebras it is sometimes convenient to consider nonunital maps. Define $\mathrm{rcp}'(\Omega, \delta)$ exactly as $\mathrm{rcp}(\Omega, \delta)$ but by requiring θ and γ to be contractive completely positive maps instead of unital ones (whether we require $\pi \colon A \to B$ to be unital or not is not important as we can always replace B by $\pi(1)B\pi(1)$), and then define $ht'(\alpha)$. We obviously have $\mathrm{rcp}'(\Omega, \delta) \leq \mathrm{rcp}(\Omega, \delta)$. But if $1 \in \Omega$, so $x = (\gamma \circ \theta)(1)$ is close to 1, we can consider the completely positive map $\tilde{\gamma} = x^{-1/2}\gamma(\cdot)x^{-1/2}$, which is close to γ. Then by A.12 we can modify θ and $\tilde{\gamma}$ to unital completely positive maps keeping $\tilde{\gamma} \circ \theta$ unchanged. This shows that if $1 \in \Omega$, then $ht'(\Omega; \alpha) = ht(\Omega; \alpha)$, so that $ht'(\alpha) = ht(\alpha)$.

(ii) Let A be a nonunital C*-algebra, α an automorphism of A. One way to define the topological entropy of α is to consider the algebra $A^\sim$ obtained by adjoining unit to A, extend α to an automorphism $\tilde{\alpha}$ of $A^\sim$, and define $ht(\alpha) = ht(\tilde{\alpha})$. Another possibility is to define $ht(\alpha)$ using rcp' from the previous remark, that is, by considering contractive completely positive maps. The result will be the same. Indeed, by A.4 any contractive completely positive maps $A \to \mathrm{Mat}_n(\mathbb{C})$ and $\mathrm{Mat}_n(\mathbb{C}) \to B$ extend to unital completely positive maps $A^\sim \to \mathrm{Mat}_n(\mathbb{C})^\sim \cong \mathrm{Mat}_n(\mathbb{C}) \oplus \mathbb{C}$ and $\mathrm{Mat}_n(\mathbb{C})^\sim \to B^\sim$.

It is then easy to check that all the main properties of the topological entropy remain true for nonunital C*-algebras.

(iii) Lemma 6.1.4(vii) shows that $\mathrm{rcp}_A(\Omega, \delta)$ depends only on the position of Ω in the C*-algebra generated by Ω. Using Arveson's extension theorem for operator systems we can conclude that it actually depends only on the position of Ω in the operator system spanned by Ω, Ω^* and 1.

More precisely, recall that an **operator system** is a closed self-adjoint subspace X of a unital C*-algebra containing the unit of the algebra. The notion of a completely positive map makes sense for operator systems. Then a unital completely positive map $\theta \colon X \to Y$ is called a **complete order embedding** if it is injective and the inverse map $\theta(X) \to X$ is completely positive. Equivalently, θ is completely isometric, that is, the map $\theta \otimes \mathrm{id} \colon X \otimes \mathrm{Mat}_n(\mathbb{C}) \to Y \otimes \mathrm{Mat}_n(\mathbb{C})$ is isometric for all $n \in \mathbb{N}$.

Let Ω be a finite subset of an operator system X. Define $\mathrm{rcp}_X(\Omega, \delta)$ by considering approximate factorizations of complete order embeddings of X into C*-algebras through finite dimensional C*-algebras. Then if $\theta \colon X \hookrightarrow A$ is a complete order embedding, we have $\mathrm{rcp}_A(\theta(\Omega), \delta) = \mathrm{rcp}_X(\Omega, \delta)$.

In particular, we get the following strengthening of Theorem 6.2.2(iii). If $\theta \colon B \to A$ is a complete order embedding of B into A, and $\alpha \circ \theta = \theta \circ \beta$, then $ht(\beta) \le ht(\alpha)$. ♦

6.3 Notes

The notion of topological entropy presented here was introduced by Voiculescu [227] for nuclear C*-algebras and extended by Brown to exact algebras [32] (we refer the reader to the book by Wassermann [232] for more information on exact C*-algebras). Most of the results in this chapter are contained in those papers. An important exception is property (v) in Lemma 6.1.4 implying Theorem 6.2.6, which was observed by Ozawa [153].

As we proved in Chap. 3, inner automorphisms of type I algebras have zero dynamical entropy. This fact and Example 6.2.10(i) suggest that the topological entropy of inner automorphisms of type I C*-algebras is zero. This is, however, known only in a few cases, e.g. for homogeneous C*-algebras [33].

As we noted in Remark 6.2.11(i), instead of unital completely positive maps we can deal with contractive completely positive maps. Pop and Smith [167] showed that we can equally well consider completely contractive maps, that is, maps $\theta \colon A \to B$ such that $\theta \otimes \mathrm{id} \colon A \otimes \mathrm{Mat}_n(\mathbb{C}) \to B \otimes \mathrm{Mat}_n(\mathbb{C})$ is contractive for any $n \in \mathbb{N}$. The reason is briefly that if we have a factorization of an algebra through a matrix algebra via completely contractive maps such that their composition almost preserves the unit, then we can construct a contractive completely positive factorization through the same matrix algebra [204]. It is, however, unknown whether it is possible instead to use completely bounded maps with a universal bound on the norms.

Another open problem is how topological entropy behaves in quotients. In other words, if α is an automorphism of A leaving a two-sided ideal $I \subset A$ invariant, and $\bar\alpha$ is the automorphism of $B = A/I$ defined by α, the question is whether the inequality $ht(\bar\alpha) \le ht(\alpha)$ holds. If there exists a unital completely positive splitting $\sigma\colon B \to A$ such that $\alpha\circ\sigma = \sigma\circ\bar\alpha$, then this is true by Remark 6.2.11(iii). It is shown in [36] that if we in addition assume that there exists an approximate unit $\{e_i\}_i$ in I such that $\alpha(e_i) = e_i$ for any i, then moreover $ht(\alpha) = \max\{ht(\bar\alpha), ht(\alpha|_I)\}$.

As in the case of dynamical entropy, the definition of topological entropy used in this chapter is not the only one possible. Similarly to the Lindblad-Alicki-Fannes entropy discussed in Notes to Chap. 3, there is an approach based on partitions of unity due to Hudetz [93], [94], [14]. Let us say that a finite set $X = \{x_i\}_i$ of operators is an operator cover if $\sum_i x_i^* x_i$ is invertible. Denote by $\bar N(X)$ the cardinality of a minimal operator subcover of X. Given two operator covers $X = \{x_i\}_i$ and $Y = \{y_j\}$ we can form a cover $X \vee Y$ consisting of the elements $y_j x_i$ (in Chap. 3 we denoted this operation by $Y \circ X$). One of the unpleasant features of this notion is that if we consider $X^{(n)} = X \vee \ldots \vee X$ (n times) then $\bar N(X^{(n)})$ can grow exponentially fast as $n \to \infty$. Still it is possible to manufacture a notion of entropy using it, see the above mentioned papers for details.

Thomsen [215] proposed a different operation of joining operator covers. He requires operator covers to be in addition self-adjoint sets and calls them partitions. Given n partitions $X_1, \ldots, X_n$ he defines $X_1 \vee \ldots \vee X_n$ as the partition consisting of the elements $x_{\sigma(1)} \ldots x_{\sigma(n)}$, where $x_i \in X_i$ and $\sigma \in S_n$. As was the case for the Lindblad-Alicki-Fannes entropy, in either approach one cannot consider all possible operator covers or partitions in a C*-algebra, since this leads to an essentially uncomputable invariant. So Hudetz introduces a notion of appropriately approximated C*-algebras, while Thomsen works with local C*-algebras.

There are a number of dynamical invariants whose definitions follow the same pattern as that of topological entropy, the main difference being the way a finite set Ω is approximated and what kind of a rank-type function is used, see e.g. [44], [110], [109]. We would like to mention one of them, which was introduced by Voiculescu [227].

Let (A, φ, α) be a C*-dynamical system. For a finite $\Omega \subset A$ and $\delta > 0$ denote by $\mathrm{rcp}_\varphi(\Omega, \delta)$ the minimal $n \in \mathbb{N}$ for which there exist unital completely positive maps $\theta\colon A \to \mathrm{Mat}_n(\mathbb{C})$ and $\gamma\colon \mathrm{Mat}_n(\mathbb{C}) \to \pi_\varphi(A)''$ such that $\varphi \circ \gamma \circ \theta = \varphi$ and $\|(\gamma \circ \theta)(a) - \pi_\varphi(a)\|_\varphi < \delta$ for any $a \in \Omega$, where $\|x\|_\varphi = \varphi(x^*x)^{1/2}$. Then we define the **completely positive approximation entropy** $hcpa_\varphi(\alpha)$ just as we defined the topological entropy $ht(\alpha)$. Most of the results on $ht(\alpha)$ can be proved in roughly the same way for $hcpa_\varphi(\alpha)$. In particular, if (B, ψ, β) is another C*-dynamical system, then

$$hcpa_{\varphi\otimes\psi}(\alpha \otimes \beta) \le hcpa_\varphi(\alpha) + hcpa_\psi(\beta).$$

Since we have the opposite inequality for the entropy $h_\varphi(\alpha)$, we get the equality

$$h_{\varphi\otimes\psi}(\alpha \otimes \beta) = h_\varphi(\alpha) + h_\psi(\beta)$$

whenever $hcpa_\varphi(\alpha) = h_\varphi(\alpha)$ and $hcpa_\psi(\beta) = h_\psi(\beta)$.

If (X,T) is a topological dynamical system, μ a T-invariant probability measure on X, $\alpha(f) = f \circ T^{-1}$ and $\varphi(f) = \int f \, d\mu$ for $f \in C(X)$, then

$$h_\mu(T) \le hcpa_\varphi(\alpha) \le h_{top}(T).$$

If the system is ergodic, then $h_\mu(T) = hcpa_\varphi(\alpha)$. This can be deduced either from the above inequalities and the Jewett-Krieger theorem, Theorem 1.2.3, or using the Shannon-McMillan-Breiman theorem, Theorem 1.2.4. Without the ergodicity assumption the equality $h_\mu(T) = hcpa_\varphi(\alpha)$ is not always true. To see this, observe that if $(A, \varphi, \alpha) = (A_1 \oplus A_2, \lambda\varphi_1 \oplus (1-\lambda)\varphi_2, \alpha_1 \oplus \alpha_2)$, $0 < \lambda < 1$, then $hcpa_\varphi(\alpha) = \max_i hcpa_{\varphi_i}(\alpha_i)$, while $h_\varphi(\alpha) = \lambda h_{\varphi_1}(\alpha_1) + (1-\lambda)h_{\varphi_2}(\alpha_2)$. To get a similar invariant which coincides with $h_\varphi(\alpha)$ for arbitrary abelian systems, we should have used $\exp(S(\varphi \circ \gamma))$ in place of n in the definition of rcp_φ.

Voiculescu's motivation for introducing completely positive approximation entropy $hcpa_\varphi(\alpha)$ was that the equality $h_\varphi(\alpha) = hcpa_\varphi(\alpha)$, when it holds, should be thought of as a weak form of the Shannon-McMillan-Breiman theorem. For a result in this direction see [142].

The idea of using some sort of approximation of finite sets of an algebra to define entropy was apparently first used in the work of Connes and Størmer [52]. Their definition is as follows. Let $M \subset B(H)$ be a von Neumann algebra, ξ a cyclic and separating tracial vector, and $\tau = \omega_\xi$ the trace on M. For $\delta > 0$ and a finite set Ω of unitaries in M define

$$h_\delta^\tau(\Omega) = \inf\{S(\varphi) \mid \varphi \in B(H)_*, \ \varphi|_M = \tau, \ S(\varphi, \varphi \circ \mathrm{Ad}\, u) < \delta \text{ for all } u \in \Omega\}.$$

Then define the entropy of a τ-preserving automorphism α of M by the same procedure as before: take $\limsup$ as $n \to \infty$ of the quantities

$$\frac{1}{n} h_\delta^\tau(\Omega \cup \alpha(\Omega) \cup \ldots \cup \alpha^{n-1}(\Omega)),$$

and then supremum over $\delta > 0$ and finite sets of unitaries $\Omega \subset M$. The only fact which is known about this entropy is that it coincides with the Kolmogorov-Sinai entropy for abelian M.

Voiculescu himself introduced another type of approximation entropy, called the **perturbation-theoretic entropy** [223], [224], [225]. To define it recall that the Macaev ideal $\mathcal{C}_\infty^-(H)$ is the space of compact operators T on H such that

$$\|T\|_\infty^- = \sum_{n=1}^\infty \frac{s_n(T)}{n} < \infty,$$

where $s_1(T) \ge s_2(T) \ge \ldots$ are the eigenvalues of $|T|$. For a finite set $\Omega = \{x_1, \ldots, x_n\}$ of operators on H put

$$k_\infty^-(\Omega) = \liminf_a \max_{1 \le i \le n} \|[a, x_i]\|_\infty^-,$$

where the lim inf is taken over the net of finite rank operators a, $0 \leq a \leq 1$, with its natural order. Now if (A, φ, α) is a C*-dynamical system, and U is the canonical unitary implementing α in the GNS-representation, we define

$$H_P(U, \pi_\varphi(A)) = \sup k_\infty^-(\Omega \cup \{U\}),$$

where the supremum is taken over all finite subsets Ω of $\pi_\varphi(A)$ generating a finite dimensional C*-algebra. It is shown in [223] that if (X, μ, T) is an ergodic dynamical system, and U_T is the unitary on $L^2(X, \mu)$ implementing T, then

$$\frac{1}{2}h(T) \leq H_P(U_T, L^\infty(X, \mu)) \leq 6h(T).$$

It is also known that for noncommutative systems this entropy is quite different from the dynamical entropy considered in Chap. 3. E.g. the dynamical entropy of a free shift to be considered in Chap. 14 is zero, while the perturbation-theoretic entropy is infinite [223].

It is interesting to note [224] that the perturbation-theoretic entropy allows us to define the entropy of an action of a finitely generated group G with a fixed set S of generators on a von Neumann algebra M. We do not have to assume that G is amenable or that the action preserves a state. Namely, if we consider M in its standard form, $M \subset B(H)$, then any automorphism of M is implemented by a canonical unitary. If u_s, $s \in S$, are the unitaries corresponding to the generators of the group, we can then define $H_P(S, M) = H_P(\{u_s\}_s, M)$. See [148] for an estimate of this entropy for the action of a free group on its boundary, which may suggest that perturbation-theoretic entropy is related to entropy of pseudogroups [71], [229].

7
Dynamics on the State Space

Throughout this chapter we fix a unital separable C*-algebra A and an automorphism α of A. Denote by $X = S(A)$ the state space of A, and by T the homeomorphism of X defined by $T\varphi = \varphi \circ \alpha^{-1}$. It would be too simple (and actually quite disappointing) if the entropy of α could be computed by looking at (X, T). Nevertheless it is a natural problem to investigate relations between entropic properties of (A, α) and (X, T). We shall see that either both systems have positive entropy or both have zero entropy.

7.1 Measure Entropy

We first consider measure-theoretic entropy.

Theorem 7.1.1. *Let notation be as above and let φ be an α-invariant state on A. Assume $h_\mu(T) = 0$ for any T-invariant probability measure μ on X with barycenter φ. Then $h_\varphi(\alpha) = 0$.*

In particular, if $h_{top}(T) = 0$ then $h_\varphi(\alpha) = 0$ for every α-invariant state φ.

This is a consequence of the inequality $h_\varphi(\alpha) \le h_\varphi^{ST}(\alpha)$ from Theorem 5.1.5 and the following result of independent interest, which in particular implies that $h_\varphi^{ST}(\alpha)$ is not larger than the supremum of $h_\mu(T)$ over all T-invariant probability measures μ on X with barycenter φ.

Proposition 7.1.2. *We have*

$$h_\varphi^{ST}(\alpha) = \sup\{H_\mu(\xi|\xi^-) - H_{\lambda,A}(\xi|\xi^-)\},$$

where the supremum is taken over all stationary canonical couplings λ of (A, φ, α) with (X, μ, T) and over all finite measurable partitions ξ of (X, μ).

For the proof we need the following **Pinsker formula**, which is a refinement of the inequality from Lemma 1.1.3.

Lemma 7.1.3. *If (Y, ν, S) is a dynamical system, then for any two finite measurable partitions ξ and ζ of (Y, ν) we have*

$$h(\xi \vee \zeta; S) = h(\zeta; S) + H(\xi | \xi^- \vee \zeta^\pm), \quad \text{where } \zeta^\pm = \vee_{-\infty}^\infty S^k \zeta.$$

Proof. Observe first that by a computation similar to (1.3) we have

$$H(\xi | \xi^- \vee \zeta^\pm) = \lim_{n \to \infty} \frac{1}{n} H(\vee_{k=0}^{n-1} S^k \xi | \zeta^\pm). \tag{7.1}$$

Denote the partitions $\vee_{k=0}^{n-1} S^k \xi$ and $\vee_{k=0}^{n-1} S^k \zeta$ by ξ_n and ζ_n, respectively. By Lemma 1.1.3,

$$h(\xi_n \vee \zeta_n; S^n) \leq h(\zeta_n; S^n) + H(\xi_n | \zeta^\pm).$$

Since $h(\xi_n \vee \zeta_n; S^n) = nh(\xi \vee \zeta; S)$ and $h(\zeta_n; S^n) = nh(\zeta; S)$, dividing the above inequality by n, letting $n \to \infty$ and using (7.1) we get the inequality $\leq$ in the lemma.

On the other hand,

$$H(\xi_n \vee \zeta_n) = H(\zeta_n) + H(\xi_n | \zeta_n) \geq H(\zeta_n) + H(\xi_n | \zeta^\pm).$$

Dividing by n, letting $n \to \infty$ and using (7.1) again, we get the inequality $\geq$ in the lemma. $\square$

Proof of Proposition 7.1.2. Let Λ be a stationary coupling of (A, φ, α) with (Y, ν, S). We can desintegrate the state Λ over (Y, ν). That is, there exists a measurable map $Y \ni y \mapsto \Lambda_y \in X$ such that

$$\Lambda(a \otimes f) = \int_Y \Lambda_y(a) f(y) \, d\nu(y) \quad \text{for } a \in A, \ f \in L^\infty(Y, \nu).$$

Existence of such a map can be deduced from C.3-C.5, by first decomposing the space of the GNS-representation defined by Λ into a direct integral of Hilbert spaces over (Y, ν), and then decomposing the representation of A and a normal extension of Λ to the algebra of decomposable operators into corresponding direct integrals (we used a similar argument in the proof of Theorem 5.2.1). Let μ be the image of ν under the map $y \mapsto \Lambda_y$. Since Λ is invariant, we have $\Lambda_{Sy} = \Lambda_y \circ \alpha^{-1}$ for a.e. $y \in Y$. Since ν is S-invariant, it follows that μ is T-invariant. Hence μ defines a stationary canonical coupling λ,

$$\lambda(a \otimes f) = \int_X \psi(a) f(\psi) d\mu(\psi) = \int_Y \Lambda_y(a) f(\Lambda_y) \, d\nu(y) \quad \text{for } a \in A, \ f \in C(X).$$

Identify $L^\infty(X, \mu)$ with a subalgebra of $L^\infty(Y, \nu)$ using the map $y \mapsto \Lambda_y$. Let ε_X be the measurable partition of Y such that $L^\infty(X) = L^\infty(Y/\varepsilon_X)$. Then λ is just the restriction of Λ to $A \otimes L^\infty(Y/\varepsilon_X)$. By Proposition 5.1.4 it suffices to show that λ is as good as Λ for computing the entropy, that is, the supremum

7.1 Measure Entropy

$$\sup\{H_\nu(\xi|\xi^-) - H_{\Lambda,A}(\xi|\xi^-)\}$$

over finite partitions ξ of Y will not change if we only consider $\xi \prec \varepsilon_X$.

Let ξ be a finite partition of Y. Since by (5.3)

$$H_\nu(\xi|\xi^-) - H_{\Lambda,A}(\xi|\xi^-) = \lim_{n\to\infty} \frac{1}{n} H_\Lambda(A; \vee_{k=0}^{n-1} S^k \xi),$$

by virtue of Proposition 5.1.3(i), we see that the expression $H_\nu(\xi|\xi^-) - H_{\Lambda,A}(\xi|\xi^-)$ is increasing in ξ, so that

$$H_\nu(\xi|\xi^-) - H_{\Lambda,A}(\xi|\xi^-) \le H_\nu(\xi \vee \vartheta|\xi^- \vee \vartheta^-) - H_{\Lambda,A}(\xi \vee \vartheta|\xi^- \vee \vartheta^-)$$

for any finite partition $\vartheta \prec \varepsilon_X$. By the Pinsker formula, Lemma 7.1.3,

$$H_\nu(\xi \vee \vartheta|\xi^- \vee \vartheta^-) = H_\nu(\vartheta|\vartheta^-) + H_\nu(\xi|\xi^- \vee \vartheta^\pm),$$

where $\vartheta^\pm = \vee_{-\infty}^\infty S^k \vartheta$. Using Proposition 5.1.3(iv),(vi) we similarly get

$$H_{\Lambda,A}(\xi \vee \vartheta|\xi^- \vee \vartheta^-) = H_{\Lambda,A}(\vartheta|\vartheta^-) + H_{\Lambda,A}(\xi|\xi^- \vee \vartheta^\pm).$$

Therefore

$$H_\nu(\xi|\xi^-) - H_{\Lambda,A}(\xi|\xi^-) \le H_\nu(\vartheta|\vartheta^-) - H_{\Lambda,A}(\vartheta|\vartheta^-)$$
$$+ H_\nu(\xi|\xi^- \vee \vartheta^\pm) - H_{\Lambda,A}(\xi|\xi^- \vee \vartheta^\pm).$$

Thus it suffices to show that $H_\nu(\xi|\xi^- \vee \vartheta^\pm) - H_{\Lambda,A}(\xi|\xi^- \vee \vartheta^\pm)$ can be made arbitrarily small for some $\vartheta \prec \varepsilon_X$. If $\{\vartheta_n\}_n$ is an increasing sequence of finite partitions such that $\vee_n \vartheta_n = \varepsilon_X$, then

$$H_\nu(\xi|\xi^- \vee \vartheta_n^\pm) \searrow H_\nu(\xi|\xi^- \vee \varepsilon_X) \text{ and } H_{\Lambda,A}(\xi|\xi^- \vee \vartheta_n^\pm) \searrow H_{\Lambda,A}(\xi|\xi^- \vee \varepsilon_X)$$

by Proposition 5.1.3(iv). Hence it is enough to prove that $H_\nu(\xi|\zeta) = H_{\Lambda,A}(\xi|\zeta)$ for any $\zeta \succ \varepsilon_X$. For this, in turn, it is enough to check that

$$S(\nu(\cdot 1_Z)|_{L^\infty(Y/\zeta)}, \nu|_{L^\infty(Y/\zeta)}) = S(\Lambda(\cdot \otimes \cdot 1_Z)|_{A\otimes L^\infty(Y/\zeta)}, \Lambda|_{A\otimes L^\infty(Y/\zeta)})$$

for any measurable $Z \subset Y$. This is indeed the case by Corollary 2.3.3, since the map

$$E: A \otimes L^\infty(Y/\zeta) \to L^\infty(Y/\zeta), \quad E(a \otimes f)(y) = \Lambda_y(a) f(y),$$

is a conditional expectation which preserves any functional of the form

$$a \otimes f \mapsto \Lambda(a \otimes f 1_Z) = \int_Z \Lambda_y(a) f(y) \, d\nu(y), \quad Z \subset Y,$$

on $A \otimes L^\infty(Y/\zeta)$. $\square$

7.2 Topological Entropy

It is natural to conjecture that if $h_\varphi^{ST}(\alpha) = 0$ for any α-invariant state φ then $h_{top}(T) = 0$. We do not know whether this is true. We have however the following result.

Theorem 7.2.1. *If $ht(\alpha) = 0$ then $h_{top}(T) = 0$.*

The proof is based on the following combinatorial statement.

Proposition 7.2.2. *For given $\varepsilon > 0$ and $a > 0$, there exist $n_0 \in \mathbb{N}$ and $b > 0$ such that if $n \geq n_0$ and $\Phi \colon \mathrm{Mat}_m(\mathbb{C})^* \to \ell_\infty^n$ is a linear contractive map such that the image of the unit ball contains at least e^{an} real elements that are ε-separated, then $m \geq e^{bn}$.*

Recall that we say that a subset of a metric space is ε-separated if the distance between any two different elements of the set is at least ε.

We shall postpone the proof of Proposition 7.2.2 and first use it to deduce Theorem 7.2.1.

Proof of Theorem 7.2.1. Assume $A \subset B(H)$. Choose a relatively compact subset Ω_0 of the unit ball of A which consists of self-adjoint elements and spans a dense subspace of A (e.g. take a sequence which converges to zero and spans a dense subspace of A). Then the weak* topology on X is defined by the metric
$$d(\varphi, \psi) = \sup_{y \in \Omega_0} |\varphi(y) - \psi(y)|.$$

Assume $h_{top}(T) > a_0 > 0$. Recalling the definition of the topological entropy in terms of separated sets from Sect. 1.2, this means that for any sufficiently small $\varepsilon > 0$ and for all n large enough there exist at least $e^{a_0 n}$ points of X that are $(n, 4\varepsilon)$-separated with respect to T^{-1}, that is, they are 4ε-separated in the metric
$$d_n(\varphi, \psi) = \max_{0 \leq k \leq n-1} d(T^{-k}\varphi, T^{-k}\psi) = \max_{0 \leq k \leq n-1} d(\varphi \circ \alpha^k, \psi \circ \alpha^k).$$

Choose a finite subset $\Omega \subset \Omega_0$ such that
$$d(\varphi, \psi) < \max_{y \in \Omega} |\varphi(y) - \psi(y)| + \varepsilon \text{ for } \varphi, \psi \in X.$$

Given $n \in \mathbb{N}$ and a unital completely positive map $\sigma \colon A \to \mathrm{Mat}_m(\mathbb{C})$, define the map
$$\Phi \colon \mathrm{Mat}_m(\mathbb{C})^* \to \ell_\infty^{n|\Omega|}, \quad \Phi(f) = \{(f \circ \sigma \circ \alpha^k)(y)\}_{0 \leq k \leq n-1, y \in \Omega},$$
where we fix a certain order on the set $\{0, \ldots, n-1\} \times \Omega$. Assume there exists a unital completely positive map $\gamma \colon \mathrm{Mat}_m(\mathbb{C}) \to B(H)$ such that
$$\|(\gamma \circ \sigma \circ \alpha^k)(y) - \alpha^k(y)\| < \varepsilon \text{ for } 0 \leq k \leq n-1, \ y \in \Omega.$$

7.2 Topological Entropy   111

Extend each state φ on A to a state $\tilde{\varphi}$ on $B(H)$. Then

$$|(\tilde{\varphi} \circ \gamma \circ \sigma \circ \alpha^k)(y) - (\varphi \circ \alpha^k)(y)| < \varepsilon.$$

If two states φ and ψ are $(n, 4\varepsilon)$-separated with respect to T^{-1}, there exists k, $0 \le k \le n-1$, such that $d(\varphi \circ \alpha^k, \psi \circ \alpha^k) \ge 4\varepsilon$. Hence there exists $y \in \Omega$ such that

$$|(\varphi \circ \alpha^k)(y) - (\psi \circ \alpha^k)(y)| > 3\varepsilon,$$

whence

$$|(\tilde{\varphi} \circ \gamma \circ \sigma \circ \alpha^k)(y) - (\tilde{\psi} \circ \gamma \circ \sigma \circ \alpha^k)(y)| > \varepsilon,$$

so that $\Phi(\tilde{\varphi} \circ \gamma)$ and $\Phi(\tilde{\psi} \circ \gamma)$ are ε-separated. Thus the image of the unit ball of $\mathrm{Mat}_m(\mathbb{C})^*$ under Φ contains at least $e^{a_0 n}$ real elements that are ε-separated. Hence by Proposition 7.2.2 applied to $a = a_0/|\Omega|$ we can find $b > 0$ such that $m \ge e^{bn}$ if n is sufficiently large. In other words,

$$\mathrm{rcp}(\Omega \cup \alpha(\Omega) \cup \ldots \alpha^{n-1}(\Omega), \varepsilon) \ge e^{bn},$$

so that $ht(\alpha) \ge b > 0$.   □

In order to prove Proposition 7.2.2 we shall compare the type 2 constants of the dual spaces. Recall that for a Banach space $\mathcal{X}$ the type 2 constant $T_2(\mathcal{X})$ is defined as the infimum of the positive numbers T such that for any $N \in \mathbb{N}$ and $x_1, \ldots, x_N \in \mathcal{X}$ we have

$$\frac{1}{2^N} \sum_{\varepsilon_1, \ldots, \varepsilon_N = \pm 1} \left\| \sum_{k=1}^{N} \varepsilon_k x_k \right\|^2 \le T^2 \sum_{k=1}^{N} \|x_k\|^2. \quad (7.2)$$

Lemma 7.2.3. *We have:*
(i) $T_2(\ell_1^n) \ge \sqrt{n}$;
(ii) there is $C > 0$ such that $T_2(\mathrm{Mat}_n(\mathbb{C})) \le C\sqrt{\log n}$ (for $n \ge 2$).

Proof. To prove the first inequality it suffices to take $N = n$ and the canonical basis of ℓ_1^n for x_k's.

To prove the second inequality consider the Schatten p-class $S_p(n)$, that is, the space $\mathrm{Mat}_n(\mathbb{C})$ with the norm $\|x\|_p = \mathrm{Tr}(|x|^p)^{1/p}$. We claim that there exists a constant C' such that

$$T_2(S_{2m}(n)) \le C'\sqrt{m} \quad \text{for any} \quad m, n \in \mathbb{N}. \quad (7.3)$$

Assuming this and using that $\|x\| \le \|x\|_{2m} \le n^{1/2m}\|x\|$, we get from (7.2)

$$T_2(\mathrm{Mat}_n(\mathbb{C})) \le C'n^{1/2m}\sqrt{m}.$$

Applying this to $m = [\log n]$, the integer part of $\log n$, we obtain the second inequality of the lemma.

7 Dynamics on the State Space

It remains to prove (7.3). We shall prove that

$$T_2(S_{2m}(n)) \le \left(\frac{(2m)!}{2^m m!}\right)^{1/2m},$$

which implies (7.3) by the Stirling formula $m! \sim (m/e)^m \sqrt{2\pi m}$.

Consider the measure with weights $(\frac{1}{2}, \frac{1}{2})$ on the two-point set $\{-1, 1\}$. Consider then the space $Y = \{-1, 1\}^N$ with the product measure μ, and let $r_k : Y \to \{\pm 1\}$ be the projection onto the k-th coordinate. Then, by Hölder's inequality,

$$\frac{1}{2^N} \sum_{\varepsilon_1, \ldots, \varepsilon_N = \pm 1} \left\| \sum_{k=1}^N \varepsilon_k x_k \right\|_{2m}^2 = \int_Y \left\| \sum_{k=1}^N r_k(t) x_k \right\|_{2m}^2 d\mu(t)$$

$$\le \left(\int_Y \left\| \sum_{k=1}^N r_k(t) x_k \right\|_{2m}^{2m} d\mu(t) \right)^{1/m}.$$

Since $\|x\|_{2m}^{2m} = \text{Tr}((x^*x)^m)$, the m-th power of the right hand side is equal to

$$\sum_{k_1, \ldots, k_{2m} = 1}^N \int_Y r_{k_1}(t) \ldots r_{k_{2m}}(t) \, d\mu(t) \text{Tr}(x_{k_1}^* x_{k_2} \ldots x_{k_{2m-1}}^* x_{k_{2m}}).$$

Since the random variables $r_1, \ldots, r_N$ are independent, and $\int_Y r_k^l d\mu$ is zero or one depending on whether l is odd or even, the integral

$$\int_Y r_{k_1}(t) \ldots r_{k_{2m}}(t) \, d\mu(t)$$

is zero unless each k, $1 \le k \le N$, appears in the sequence $k_1, \ldots, k_{2m}$ an even number of times, and then the integral is 1. Denoting by $\sum'$ the summation over the indices satisfying the latter property, we thus have to show that

$$\sideset{}{'}\sum_{k_1, \ldots, k_{2m}} \text{Tr}(x_{k_1}^* x_{k_2} \ldots x_{k_{2m-1}}^* x_{k_{2m}}) \le \frac{(2m)!}{2^m m!} \left(\sum_k \|x_k\|_{2m}^2 \right)^m. \quad (7.4)$$

Apply the generalized **Horn inequality**

$$\text{Tr}(|y_1 \ldots y_p|) \le \sum_{j=1}^n s_j(y_1) \ldots s_j(y_p)$$

(see e.g. [198, Theorem 1.15] for $p = 2$, the same proof works in general), where $s_j(y_k)$ is the j-th s-number of $y_k \in \text{Mat}_n(\mathbb{C})$, i.e., $s_1(y_k) \ge \ldots \ge s_n(y_k)$ are the eigenvalues of $|y_k|$. Using that $s_j(x_k) = s_j(x_k^*)$, we then see that the left hand side of (7.4) is not larger than

$$\sum_j \sum_{k_1,\dots,k_{2m}}{}' s_j(x_{k_1})\dots s_j(x_{k_{2m}}).$$

Since the number of multi-indices $(k_1,\dots,k_{2m})$ containing each k, $1\le k\le N$, $2m_k$ times is equal to $(2m)!((2m_1)!\dots(2m_N)!)^{-1}$, the above expression equals

$$\sum_j \sum_{\substack{m_1,\dots,m_N \\ m_1+\dots+m_N=m}} \frac{(2m)!}{(2m_1)!\dots(2m_N)!} s_j(x_1)^{2m_1}\dots s_j(x_N)^{2m_N}.$$

By virtue of the inequality $(2m_k)!\ge 2^{m_k} m_k!$, the last expression is not larger than

$$\frac{(2m)!}{2^m m!}\sum_j \sum_{\substack{m_1,\dots,m_N \\ m_1+\dots+m_N=m}} \frac{m!}{m_1!\dots m_N!} s_j(x_1)^{2m_1}\dots s_j(x_N)^{2m_N}$$

$$= \frac{(2m)!}{2^m m!}\sum_j \left(\sum_k s_j(x_k)^2\right)^m$$

$$\le \frac{(2m)!}{2^m m!}\left(\sum_k\left(\sum_j s_j(x_k)^{2m}\right)^{1/m}\right)^m$$

(by the triangle inequality for ℓ_m^N)

$$= \frac{(2m)!}{2^m m!}\left(\sum_k \|x_k\|_{2m}^2\right)^m,$$

which establishes (7.4) and thus completes the proof of the lemma. □

We shall also need the following form of the combinatorial Sauer-Shelah lemma. For a set I we denote by 2^I the set of all subsets of I. If $J\subset I$ and $\mathcal{A}\subset 2^I$, we denote by $\mathcal{A}(J)$ the subset of 2^J consisting of all sets of the form $H\cap J$, $H\in\mathcal{A}$.

Lemma 7.2.4. *For every $c>0$ there exist $d>0$ and $n_0\in\mathbb{N}$ such that if I is a finite set, $|I|=n\ge n_0$, and $\mathcal{A}\subset 2^I$ satisfies $|\mathcal{A}|\ge e^{cn}$, then there exists $J\subset I$ such that $|J|\ge dn$ and $\mathcal{A}(J)=2^J$.*

Proof. We shall prove the following more precise result. If $m\le n$ is such that

$$|\mathcal{A}|>\sum_{k=0}^{m-1}\binom{n}{k},$$

then there exists J such that $|J|=m$ and $\mathcal{A}(J)=2^J$. To show that this is indeed enough, it suffices to check that there exists $d>0$ such that the above sum is smaller than e^{cn} for sufficiently large n and $m=[dn]+1$. Since $\binom{n}{k}\le\binom{n}{m}\binom{m}{k}$ for $k\le m$, the above sum is smaller than $2^m\binom{n}{m}$, and by the

Stirling formula we get $2^{dn+1}\binom{n}{[dn]+1} \leq 2^{c(d)n}$ for sufficiently large n, with $c(d) \to 0$ as $d \to 0$. So we just have to choose d such that $c(d) \leq c$.

The proof of the above statement is by induction on n. Since the result is obvious for $m = 1$ and $m = n$, assume $1 < m < n$. Fix any point $p \in I$, and set
$$I_p = I \setminus \{p\}, \quad \mathcal{A}_p = \mathcal{A}(I_p).$$
If $|\mathcal{A}_p| > \sum_{k=0}^{m-1}\binom{n-1}{k}$, the result is true by induction. So assume
$$|\mathcal{A}_p| \leq \sum_{k=0}^{m-1}\binom{n-1}{k}.$$

Define $\mathcal{B} = \{H \subset I_p \mid H \in \mathcal{A}, \; H \cup \{p\} \in \mathcal{A}\}$. Then every element in $\mathcal{B}$ has two preimages in $\mathcal{A}$ under the map $\mathcal{A} \ni H \mapsto H \cap I_p \in \mathcal{A}_p$, while every element in $\mathcal{A}_p \setminus \mathcal{B}$ has one preimage. Therefore $|\mathcal{A}| = |\mathcal{A}_p \setminus \mathcal{B}| + 2|\mathcal{B}| = |\mathcal{A}_p| + |\mathcal{B}|$. Using that $\binom{n}{k} - \binom{n-1}{k} = \binom{n-1}{k-1}$ for $k \geq 1$, we then get
$$|\mathcal{B}| = |\mathcal{A}| - |\mathcal{A}_p| > \sum_{k=0}^{m-1}\binom{n}{k} - \sum_{k=0}^{m-1}\binom{n-1}{k} = \sum_{k=0}^{m-2}\binom{n-1}{k}.$$

By induction there exists $J_p \subset I_p$ such that $|J_p| = m-1$ and $\mathcal{B}(J_p) = 2^{J_p}$. Put $J = J_p \cup \{p\}$. Then by definition of $\mathcal{B}$, for any $H \subset J_p$ there exists $H' \subset I_p$ such that $H' \cap J_p = H$ and both H' and $H' \cup \{p\}$ belong to $\mathcal{A}$. This shows that $\mathcal{A}(J) = 2^J$. □

This lemma allows us to replace the assumption on the number of ε-separated elements in Proposition 7.2.2 by the assumption that the image of the unit ball under the map Φ in the proposition contains a ball of a fixed radius.

Lemma 7.2.5. *For given $\varepsilon > 0$ and $a > 0$, there exist $n_0 \in \mathbb{N}$, $d > 0$ and $\delta > 0$ such that if $n \geq n_0$ and Ω is a symmetric and convex subset of the unit ball of $\ell_\infty^n(\mathbb{R})$ containing at least e^{an} elements that are ε-separated, then there exists a subset $J \subset \{1, \ldots, n\}$ such that $|J| \geq dn$ and the image of Ω under the projection $\pi_J \colon \ell_\infty^n(\mathbb{R}) \to \ell_\infty(J; \mathbb{R})$ contains the ball of radius δ.*

Proof. Let $Y \subset \Omega$ be an ε-separated set, $|Y| \geq e^{an}$. Fix $N \in \mathbb{N}$ to be specified later, and for each k, $0 \leq k < N$, consider the set $U_k \subset \mathbb{R}$ defined by
$$U_k = \bigcup_{m=-M}^{M}[m\varepsilon + k\delta_0, m\varepsilon + (k+1)\delta_0),$$
where $M = 1 + [1/\varepsilon]$ and $\delta_0 = \varepsilon/N$. Let Y_k be the subset of Y consisting of elements $y = (y_1, \ldots, y_n)$ such that the set $J_k(y) = \{j \mid y_j \in U_k\}$ contains no more than n/N elements. Since the sets U_k are disjoint and cover the

interval $[-1, 1]$, the sets $J_k(y)$, $0 \leq k \leq N-1$, are disjoint, and their union is $\{1, \ldots, n\}$. Hence $|J_k(y)| \leq n/N$ for some k. Therefore $\cup_{k=0}^{N-1} Y_k = Y$. Fix k such that $|Y_k| \geq |Y|/N$.

For all m, $|m| \leq M$, and $I \subset \{1, \ldots, n\}$ with $|I| > n(N-1)/N$ consider $\mathcal{A}_{m,I} \subset 2^I$ defined as follows. A subset $H \subset I$ is in $\mathcal{A}_{m,I}$ if there exists $y \in Y_k$ such that

$$y_j \geq m\varepsilon + (k+1)\delta_0, \quad \text{if } j \in H,$$
$$y_j < m\varepsilon + k\delta_0, \quad \text{if } j \in I \backslash H,$$
$$y_j \in [m\varepsilon + k\delta_0, m\varepsilon + (k+1)\delta_0) \subset U_k, \quad \text{if } j \in \{1, \ldots, n\} \backslash I.$$

For any m and $y \in Y_k$ there exists exactly one set I such that y defines an element of $\mathcal{A}_{m,I}$. In other words, for each m we have a map from Y_k into the disjoint union $\sqcup_I \mathcal{A}_{m,I}$. If two points $y, z \in Y_k$ have the same images under these maps, that is, they define the same element of 2^{I_m} for each m (for some uniquely defined I_m), then they cannot be ε-separated, and hence $y = z$. It follows that

$$|Y_k| \leq \prod_m \left(\sum_I |\mathcal{A}_{m,I}| \right).$$

Let $c_n(N)$ be the number of I's such that $|I| > n(N-1)/N$. Equivalently, $c_n(N)$ is the number of J's with $|J| \leq n/N$, so $c_n(N) = \sum_{j \leq n/N} \binom{n}{j}$. Then

$$\prod_{m=-M}^{M} \left(\sum_I |\mathcal{A}_{m,I}| \right) \leq \left(c_n(N) \max_{m,I} |\mathcal{A}_{m,I}| \right)^{2M+1}.$$

Since $|Y_k| \geq |Y|/N$, we get

$$\max_{m,I} |\mathcal{A}_{m,I}| \geq c_n(N)^{-1} \left(\frac{|Y|}{N} \right)^{1/(2M+1)} \geq c_n(N)^{-1} N^{-1/(2M+1)} e^{an/(2M+1)}.$$
(7.5)

We have already remarked in the proof of Lemma 7.2.4 that $c_n(N)$ does not grow faster than $e^{c(N)n}$ as $n \to \infty$, with $c(N) \to 0$ as $N \to \infty$. Thus we can choose N and $c > 0$ such that the right hand side of (7.5) is not smaller than e^{cn} for sufficiently large n. Fix m and I, $|I| > n(N-1)/N$, such that $|\mathcal{A}_{m,I}| \geq e^{cn}$.

By Lemma 7.2.4 there exists $J \subset I$ such that $|J| \geq dn$ and $\mathcal{A}_{m,I}(J) = 2^J$. In other words, for any $H \subset J$ there is $y \in Y$ satisfying

$$y_j \geq m\varepsilon + (k+1)\delta_0, \quad \text{if } j \in H,$$
$$y_j < m\varepsilon + k\delta_0, \quad \text{if } j \in J \backslash H.$$

Since the set $\pi_J(\Omega)$ is convex, this implies (e.g. by the Hahn-Banach theorem) that it contains all elements $y \in \ell_\infty(J; \mathbb{R})$ such that

$$m\varepsilon + k\delta_0 \leq y_j \leq m\varepsilon + (k+1)\delta_0, \quad j \in J.$$

In other words, it contains a translation of the ball of radius $\delta_0/2$ in $\ell_\infty(J;\mathbb{R})$, namely, $B_{\delta_0/2}(y) = y + B_{\delta_0/2}(0)$ with $y_j = m\varepsilon + k\delta_0 + \delta_0/2$ for $j \in J$. Since $\pi_J(\Omega)$ is also symmetric, it follows that it contains $\frac{1}{2}(y + B_{\delta_0/2}(0) - y) = B_{\delta_0/4}(0)$, the ball of radius $\delta = \delta_0/4$. □

Proof of Proposition 7.2.2. By Lemma 7.2.5 we can find $J \subset \{1,\ldots,n\}$ such that $|J| \geq dn$ and the image of the unit ball of $\mathrm{Mat}_m(\mathbb{C})^*$ under the map $\pi_J \circ \Phi$ contains the ball of radius δ in $\ell_\infty(J;\mathbb{R})$. Decomposing elements of $\ell_\infty(J)$ into real and imaginary parts, we conclude that the image of the unit ball contains the ball of radius $\delta_0 = \delta/2$ in $\ell_\infty(J)$.

Then the dual map

$$S = (\pi_J \circ \Phi)^* \colon \ell_1(J) \to \mathrm{Mat}_m(\mathbb{C})$$

has the property $\delta_0 \|y\|_1 \leq \|Sy\| \leq \|y\|_1$. Hence the type 2 constants satisfy

$$T_2(\mathrm{Mat}_m(\mathbb{C})) \geq \delta_0 T_2(\ell_1(J)).$$

By Lemma 7.2.3 this gives

$$C\sqrt{\log m} \geq \delta_0\sqrt{|J|} \geq \delta_0\sqrt{dn},$$

so that $m \geq e^{(\delta_0/C)^2 dn}$. □

Although we have proved that $ht(\alpha)$ is positive as soon as $h_{top}(T)$ is positive, our proof does not provide any estimates for $ht(\alpha)$. This is not a drawback of the proof, as in fact $h_{top}(T)$ can take only two values, zero and infinity.

Proposition 7.2.6. *We have either $h_{top}(T) = 0$ or $h_{top}(T) = \infty$.*

Proof. As in the proof of Theorem 7.2.1, let the metric on X be given by

$$d(\varphi, \psi) = \sup_{y \in \Omega_0} |\varphi(y) - \psi(y)|.$$

If $h_{top}(T) > a > 0$, there exists $\varepsilon > 0$ such that for all n sufficiently large there exists e^{an} points that are (n,ε)-separated with respect to T^{-1}.

Fix $r \in \mathbb{N}$ and choose $\varepsilon' > 0$ and $\lambda_1, \ldots, \lambda_r > 0$ such that $\sum_j \lambda_j = 1$ and

$$\varepsilon\lambda_i - 2\sum_{j=i+1}^{r} \lambda_j > \varepsilon' \text{ for } i = 1,\ldots,r-1.$$

Let E_n be an (n,ε)-separated set. Consider the set E'_n consisting of states of the form

$$\sum_{j=1}^{r} \lambda_j \varphi_j, \text{ where } \varphi_j \in E_n \text{ for } j = 1,\ldots,r.$$

We claim that $|E'_n| = |E_n|^r$ and the set E'_n is (n,ε')-separated. This would imply that $h_{top}(T) \geq ra$, so that $h_{top}(T) = \infty$, as r was arbitrary.

Let $(\varphi_1, \ldots, \varphi_r)$ and $(\psi_1, \ldots, \psi_r)$ be different r-tuples in E_n^r, and let i be the minimal number such that $\varphi_i \neq \psi_i$. As φ_i and ψ_i are (n, ε)-separated, there exist $y \in \Omega_0$ and k, $0 \leq k \leq n-1$, such that

$$|(\varphi_i \circ \alpha^k)(y) - (\psi_i \circ \alpha^k)(y)| \geq \varepsilon.$$

Then, since $\varphi_j = \psi_j$ for $j < i$, the distance between $T^{-k}(\sum_j \lambda_j \varphi_j)$ and $T^{-k}(\sum_j \lambda_j \psi_j)$ is not smaller than

$$\left| \sum_j \lambda_j \left((\varphi_j \circ \alpha^k)(y) - (\psi_j \circ \alpha^k)(y) \right) \right|$$

$$\geq \lambda_i |(\varphi_i \circ \alpha^k)(y) - (\psi_i \circ \alpha^k)(y)| - \sum_{j=i+1}^{r} \lambda_j (|(\varphi_j \circ \alpha^k)(y)| + |(\psi_j \circ \alpha^k)(y)|)$$

$$\geq \varepsilon \lambda_i - 2 \sum_{j=i+1}^{r} \lambda_j > \varepsilon',$$

and the proof is complete. $\square$

Remark 7.2.7. The above proof actually shows that the topological entropy of any affine homeomorphism of a convex compact subset of a locally convex space is either zero or infinity. ◆

We finish this chapter with the following sufficient condition for positivity of entropy, which is implicit in the proof of Theorem 7.2.1. Recall first that a subset I of $\mathbb{N}$ is said to have positive upper density if

$$\limsup_{n \to \infty} \frac{|I \cap \{1, \ldots, n\}|}{n} > 0.$$

Proposition 7.2.8. *Suppose there exist a self-adjoint element $a \in A$ and a subset $I = \{n_1 < n_2 < \ldots\} \subset \mathbb{N}$ of positive upper density such that the map $\Gamma: \ell_1 \to A$ defined by $\Gamma(e_k) = \alpha^{n_k}(a)$, $k \in \mathbb{N}$, where $\{e_k\}_k$ denotes the canonical basis of ℓ_1, is an isomorphism of ℓ_1 onto a closed subspace of A. Then $ht(\alpha) > 0$.*

Proof. Assume $A \subset B(H)$. We may also assume that $\|a\| \leq 1$, so that Γ is a contraction. Let $\delta > 0$ be such that $\|\Gamma(x)\| \geq \delta \|x\|_1$ for $x \in \ell_1$.

For each $n \in \mathbb{N}$ set $I_n = I \cap \{1, \ldots, n\}$. Denote by Γ_n the restriction of Γ to $\ell_1^{|I_n|}$. If unital completely positive maps $\sigma: A \to \text{Mat}_m(\mathbb{C})$ and $\gamma: \text{Mat}_m(\mathbb{C}) \to B(H)$ are such that $\|(\gamma \circ \sigma \circ \alpha^k)(a) - \alpha^k(a)\| < \delta/2$ for $k = 1, \ldots, n$, then $\|\Gamma_n - \gamma \circ \sigma \circ \Gamma_n\| < \delta/2$. It follows that

$$\|(\gamma \circ \sigma \circ \Gamma_n)(x)\| \geq \frac{\delta}{2} \|x\|_1 \quad \text{for} \quad x \in \ell_1^{|I_n|}.$$

Since γ is a contraction, the same inequality holds for the contraction $\sigma \circ \Gamma_n \colon \ell_1^{|I_n|} \to \mathrm{Mat}_m(\mathbb{C})$. The same argument as in the proof of Proposition 7.2.2 then shows that

$$C\sqrt{\log m} \geq \frac{\delta}{2}\sqrt{|I_n|},$$

so $m \geq e^{(\delta/2C)^2 |I_n|}$. Since $|I_n| \geq dn$ for certain sufficiently large n's and $d > 0$ by assumption, we conclude that

$$ht(\{a\}, \delta/2; \alpha) \geq \left(\frac{\delta}{2C}\right)^2 d > 0,$$

and the proof is complete. $\square$

If $\{a_n\}_{n=1}^{\infty}$ is a sequence of self-adjoint elements in the unit ball of a C*-algebra A, then in order to check that the map $\ell_1 \to A$, $e_n \mapsto a_n$, is an isomorphism onto a closed subspace it suffices to find $\varepsilon > 0$ such that for any $n \in \mathbb{N}$ and $\varepsilon_1, \ldots, \varepsilon_n = \pm 1$ there exists a state $\varphi_{\varepsilon_1 \ldots \varepsilon_n}$ satisfying

$$\varepsilon_k \varphi_{\varepsilon_1 \ldots \varepsilon_n}(a_k) \geq \varepsilon, \quad k = 1, \ldots, n.$$

Indeed, if $\lambda_1, \ldots, \lambda_n \in \mathbb{R}$ and ε_k is the sign of λ_k, then

$$\varphi_{\varepsilon_1 \ldots \varepsilon_n}(\lambda_1 a_1 + \ldots + \lambda_n a_n) \geq \varepsilon(|\lambda_1| + \ldots + |\lambda_n|),$$

so that $\|\lambda_1 a_1 + \ldots + \lambda_n a_n\| \geq \varepsilon \|(\lambda_1, \ldots, \lambda_n)\|_1$.

By the proof of the above proposition it then follows that there exist $\delta > 0$ and $d > 0$ such that

$$\mathrm{rcp}(\{a_1, \ldots, a_n\}, \delta) \geq e^{dn}$$

for all sufficiently large $n \in \mathbb{N}$. Note however that in this case the same conclusion can be made without relying on the estimates of the type 2 constants from Lemma 7.2.3. Indeed, consider the channels $\gamma_k \colon \mathbb{C}^2 \to A$ defined by

$$\gamma_k(p_1) = \frac{1}{2}(1 + a_k), \quad \gamma_k(p_2) = \frac{1}{2}(1 - a_k),$$

where $p_1 = (1, 0)$, $p_2 = (0, 1)$. For $n \in \mathbb{N}$ choose $\varphi_{\varepsilon_1 \ldots \varepsilon_n}$ as above and put

$$\varphi = \frac{1}{2^n} \sum_{\varepsilon_1, \ldots, \varepsilon_n = \pm 1} \varphi_{\varepsilon_1 \ldots \varepsilon_n}.$$

Then

$$H_\varphi(\gamma_1, \ldots, \gamma_n; \{2^{-n}\varphi_{\varepsilon_1 \ldots \varepsilon_n}\}) = \sum_{k=1}^{n} H_\varphi(\gamma_k; \{2^{-n}\varphi_{\varepsilon_k}^{(k)}\}).$$

By the choice of the $\varphi_{\varepsilon_1 \ldots \varepsilon_n}$'s,

$$2^{-n+1}(\varphi_1^{(k)} \circ \gamma_k)(p_1) \geq \frac{1}{2}(1+\varepsilon) \quad \text{and} \quad 2^{-n+1}(\varphi_{-1}^{(k)} \circ \gamma_k)(p_1) \leq \frac{1}{2}(1-\varepsilon).$$

Note that if μ_1 and μ_2 are two states on $\mathbb{C}^2$ such that $\mu_1(p_1) \geq (1+\varepsilon)/2$ and $\mu_2(p_1) \leq (1-\varepsilon)/2$, and $\mu = (\mu_1 + \mu_2)/2$, then

$$\|\mu - \mu_i\| \geq |\mu(p_1) - \mu_i(p_1)| = \frac{1}{2}|\mu_1(p_1) - \mu_2(p_1)| \geq \frac{\varepsilon}{2}.$$

Hence there exists $\varepsilon' > 0$ depending only on ε such that

$$\frac{1}{2}S(\mu_1,\mu) + \frac{1}{2}S(\mu_2,\mu) \geq \varepsilon'.$$

It follows that $H_\varphi(\gamma_1,\ldots,\gamma_n) \geq n\varepsilon'$. Then by an argument similar to that in the proof of Proposition 6.2.7, for any $d < \varepsilon'$ we can find $\delta > 0$ such that

$$\mathrm{rcp}(\{a_1,\ldots,a_n\},\delta) \geq e^{dn}$$

for any $n \in \mathbb{N}$.

Example 7.2.9. Let u be the shift to the right on $\ell_2(\mathbb{Z})$. Denote by α the automorphism $\mathrm{Ad}\, u$ of $B(H)$. Let $a \in B(H)$ be a self-adjoint operator such that for any $n \in \mathbb{N}$ and $\varepsilon_1,\ldots,\varepsilon_n = \pm 1$ there exists $m \in \mathbb{Z}$ such that

$$(ae_{m-k}, e_{m-k}) = (\alpha^k(a)e_m, e_m) = \varepsilon_k \quad \text{for} \quad k = 1,\ldots,n,$$

where $\{e_k\}_{k \in \mathbb{Z}}$ is the canonical basis in $\ell_2(\mathbb{Z})$. Then for any α-invariant C*-subalgebra A of $B(H)$ containing a we have $ht(\alpha|_A) > 0$. ♦

7.3 Notes

Proposition 7.1.2, of which Theorem 7.1.1 is an obvious consequence, was proved by Sauvageot [190]. The Pinsker formula, Lemma 7.1.3, appeared in [164].

In the abelian case Theorem 7.2.1 was proved by Glasner and Weiss [73]. They gave two different proofs of the result, one measure-theoretic, and one using geometry of the Banach spaces ℓ_1 and ℓ_∞. The second proof was extended to noncommutative systems by Kerr [107]. The estimate of the type 2 constants of matrix algebras in Lemma 7.2.3 is attributed in [107] to Tomczak-Jaegermann, who obtained the estimate of the type 2 constants of the Schatten ideals in [216]. Inequality (7.3) can be extended to noninteger $m \geq 1$ by interpolation [216]. Lemma 7.2.4, or more precisely the statement established in the course of its proof, is a famous result of Perles and Sauer [187], Shelah [196] and Vapnik and Červonenkis [219]. Lemma 7.2.5 is apparently well-known to specialists on geometric Banach space theory. We follow Glasner and Weiss [73] in its proof. Proposition 7.2.6 was proved by Sigmund [197] in the abelian case. As was observed in [107], the same proof works for noncommutative systems. Proposition 7.2.8 is due to Kerr and Li [108]. In the abelian case the converse statement is also true [108], which extends a result of Glasner and Weiss [73]. Example 7.2.9 is from [107].

In the abelian case the converse to Theorem 7.2.1 clearly holds, since the spectrum of an abelian algebra can be identified with the pure state space. To put it differently, if A is a C*-algebra and $X = S(A)$ is its state space, then we can define an injective unital completely positive map $\theta \colon A \to C(X)$ by $\theta(a)(\varphi) = \varphi(a)$. If α is an automorphism of A, $T\varphi = \varphi \circ \alpha^{-1}$, and $\hat{\alpha}$ is the automorphism of $C(X)$ defined by T, $\hat{\alpha}(f) = f \circ T^{-1}$, then $\hat{\alpha} \circ \theta = \theta \circ \alpha$. If A is abelian then θ is a complete order embedding, so by Remark 6.2.11(iii) we have $ht(\alpha) \leq ht(\hat{\alpha}) = h_{top}(T)$. For general A the map θ is far from being a complete order embedding. In fact, the inverse map $\theta^{-1} \colon \theta(A) \to A$ is not even completely bounded unless A is subhomogeneous, that is, a subalgebra of $C(Y) \otimes \text{Mat}_n(\mathbb{C})$ for a compact set Y and $n \in \mathbb{N}$ [95], [203].

Kerr and Li [108] introduced the contractive approximation entropy $hc(\alpha)$, where they use isometric embeddings instead of injective homomorphisms or complete order embeddings, and contractive factorizations through abelian finite dimensional C*-algebras instead of contractive completely positive factorizations through matrix algebras. One of the advantages of this entropy is that the converse to Theorem 7.2.1 holds [108], as one can see using the isometric embedding of A into $C(\tilde{X})$, where $\tilde{X}$ is the unit ball in A^*, and the fact that the topological entropy of the homeomorphism of $\tilde{X}$ defined by α is zero as soon as $h_{top}(T) = 0$ [107]. The price one pays is that for noncommutative systems contractive approximation entropy behaves very differently from the Voiculescu-Brown topological entropy. E.g. if α is the shift on $\text{Mat}_m(\mathbb{C})^{\otimes \mathbb{Z}}$ then, as we know, $ht(\alpha) = \log m$, while $hc(\alpha) = +\infty$ [108].

It remains an open question whether the converse to Theorem 7.2.1 is true for exact or nuclear algebras. In this respect it would be interesting to investigate automorphisms of noncommutative tori, see Chap. 11, where we have positive topological entropy, but generically only one invariant state with zero dynamical entropy.

8
Crossed Products

In this chapter we shall prove that canonical extensions of automorphisms to crossed products have the same entropy as the original systems. This is interesting and nontrivial already for crossed products by $\mathbb{Z}$, but we shall work with larger classes of groups and with twisted crossed products as this does not require significant additional efforts.

8.1 Crossed Products by Discrete Amenable Groups

Let G be a discrete group, and $\beta\colon G \to \mathrm{Aut}(A)$ an action of G on a unital C*-algebra A. Let also ω be a $\mathbb{T}$-valued **2-cocycle** on G, so

$$\omega(h,k)\overline{\omega(gh,k)}\omega(g,hk)\overline{\omega(g,h)} = 1 \quad \text{for} \quad g,h,k \in G.$$

Note that if $e \in G$ is the unit element then the cocycle identity implies that $\omega(g,e) = \omega(e,e) = \omega(e,k)$ for any $g,k \in G$ (put $h = k = e$ and $g = h = e$), and $\omega(g, g^{-1}) = \omega(g^{-1}, g)$ for any $g \in G$ (put $h = g^{-1}$ and $k = g$).

Consider the **full crossed product** $A \rtimes_{\beta,\omega} G$, which is the universal unital C*-algebra generated by a copy of A and by unitaries u_g, $g \in G$, such that

$$u_g a = \beta_g(a) u_g \quad \text{and} \quad u_g u_h = \omega(g,h) u_{gh}.$$

Note that $u_e = \omega(e,e)1$ and $u_g^* = \overline{\omega(g,g^{-1})\omega(e,e)} u_{g^{-1}}$ for any $g \in G$.

Assume $A \subset B(H)$. Then we can define a representation π of $A \rtimes_{\beta,\omega} G$ on $H \otimes \ell_2(G)$ as follows:

$$\pi(a)(\xi \otimes \delta_h) = \beta_{h^{-1}}(a)\xi \otimes \delta_h, \quad \pi(u_g)(\xi \otimes \delta_h) = \omega(g,h)\xi \otimes \delta_{gh}, \quad (8.1)$$

where $\delta_g \in \ell_2(G)$ is defined by $\delta_g(h) = 0$ for $h \neq g$ and $\delta_g(g) = 1$. The **reduced crossed product** $A \rtimes_{r,\beta,\omega} G$ is by definition the quotient of the full crossed product by the kernel of π, which is independent of the embedding of A into $B(H)$. Recall that if G is amenable then the full and reduced crossed product coincide, which we shall actually prove as a byproduct. If $\omega \equiv 1$ we write $A \rtimes_\beta G$ and $A \rtimes_{r,\beta} G$ for the crossed products.

If $G = \mathbb{Z}$ then the action of G on A is given by one automorphism β of A, and the crossed product $A \rtimes_\beta \mathbb{Z}$ is generated by A and a unitary u such that $ua = \beta(a)u$. We say that $u \in A \rtimes_\beta \mathbb{Z}$ is the canonical unitary implementing β.

Let α be an automorphism of A commuting with β_g for every $g \in G$. Then α extends to an automorphism $\bar\alpha$ of $A \rtimes_{r,\beta,\omega} G$ such that $\bar\alpha(u_g) = u_g$.

Using the canonical conditional expectation $E \colon A \rtimes_{r,\beta,\omega} G \to A$ defined by $E(au_g) = 0$ for $g \neq e$, we can also extend any state φ on A to a state $\bar\varphi$ on the crossed product. If φ is α-invariant, then $\bar\varphi$ is $\bar\alpha$-invariant.

Theorem 8.1.1. *Let G be a discrete amenable group, $\omega \in Z^2(G; \mathbb{T})$ a 2-cocycle, $\beta \colon G \to \mathrm{Aut}(A)$ an action of G on A, and α an automorphism of A commuting with β. Then for the automorphism $\bar\alpha$ of $A \rtimes_{\beta,\omega} G$ we have*

(i) $ht(\bar\alpha) = ht(\alpha)$;

(ii) *if φ is an α- and β-invariant state on A such that there exists a φ-approximating net, then $h_{\bar\varphi}(\bar\alpha) = h_\varphi(\alpha)$.*

If $A \subset B(H)$ is a von Neumann algebra, we can also define the von Neumann algebra crossed product of A by G, which is by definition the von Neumann algebra generated by $A \rtimes_{r,\beta,\omega} G$ in $B(H \otimes \ell_2(G))$. Since the restriction of an automorphism of a von Neumann algebra to a dense C*-subalgebra gives the same dynamical entropy, part (ii) is of course also true for W*-dynamical systems and von Neumann algebra crossed products.

The most interesting particular case of the above theorem is formulated in the following corollary.

Corollary 8.1.2. *Let α be an automorphism of a unital C*-algebra A. Let $u \in A \rtimes_\alpha \mathbb{Z}$ be the canonical unitary implementing α. Then*

(i) $ht(\mathrm{Ad}\, u) = ht(\alpha)$;

(ii) *if φ is an α-invariant state on A such that there exists a φ-approximating net, then $h_{\bar\varphi}(\mathrm{Ad}\, u) = h_\varphi(\alpha)$.* □

To prove Theorem 8.1.1 we shall approximate the system $(A \rtimes_{r,\beta,\omega} G, \bar\alpha)$ by amplifications of the system (A, α).

Let f be a positive finitely supported function on G such that $\|f\|_1 = \sum_h f(h) = 1$. Denote by F the support of f. We shall construct two unital completely positive maps

$$\Phi_f \colon A \rtimes_{r,\beta,\omega} G \to A \otimes B(\ell_2(F)) \quad \text{and} \quad \Psi_f \colon A \otimes B(\ell_2(F)) \to A \rtimes_{r,\beta,\omega} G.$$

To define Φ_f assume $A \subset B(H)$, identify $\ell_2(F)$ with a subspace of $\ell_2(G)$, and let $P_F \colon H \otimes \ell_2(G) \to H \otimes \ell_2(F)$ be the orthogonal projection. Then consider the representation π of $A \rtimes_{r,\beta,\omega} G$ on $H \otimes \ell_2(G)$ defined above, and set

8.1 Crossed Products by Discrete Amenable Groups 123

$$\Phi_f(x) = P_F\pi(x)P_F \quad \text{for } x \in A \rtimes_{r,\beta,\omega} G.$$

A priori we have $\Phi_f(x) \in B(H) \otimes B(\ell_2(F))$, but the equality

$$P_F\pi(au_g)(\xi \otimes \delta_h) = \begin{cases} \omega(g,h)(\beta_{h^{-1}g^{-1}}(a)\xi \otimes \delta_{gh}), & \text{if } gh \in F, \\ 0, & \text{otherwise,} \end{cases}$$

see (8.1), shows that $\Phi_f(x) \in A \otimes B(\ell_2(F))$. More precisely, we have

$$\Phi_f(au_g) = \sum_{h \in F \cap g^{-1}F} \omega(g,h)\beta_{h^{-1}g^{-1}}(a) \otimes e_{gh,h}, \tag{8.2}$$

where $\{e_{s,t}\}_{s,t \in F}$ is the set of matrix units in $B(\ell_2(F))$ corresponding to the basis $\{\delta_h\}_{h \in F}$. In particular,

$$\Phi_f(a) = \sum_{h \in F} \beta_{h^{-1}}(a) \otimes e_{h,h}.$$

To define Ψ_f note that for any elements $x_s \in A \rtimes_{r,\beta,\omega} G$, $s \in F$, we have a completely positive map $A \otimes B(\ell_2(F)) \to A \rtimes_{r,\beta,\omega} G$, $a \otimes e_{s,t} \mapsto x_s a x_t^*$. This is proved similarly to A.9. Namely, identify $A \rtimes_{r,\beta,\omega} G$ with

$$(A \rtimes_{r,\beta,\omega} G) \otimes \mathbb{C}e_{h,h} \subset A \otimes B(\ell_2(F))$$

for some fixed $h \in F$, and write the map as

$$A \otimes B(\ell_2(F)) \ni x \mapsto VxV^* \in (A \rtimes_{r,\beta,\omega} G) \otimes \mathbb{C}e_{h,h},$$

where $V = \sum_{s \in F} x_s \otimes e_{h,s}$.

Apply this construction to $x_s = f(s)^{1/2}u_s$, $s \in F$, to get Ψ_f. Thus

$$\Psi_f(a \otimes e_{s,t}) = f(s)^{1/2}u_s a(f(t)^{1/2}u_t)^* = f(s)^{1/2}f(t)^{1/2}\beta_s(a)u_s u_t^*. \tag{8.3}$$

Since $\sum_h f(h) = 1$, the map Ψ_f is unital.

Proposition 8.1.3. *For every positive finitely supported function f on G such that $\|f\|_1 = 1$, the unital completely positive maps Φ_f and Ψ_f have the following properties:*
(i) $\Phi_f \circ \bar{\alpha} = (\alpha \otimes \mathrm{id}) \circ \Phi_f$ and $\Psi_f \circ (\alpha \otimes \mathrm{id}) = \bar{\alpha} \circ \Psi_f$;
(ii) *if φ is a G-invariant state, then $(\varphi \otimes \varphi_f) \circ \Phi_f = \bar{\varphi}$ and $\bar{\varphi} \circ \Psi_f = \varphi \otimes \varphi_f$, where φ_f is the state on $B(\ell_2(F))$ given by $\varphi_f(e_{s,t}) = \delta_{s,t}f(s)$;*
(iii) $(\Psi_f \circ \Phi_f)(au_g) = (\sum_{h \in G} f(gh)^{1/2}f(h)^{1/2})au_g$, *so that*

$$\|(\Psi_f \circ \Phi_f)(au_g) - au_g\| \leq \|_g f - f\|_1^{1/2}\|a\|,$$

where $_g f(h) = f(g^{-1}h)$.

Proof. Parts (i) and (ii) are immediate consequences of (8.2) and (8.3). To check (iii) we compute

$$(\Psi_f \circ \Phi_f)(au_g) = \Psi_f \left(\sum_{h \in F \cap g^{-1}F} \omega(g,h)\beta_{h^{-1}g^{-1}}(a) \otimes e_{gh,h} \right)$$

$$= \sum_{h \in F \cap g^{-1}F} f(gh)^{1/2} f(h)^{1/2} \omega(g,h) a u_{gh} u_h^*$$

$$= \left(\sum_{h \in G} f(gh)^{1/2} f(h)^{1/2} \right) a u_g.$$

Then the inequality in (iii) follows from

$$\left| \sum_{h \in G} f(gh)^{1/2} f(h)^{1/2} - 1 \right| = |(f^{1/2}, {_g}f^{1/2}) - (f^{1/2}, f^{1/2})|$$

$$\leq \|{_g}f^{1/2} - f^{1/2}\|_2$$

$$\leq \|{_g}f - f\|_1^{1/2},$$

where the last inequality holds, as $(x-y)^2 \leq |x^2 - y^2|$ for any positive numbers x and y. $\square$

Since the group G is amenable, there exists a net $\{f_i\}_i$ of positive finitely supported functions on G such that $\|f_i\|_1 = 1$ and $\|{_g}f_i - f_i\|_1 \to 0$ for any $g \in G$. Then $(\Psi_{f_i} \circ \Phi_{f_i})(x) \to x$ in norm for any $x \in A \rtimes_{r,\beta,\omega} G$.

Note that for full crossed products we can define a map $\bar{\Phi}_f$ from the full crossed into $A \otimes B(\ell_2(F))$ by composing Φ_f with the quotient map $A \rtimes_{\beta,\omega} G \to A \rtimes_{r,\beta,\omega} G$, and then define a map $\bar{\Psi}_f$ into the full crossed product by the same formula as for Ψ_f. Then we get $\bar{\Psi}_{f_i} \circ \bar{\Phi}_{f_i} \to$ id pointwise. Since by definition $\bar{\Phi}_f$ factorizes through the reduced crossed product, it follows that full and reduced crossed products by amenable groups coincide.

To complete the proof of the theorem we need the following approximation result similar to Proposition 3.2.4.

Lemma 8.1.4. *Let α be an automorphism of a C^*-algebra B, $\{B_i\}_i$ a net of C^*-algebras together with automorphisms $\alpha_i \in \mathrm{Aut}(B_i)$, and $\{\Phi_i \colon B \to B_i\}_i$ and $\{\Psi_i \colon B_i \to B\}_i$ two nets of equivariant (that is, $\Phi_i \circ \alpha = \alpha_i \circ \Phi_i$ and $\alpha \circ \Psi_i = \Psi_i \circ \alpha_i$) unital completely positive maps such that $(\Psi_i \circ \Phi_i)(b) \to b$ in norm for any $b \in B$. Then*

(i) $ht(\alpha) \leq \liminf_i ht(\alpha_i)$;

(ii) *if φ is an α-invariant state on B such that $\varphi \circ \Psi_i \circ \Phi_i = \varphi$ for any i, then $h_\varphi(\alpha) \leq \liminf_i h_{\varphi \circ \Psi_i}(\alpha_i)$.*

Proof. Let $B \subset B(H)$, $B_i \subset B(H_i)$. By Arveson's extension theorem, A.8, we can extend $\Psi_i \colon B_i \to B$ to a unital completely positive map $\bar{\Psi}_i \colon B(H_i) \to$

$B(H)$. Then given unital completely positive maps $\gamma\colon B_i \to \mathrm{Mat}_n(\mathbb{C})$ and $\theta\colon \mathrm{Mat}_n(\mathbb{C}) \to B(H_i)$ we can consider

$$\gamma \circ \Phi_i \colon B \to \mathrm{Mat}_n(\mathbb{C}) \quad \text{and} \quad \bar{\Psi}_i \circ \theta \colon \mathrm{Mat}_n(\mathbb{C}) \to B(H),$$

and estimate

$$\|(\bar{\Psi}_i \circ \theta \circ \gamma \circ \Phi_i)(b) - b\| \le \|\bar{\Psi}_i((\theta \circ \gamma \circ \Phi_i)(b) - \Phi_i(b))\| + \|(\Psi_i \circ \Phi_i)(b) - b\|$$
$$\le \|(\theta \circ \gamma)(\Phi_i(b)) - \Phi_i(b)\| + \|(\Psi_i \circ \Phi_i)(b) - b\|.$$

It follows that if $\Omega \subset B$ is a finite subset, then

$$ht(\Omega, 2\varepsilon; \alpha) \le ht(\Phi_i(\Omega), \varepsilon; \alpha_i) \le ht(\alpha_i)$$

as soon as $\|(\Psi_i \circ \Phi_i)(b) - b\| < \varepsilon$ for $b \in \Omega$. Thus (i) is proved.

Similarly, if $\gamma\colon A \to B$ is a channel, then by Proposition 3.1.3(ii)

$$h_\varphi(\Psi_i \circ \Phi_i \circ \gamma; \alpha) \le h_{\varphi \circ \Psi_i}(\Phi_i \circ \gamma; \alpha_i) \le h_{\varphi \circ \Psi_i}(\alpha_i).$$

On the other hand, by Proposition 3.1.11

$$h_\varphi(\gamma; \alpha) \le h_\varphi(\Psi_i \circ \Phi_i \circ \gamma; \alpha) + \varepsilon$$

if $\|\Psi_i \circ \Phi_i \circ \gamma - \gamma\|$ is sufficiently small. This proves (ii). □

Proof of Theorem 8.1.1. If $\{f_i\}_i$ is a net of positive finitely supported functions on G such that $\|f_i\|_1 = 1$ and $\|gf_i - f_i\|_1 \to 0$ for any $g \in G$, then by Proposition 8.1.3, the discussion following it and Lemma 8.1.4(i) we have

$$ht(\bar{\alpha}) \le \liminf_i ht(\alpha \otimes \mathrm{id}_{B(\ell_2(F_i))}),$$

where F_i is the support of f_i. By Theorem 6.2.2(iv), $ht(\alpha \otimes \mathrm{id}) = ht(\alpha)$. Thus we get $ht(\bar{\alpha}) \le ht(\alpha)$. On the other hand, the opposite inequality holds by Theorem 6.2.2(iii).

The case of dynamical entropy is similar. The inequality $h_{\bar\varphi}(\bar\alpha) \ge h_\varphi(\alpha)$ holds by Theorem 3.2.2(v). In order to get the opposite inequality we just have to recall that by Theorem 3.2.5(iv), if there exists a φ-approximating net, then $h_{\varphi \otimes \psi}(\alpha \otimes \mathrm{id}) = h_\varphi(\alpha)$ for any state ψ on a finite dimensional C*-algebra. □

8.2 Generalizations

In this section we shall discuss several generalizations of Theorem 8.1.1.

First of all, what are the minimal assumptions on G for the equality $ht(\bar\alpha) = ht(\alpha)$ to be true? Since by our definition of topological entropy an algebra is exact if and only if the entropy of the identity automorphism is zero, for $\omega \equiv 1$ the equality implies that $A \rtimes_{r,\beta} G$ is exact as soon as A is exact. But then the reduced group C*-algebra $C_r^*(G) \subset A \rtimes_{r,\beta} G$ is also exact. Groups with this property are called **exact**. We shall use an equivalent definition of exactness saying that there exists an **amenable action** of G on a compact space. That is, there are an action $\gamma \colon G \to \operatorname{Aut}(B)$ of G on an abelian unital C*-algebra B (in fact, we can always take the action by translations on $\ell_\infty(G)$) and a net $\{f_i\}_i$ of positive finitely supported functions $G \to B$ such that $\sum_h f_i(h) = 1$ and

$$\lim_i \left\| \sum_{h \in G} |f_i(gh) - \gamma_g(f_i(h))| \right\| = 0 \text{ for any } g \in G.$$

Note that if $B = \mathbb{C}1$, then we get the usual definition of amenability of G. The class of exact groups is however much bigger than the class of amenable groups. For example, free groups are exact (in Chap. 14 we shall prove a more general result).

Theorem 8.2.1. *Let G be a discrete exact group, $\omega \in Z^2(G; \mathbb{T})$ a 2-cocycle, $\beta \colon G \to \operatorname{Aut}(A)$ an action of G on A, and α an automorphism of A commuting with β. Then for the automorphism $\bar\alpha$ of $A \rtimes_{r,\beta,\omega} G$ we have $ht(\bar\alpha) = ht(\alpha)$.*

Proof. The proof is a slight modification of that of Theorem 8.1.1, so we will be sketchy.

Assume first that there exists a unital C*-subalgebra B of the center of A such that the action $\beta|_B$ of G on B is amenable and $\alpha|_B$ is trivial. Given a positive finitely supported function $f \colon G \to B$ such that $\sum_h f(h) = 1$ we can define unital completely positive maps Φ_f and Ψ_f by formulas (8.2) and (8.3), so, with $F = \operatorname{supp} f$,

$$\Phi_f(au_g) = \sum_{h \in F \cap g^{-1}F} \omega(g,h) \beta_{h^{-1}g^{-1}}(a) \otimes e_{gh,h},$$

$$\Psi_f(a \otimes e_{s,t}) = f(s)^{1/2} u_s a (f(t)^{1/2} u_t)^* = f(s)^{1/2} \beta_{st^{-1}}(f(t)^{1/2}) \beta_s(a) u_s u_t^*.$$

Then

$$(\Psi_f \circ \Phi_f)(au_g) = \sum_h f(gh)^{1/2} \beta_g(f(h)^{1/2}) au_g,$$

and as

$$\left\| \sum_h f(gh)^{1/2} \beta_g(f(h)^{1/2}) - 1 \right\| = \left\| \sum_h (f(gh)^{1/2} - \beta_g(f(h))^{1/2}) \beta_g(f(h))^{1/2} \right\|$$

$$\leq \left\| \sum_h |f(gh) - \beta_g(f(h))| \right\|^{1/2},$$

we see that the maps Φ_f and Ψ_f have properties similar to those in Proposition 8.1.3(i),(iii). Hence $ht(\bar\alpha) = ht(\alpha)$.

In the general case, given an amenable action γ of G on a unital abelian C*-algebra B, we can conclude that the extension of the automorphism $\mathrm{id} \otimes \alpha$ of $B \otimes A$ to $(B \otimes A) \rtimes_{r, \gamma \otimes \beta, \omega} G$ has entropy not greater than $ht(\mathrm{id} \otimes \alpha) = ht(\alpha)$. As $A \rtimes_{r, \beta, \omega} G$ is a subalgebra of $(B \otimes A) \rtimes_{r, \gamma \otimes \beta, \omega} G$, we get $ht(\bar\alpha) \le ht(\alpha)$, whereas the opposite inequality always holds. □

Note that the previous argument does not allow us to prove an analogue of Theorem 8.1.1(ii) for exact groups. The problem is that if a group acts amenably and preserves a state, then it is amenable. Thus if G is exact but nonamenable, there is no G-invariant state on B. In particular, there is no canonical extension of a state on A to a state on $(B \otimes A) \rtimes_{r, \gamma \otimes \beta} G$. Moreover, there exists no conditional expectation $(B \otimes A) \rtimes_{r, \gamma \otimes \beta} G \to A \rtimes_{r, \beta} G$, since otherwise the restriction to B of the composition of $\bar\varphi$ with such a conditional expectation would be a G-invariant state. So even if we manage to get an estimate of the dynamical entropy of the extension of $\mathrm{id}_B \otimes \alpha$ to the crossed product with respect to some invariant state, it will not give us an immediate estimate of the entropy of $\bar\alpha$.

We next turn to nondiscrete groups. Then the crossed products are nonunital, but the notion of topological entropy still makes sense by Remark 6.2.11(ii).

Recall the definition of a crossed product. Let G be a locally compact group, $\omega: G \times G \to \mathbb{T}$ a continuous 2-cocycle, $\beta: G \to \mathrm{Aut}(A)$ a strongly continuous action of G on a C*-algebra A. Fix a left-invariant Haar measure on G. Consider the Banach $*$-algebra $L^1_\omega(G, A)$ consisting of functions $f: G \to A$ such that $\int_G \|f(g)\| dg < \infty$, with product

$$(f_1 *_\omega f_2)(g) = \int_G \omega(h, h^{-1}g) f_1(h) \beta_h(f_2(h^{-1}g)) dh$$

and involution

$$f^*(g) = \overline{\omega(g, g^{-1}) \omega(e, e)} \Delta(g^{-1}) \beta_g(f(g^{-1})^*),$$

where Δ is the modular function of G. Then the full crossed product algebra $A \rtimes_{\beta, \omega} G$ is the enveloping C*-algebra of $L^1_\omega(G, A)$. The algebra A is in general a subalgebra not of the crossed product, but of its multiplier algebra. Namely, for $a \in A$ and $f \in L^1_\omega(G, A)$ we define $(af)(g) = af(g)$ and $(fa)(g) = f(g) \beta_g(a)$. Similarly we have unitary elements u_g, $g \in G$, in the multiplier algebra defined by $(u_g f)(h) = \omega(g, g^{-1}h) \beta_g(f(g^{-1}h))$ and $(fu_g)(h) = \omega(hg^{-1}, g) \Delta(g^{-1}) f(hg^{-1})$.

If $A \subset B(H)$ then similarly to (8.1) define a representation π of $A \rtimes_{\beta, \omega} G$ on $L^2(G, H) \cong H \otimes L^2(G)$. Its extension to the multiplier algebra is given by

$$(\pi(a)\xi)(h) = \beta_{h^{-1}}(a) \xi(h), \quad (\pi(u_g)\xi)(h) = \omega(g, g^{-1}h) \xi(g^{-1}h), \quad (8.4)$$

so that for $f \in L^1_\omega(G, A)$ we have

$$\pi(f) = \int_G \pi(f(g))\pi(u_g)dg.$$

Then $A \rtimes_{r,\beta,\omega} G$ is the quotient of the full crossed product by the kernel of this representation. Again, this kernel is independent of the choice of the embedding of A into $B(H)$, and if G is amenable then the kernel is trivial.

Finally, if α is an automorphism of A commuting with the automorphisms β_g, $g \in G$, we can define an automorphism $\bar\alpha$ of $A \rtimes_{r,\beta,\omega} G$ such that $(\bar\alpha(f))(g) = \alpha(f(g))$ for $f \in L^1_\omega(G, A)$.

Theorem 8.2.2. *Let G be a locally compact amenable group, $\omega \in Z^2(G; \mathbb{T})$ a continuous 2-cocycle, $\beta: G \to \mathrm{Aut}(A)$ a strongly continuous action of G on a C^*-algebra A, and α an automorphism of A commuting with β. Then for the automorphism $\bar\alpha$ of $A \rtimes_{\beta,\omega} G$ we have $ht(\bar\alpha) = ht(\alpha)$.*

Proof. The proof is again similar to that of Theorem 8.1.1, but the details become a bit more involved. This time the maps Φ_f and Ψ_f will be defined for positive compactly supported functions f on G, so that

$$\Phi_f: A \rtimes_{r,\beta,\omega} G \to A \otimes K(L^2(F)) \quad \text{and} \quad \Psi_f: A \otimes K(L^2(F)) \to A \rtimes_{r,\beta,\omega} G,$$

where $F = \mathrm{supp}\, f$, and $K(L^2(F))$ denotes the algebra of compact operators on $L^2(F)$. To define these maps represent A faithfully on a Hilbert H, and consider the representation of the crossed product on $H_0 = L^2(G, H)$ as above. Let $P_F: H_0 \to L^2(F, H)$ be the orthogonal projection, and set $\Phi_f(x) = P_F \pi(x) P_F$. Then, using (8.4), similarly to (8.2) we have

$$(\Phi_f(au_g)\xi)(h) = \omega(g, g^{-1}h)\beta_{h^{-1}}(a)\xi(g^{-1}h) \quad \text{for } \xi \in L^2(F, H). \tag{8.5}$$

To define Ψ_f we shall consider $A \otimes K(L^2(F))$ acting on $H_0 \otimes L^2(F) = L^2(F, H_0)$ rather than on $H \otimes L^2(F)$, and define a map $B(L^2(F, H_0)) \to B(H_0)$. Consider the operator

$$V_f: H_0 \to L^2(F, H_0), \quad (V_f\xi)(s) = f(s)^{1/2}\pi(u_s)^*\xi.$$

The adjoint operator is given by

$$V_f^*\zeta = \int_F f(s)^{1/2}\pi(u_s)\zeta(s)\,ds.$$

In particular, $V_f^*V_f$ is just the scalar operator $\|f\|_1$, so that if $\|f\|_1 = 1$ we get a unital completely positive map

$$\tilde\Psi_f: B(L^2(F, H_0)) \to B(H_0), \quad \tilde\Psi_f(x) = V_f^*xV_f.$$

To see that $\tilde\Psi_f(\pi(A) \otimes K(L^2(F))) \subset \pi(A \rtimes_{r,\beta,\omega} G)$, consider an integral operator T on $L^2(F)$ with continuous kernel $k \in C(F \times F)$, in other words

8.2 Generalizations 129

$$(T\xi)(s) = \int_F k(s,t)\xi(t)\,dt.$$

Then a simple computation yields (compare with (8.3))

$$\tilde{\Psi}_f(\pi(a)\otimes T) = \iint_{F\times F} f(s)^{1/2}f(t)^{1/2}k(s,t)\pi(\beta_s(a)u_s u_t^*)ds\,dt \in \pi(A\rtimes_{r,\beta,\omega}G).$$

Since integral operators are dense in the algebra of compact operators, we conclude that $\tilde{\Psi}_f(\pi(A) \otimes K(L^2(F))) \subset \pi(A \rtimes_{r,\beta,\omega} G)$. Hence there exists a contractive completely positive map $\Psi_f\colon A \otimes K(L^2(F)) \to A \rtimes_{r,\beta,\omega} G$ such that $(\pi \circ \Psi_f)(a \otimes T) = \tilde{\Psi}_f(\pi(a) \otimes T)$ for $a \in A$ and $T \in K(L^2(F))$.

Consider the representation ρ of $A \otimes K(L^2(F))$ on $L^2(F, H_0) = H_0 \otimes L^2(F)$ defined by $\rho(a \otimes T) = \pi(a) \otimes T$. Then, denoting the extension of ρ to the multiplier algebra by the same letter, (8.5) can be written as

$$(\rho(\Phi_f(au_g))\zeta)(s) = \omega(g,g^{-1}s)\pi(\beta_{s^{-1}}(a))\zeta(g^{-1}s) \text{ for } \zeta \in L^2(F,H_0),\ s \in F.$$

For $\xi \in H_0$ we then compute

$$\pi((\Psi_f \circ \Phi_f)(au_g))\xi = V_f^* \rho(\Phi_f(au_g))V_f\xi$$
$$= \int_F f(s)^{1/2}\pi(u_s)(\rho(\Phi_f(au_g))V_f\xi)(s)\,ds$$
$$= \int_F f(s)^{1/2}\omega(g,g^{-1}s)\pi(u_s)\pi(\beta_{s^{-1}}(a))(V_f\xi)(g^{-1}s)\,ds$$
$$= \int_F f(s)^{1/2}f(g^{-1}s)^{1/2}\omega(g,g^{-1}s)\pi(u_s\beta_{s^{-1}}(a)u_{g^{-1}s}^*)\xi\,ds$$
$$= \int_F f(s)^{1/2}f(g^{-1}s)^{1/2}\pi(au_g)\xi\,ds,$$

so that exactly as in Proposition 8.1.3(iii)

$$(\Psi_f \circ \Phi_f)(au_g) = \left(\int_F f(s)^{1/2}f(g^{-1}s)^{1/2}\,ds\right)au_g.$$

By an analogue of Lemma 8.1.4(i) for contractive completely positive maps instead of unital ones, this allows us to conclude that $ht(\bar{\alpha}) \leq ht(\alpha)$.

If G is nondiscrete, A is not a subalgebra of the crossed product, so we need an additional argument for $ht(\alpha) \leq ht(\bar{\alpha})$. This has nothing to do with amenability and is done as follows. It suffices to construct two nets $\{\varphi_i\colon A \to A \rtimes_{r,\beta,\omega} G\}_i$ and $\{\psi_i\colon A \rtimes_{r,\beta,\omega} G \to A\}_i$ of equivariant completely positive contractive maps such that $\|(\psi_i \circ \varphi_i)(a) - a\| \to 0$ for any $a \in A$.

Choose a net $\{f_i\}_i \subset L^1(G)$ of functions such that $\|f_i\|_1 = 1$, and the support of f_i is eventually contained in an arbitrarily small neighbourhood of the unit element $e \in G$. Put $x_i = \int_G f_i(g)u_g dg$. Then set $\varphi_i(a) = x_i a x_i^*$.

For $\zeta \in L^2(G)$, $\|\zeta\|_2 = 1$, consider the Fubini map

$$\psi_\zeta = \mathrm{id} \otimes \omega_\zeta \colon B(H \otimes L^2(G)) \to B(H),$$

where ω_ζ is the state on $B(L^2(G))$ given by $\omega_\zeta(T) = (T\zeta, \zeta)$. We shall see in a moment that $\psi_\zeta(\pi(A \rtimes_{r,\beta,\omega} G)) \subset A$. We claim that if $\|\beta_{h^{-1}}(a) - a\| < \varepsilon$ for $h \in \mathrm{supp}\,\zeta$, then

$$\|(\psi_\zeta \circ \pi \circ \varphi_i)(a) - a\| < \varepsilon$$

for sufficiently large i. Thus, replacing $\{\varphi_i\}_i$ by a subnet, an appropriate net $\{\psi_i\}_i$ can be chosen from the maps $\psi_\zeta \circ \pi$, $\zeta \in L^2(G)$.

To prove our claim consider the C*-subalgebra $C_r^*(G,\omega) = \mathbb{C}1 \rtimes_{r,\omega} G$ of the multiplier algebra of $A \rtimes_{r,\beta,\omega} G$. Consider the representation $\bar\pi$ of $C_r^*(G,\omega)$ on $L^2(G)$ given by $(\bar\pi(u_g)\vartheta)(h) = \omega(g, g^{-1}h)\vartheta(g^{-1}h)$. Thus $\pi(x) = 1 \otimes \bar\pi(x)$ for $x \in C_r^*(G,\omega)$. The equality

$$\psi_\zeta(\pi(au_g)) = \int \omega(g, g^{-1}h)\zeta(g^{-1}h)\overline{\zeta(h)}\beta_{h^{-1}}(a)dh$$

shows that if $\|\beta_{h^{-1}}(a) - a\| < \varepsilon$ for $h \in \mathrm{supp}\,\zeta$, then

$$\|\psi_\zeta(\pi(au_g)) - \omega_\zeta(\bar\pi(u_g))a\| < \varepsilon,$$

so that $\|\psi_\zeta(\pi(ax_ix_i^*)) - \omega_\zeta(\bar\pi(x_ix_i^*))a\| < \varepsilon$. Hence $\|\psi_\zeta(\pi(ax_ix_i^*)) - a\| < \varepsilon$ for i large enough, since $\bar\pi(x_ix_i^*)\zeta \to \zeta$. On the other hand,

$$[x_i, a] = \int_G f_i(g)(\beta_g(a) - a)u_g dg \to 0,$$

so $ax_ix_i^*$ becomes arbitrarily close to $x_iax_i^* = \varphi_i(a)$. This proves our claim and thus completes the proof of the theorem. □

Note that similarly to the case of a discrete group the maps Φ_f and Ψ_f can be defined for full crossed products. This is clearly true for Φ_f. To construct Ψ_f it suffices to take a faithful representation of the full crossed product on a Hilbert space H_0 and use the same formulas as in the proof of the theorem. Thus, exactly as for discrete groups, we may conclude that if G is amenable then the full and reduced crossed products coincide. Moreover, then the identity map of $A \rtimes_{\beta,\omega} G$ approximately factorizes through the algebras $A \otimes K(H)$, where $K(H)$ is the algebra of compact operators on H. So if in addition A is nuclear then $A \rtimes_{\beta,\omega} G$ is also nuclear, and if A is exact then by an argument similar to the proof of Lemma 6.1.2 the algebra $A \rtimes_{\beta,\omega} G$ is exact.

8.3 Notes

The question whether for an ergodic dynamical system (X, μ, T) the dynamical entropy of the inner automorphism of $L^\infty(X, \mu) \rtimes \mathbb{Z}$ defined by the canonical unitary implementing T coincides with $h_\mu(T)$, was raised by Størmer [208].

Voiculescu [227] answered this question positively by constructing the maps Φ_f and Ψ_f from Sect. 8.1. These maps were also independently introduced by Sinclair and Smith [201] for different purposes. The observation made in the proof of Theorem 8.2.2 that a similar construction works for crossed products by arbitrary locally compact groups, seems to be new.

Remark that the composition of the maps Ψ_f and Φ_f is a map of the form $au_g \mapsto \varphi(g)au_g$, where φ is a finitely supported positive definite function. Such maps were considered already by Zeller-Meier [237]. But the fact that they have nice factorizations through tensor products of the algebra and matrix algebras had apparently not been observed before Voiculescu [227] and Sinclair and Smith [201].

The result of Voiculescu was independently extended to general C*-dynamical systems by Brown [32] in the case of topological entropy and by Golodets and Neshveyev [79] in the case of dynamical entropy. A similar result is true for other types of dynamical invariants, see e.g. [35], [110]. Essentially, one has to check that such invariants (i) have the approximation property stated in Lemma 8.1.4, (ii) do not change if we tensor an automorphism with the trivial automorphism of a full matrix algebra.

For equivalence of various notions of exactness for discrete groups we refer the reader to the work of Anantharaman-Delaroche [6]. Theorem 8.2.1 was obtained by Choda [47] and Germain [70]. Theorem 8.2.2, in the case of the trivial cocycle, is due to Smith and Pop [167], whose proof is based on completely contractive factorizations of crossed products and the Takesaki-Takai duality. Smith and Pop prove, in fact, a slightly stronger result. Namely, assume that α does not commute with β, but there exists a continuous automorphism γ of G such that $\alpha \circ \beta_g = \beta_{\gamma(g)} \circ \alpha$. Assume also that the cocycle ω is γ-invariant. Then there exists an automorphism $\bar\alpha$ of $A \rtimes_{r,\beta,\omega} G$ such that $\bar\alpha(au_g) = \alpha(a)u_{\gamma(g)}$. Then, assuming that the group $\{\gamma^n\}_{n \in \mathbb{Z}}$ is relatively compact in $\mathrm{Aut}(G)$, we have $ht(\bar\alpha) = ht(\alpha)$. In order to show this one just has to use γ-invariant functions in the constructions of the maps involved, which is possible by the relative compactness assumption, and instead of the automorphism $\alpha \otimes \mathrm{id}$ of $A \otimes K(L^2(F))$ use the automorphism $\alpha \otimes \mathrm{Ad}\, U$, where U is the unitary on $L^2(G)$ defined by γ, so $(U\xi)(h) = \xi(\gamma^{-1}(h))$.

Without the assumption that $\{\gamma^n\}_{n \in \mathbb{Z}}$ is relatively compact it is natural to ask whether the inequality

$$ht(\bar\alpha) \le ht(\alpha) + ht(\alpha_\gamma)$$

holds, where α_γ is the automorphism of $C_r^*(G,\omega)$ defined by γ. Instead of $ht(\alpha_\gamma)$ one can try to use approximation entropies of automorphisms of discrete groups, e.g. the ones introduced by Choda [45] and Brown and Germain [37]. Very little is known about this problem, see [79], [45], [75] for some examples and partial results. In Chap. 11 we shall see that the inequality above can be strict already for discrete abelian groups and $\omega \equiv 1$.

The results of the present chapter show that the entropy of inner automorphisms is a computable invariant. Moreover, this invariant tells something about the position of the algebra generated by the unitary defining the automorphism. E.g. it is shown in [207] and [141] that if $h_\varphi(\operatorname{Ad} u) > 0$ for a unitary u in a hyperfinite factor, then u is not contained in any Cartan subalgebra. See [33], [143], [144] for some other results in this direction. It thus seems natural to pose the following question. Given a von Neumann algebra M, a normal state φ on M, and a von Neumann subalgebra N of the centralizer M_φ of the state φ, is it possible to define a computable entropic invariant of the action of the unitary group of N on M?

9
Variational Principle

The variational principle formulated in Chap. 1 states that for topological dynamical systems the topological entropy is the supremum of the measure entropies over all invariant measures. More generally, given a continuous function f one introduces the notion of pressure and proves that it can be obtained as the supremum of the quantities $h_\mu(T) + \int f\, d\mu$. The measures where the supremum is attained are called equilibrium, as they indeed are equilibrium states in gas lattice models. In this form, using mean entropy and a formulation of equilibrium in terms of the KMS condition, the variational principle was extended to quantum spin lattice system. The notion of mean entropy depends on the choice of a sequence of finite subsystems and so does not make sense in general. It is natural to replace it by dynamical entropy, and our goal is to prove a variational principle in this setting for certain asymptotically abelian systems. In particular, we shall prove that under certain assumptions the topological entropy is the supremum of the dynamical entropies, and if the supremum is attained at some state, then this state must be tracial.

9.1 Pressure

In order to define pressure we follow the setup of Chap. 6 of the definition of topological entropy.

Let A be a unital C*-algebra, α an automorphism of A, $H \in A^{sa}$ a self-adjoint element. For $\delta > 0$ and a finite subset $\Omega \subset A$ denote by $P(H, \Omega; \delta)$ the infimum of the quantities $\log \mathrm{Tr}_B(e^{-\theta(H)})$, where B is a finite dimensional C*-algebra and $\theta\colon A \to B$ is a unital completely positive map such that there exist a unital injective homomorphism $\pi\colon A \to C$ for some C*-algebra C and a unital completely positive map $\gamma\colon B \to C$ with $\|(\gamma \circ \theta)(a) - \pi(a)\| < \delta$ for any $a \in \Omega$. Put $P(H, \Omega; \delta) = +\infty$ if no such B and θ exist. As before, Tr_B denotes the canonical trace on B, i.e., the trace which takes the value 1 on each minimal projection.

By the argument in the proof of Lemma 6.1.2 it suffices to consider one fixed faithful representation $\pi\colon A \to B(H)$. If A is nuclear we may also take $C = A$ and π the identity map. Note also that if $H = 0$ then $P(H,\Omega;\delta) = \log \operatorname{rcp}(\Omega,\delta)$. Now put

$$P_\alpha(H,\Omega;\delta) = \limsup_{n\to\infty} \frac{1}{n} P\left(\sum_{j=0}^{n-1} \alpha^j(H), \bigcup_{j=0}^{n-1} \alpha^j(\Omega); \delta\right),$$

$$P_\alpha(H,\Omega) = \sup_{\delta>0} P_\alpha(H,\Omega;\delta).$$

Definition 9.1.1. *The pressure of α at H is*

$$P_\alpha(H) = \sup_\Omega P_\alpha(H,\Omega),$$

where the supremum is taken over all finite subsets of $\Omega \subset A$.

Note that it suffices to consider finite subsets Ω of a set spanning a dense subspace of A. It is clear also that if $H = 0$ then the pressure coincides with the topological entropy $ht(\alpha)$.

Proposition 9.1.2. *For any α-invariant state φ of A we have*

$$P_\alpha(H) \geq h_\varphi(\alpha) - \varphi(H).$$

Proof. Assume $A \subset B(H)$. Let N be a finite dimensional C*-algebra and $\gamma\colon N \to A$ be a channel. Fix $\delta > 0$. Choose a finite subset Ω_0 of N such that its convex hull contains the unit ball. Put $\Omega = \{H\} \cup \gamma(\Omega_0)$.

Let B be a finite dimensional C*-algebra, and $\theta\colon A \to B$ and $\psi\colon B \to B(H)$ be unital completely positive maps such that

$$\|(\psi \circ \theta)(a) - a\| < \delta \text{ for } a \in \alpha^k(\Omega), \ 0 \leq k \leq n-1.$$

Then $\|\psi \circ \theta \circ \alpha^k \circ \gamma - \alpha^k \circ \gamma\| < \delta$ for $0 \leq k \leq n-1$.

Let $\varphi = \sum_{i_1,\ldots,i_n} \varphi_{i_1\ldots i_n}$ be a finite decomposition. Extend each $\varphi_{i_1\ldots i_n}$ to a positive linear functional $\tilde\varphi_{i_1\ldots i_n}$ on $B(H)$, and set $\tilde\varphi = \sum_{i_1,\ldots,i_n} \tilde\varphi_{i_1\ldots i_n}$. We obviously have

$$H_\varphi(\gamma,\ldots,\alpha^{n-1}\circ\gamma;\{\varphi_{i_1\ldots i_n}\}) = H_{\tilde\varphi}(\gamma,\ldots,\alpha^{n-1}\circ\gamma;\{\tilde\varphi_{i_1\ldots i_n}\}).$$

On the other hand, by Proposition 3.1.11

$$|H_{\tilde\varphi}(\gamma,\ldots,\alpha^{n-1}\circ\gamma) - H_{\tilde\varphi}(\psi\circ\theta\circ\gamma,\ldots,\psi\circ\theta\circ\alpha^{n-1}\circ\gamma)| < n\varepsilon$$

for some ε depending only on δ and $\dim N$. Since

$$H_{\tilde\varphi}(\psi\circ\theta\circ\gamma,\ldots,\psi\circ\theta\circ\alpha^{n-1}\circ\gamma) \leq S(\tilde\varphi\circ\psi)$$

by Proposition 3.1.3(i),(iv) and Lemma 3.1.4, we get

9.1 Pressure

$$H_\varphi(\gamma, \ldots, \alpha^{n-1} \circ \gamma; \{\varphi_{i_1 \ldots i_n}\}) \leq S(\tilde{\varphi} \circ \psi) + n\varepsilon.$$

By Proposition 2.2.5 we have $\log \operatorname{Tr}_B(e^{-K}) \geq S(\omega) - \omega(K)$ for any state ω and any self-adjoint K. Therefore

$$S(\tilde{\varphi} \circ \psi) \leq \log \operatorname{Tr}_B\left(e^{-\theta\left(\sum_{j=0}^{n-1} \alpha^j(H)\right)}\right) + \tilde{\varphi}\left((\psi \circ \theta)\left(\sum_{j=0}^{n-1} \alpha^j(H)\right)\right)$$

$$\leq \log \operatorname{Tr}_B\left(e^{-\theta\left(\sum_{j=0}^{n-1} \alpha^j(H)\right)}\right) + \varphi\left(\sum_{j=0}^{n-1} \alpha^j(H)\right) + n\delta$$

$$= \log \operatorname{Tr}_B\left(e^{-\theta\left(\sum_{j=0}^{n-1} \alpha^j(H)\right)}\right) + n\varphi(H) + n\delta.$$

Thus

$$\frac{1}{n} H_\varphi(\{\alpha^j \circ \gamma\}_{0 \leq j \leq n-1}) \leq \frac{1}{n} P\left(\sum_{j=0}^{n-1} \alpha^j(H), \bigcup_{j=0}^{n-1} \alpha^j(\Omega); \delta\right) + \varphi(H) + \delta + \varepsilon.$$

It follows that $h_\varphi(\gamma; \alpha) \leq P_\alpha(H, \Omega) + \varphi(H)$, hence $h_\varphi(\alpha) \leq P_\alpha(H) + \varphi(H)$.
$\square$

Remark 9.1.3. We have chosen the minus sign $e^{-\theta(H)}$ in the definition of $P(H, \Omega; \delta)$ because of its use in physical applications [30], rather than the plus sign used in ergodic theory [231]. If A is abelian, $A = C(X)$, and $P_\alpha^{cl}(H)$ denotes the pressure as defined in [231], see also Notes at the end of the chapter, then $P_\alpha(H) = P_\alpha^{cl}(-H)$. The inequality $\leq$ can be proved similarly to the proof of Lemma 6.1.3. The converse inequality follows from Proposition 9.1.2 and the classical variational principle. $\blacklozenge$

We list some properties of the function $H \mapsto P_\alpha(H)$ on A^{sa}.

Proposition 9.1.4. *We have:*
(i) *if $H \leq K$ then $P_\alpha(H) \geq P_\alpha(K)$;*
(ii) $P_\alpha(H + c1) = P_\alpha(H) - c$ *for $c \in \mathbb{R}$;*
(iii) $P_\alpha(H)$ *is either infinite for all H or is finite valued;*
(iv) *if P_α is finite valued then $|P_\alpha(H) - P_\alpha(K)| \leq \|H - K\|$;*
(v) *for $k \in \mathbb{N}$, $P_{\alpha^k}(\sum_{j=0}^{k-1} \alpha^j(H)) = k P_\alpha(H)$;*
(vi) $P_\alpha(H + \alpha(K) - K) = P_\alpha(H)$.

Proof. Let $\theta: A \to D$ be a unital completely positive map.
If $H \leq K$, then by the Peierls-Bogoliubov inequality, Corollary 2.3.11,

$$\log \operatorname{Tr}_B\left(e^{-\theta\left(\sum_{j=0}^{n-1} \alpha^j(H)\right)}\right) \geq \log \operatorname{Tr}_B\left(e^{-\theta\left(\sum_{j=0}^{n-1} \alpha^j(K)\right)}\right).$$

Thus (i) follows.
Since

$$\log \mathrm{Tr}_B\left(e^{-\theta\left(\sum_{j=0}^{n-1} \alpha^j(H+c1)\right)}\right) = \log \mathrm{Tr}_B\left(e^{-\theta\left(\sum_{j=0}^{n-1} \alpha^j(H)\right)}\right) - nc,$$

we get (ii).

By (i) and (ii) we have

$$P_\alpha(H) \geq P_\alpha(\|H\|) = P_\alpha(0) - \|H\| = ht(\alpha) - \|H\|,$$

and similarly $P_\alpha(H) \leq ht(\alpha) + \|H\|$. This shows (iii).

By the Peierls-Bogoliubov inequality, Corollary 2.3.11,

$$\left|\frac{1}{n}\log \mathrm{Tr}_B\left(e^{-\theta\left(\sum_{j=0}^{n-1} \alpha^j(H)\right)}\right) - \frac{1}{n}\log \mathrm{Tr}_B\left(e^{-\theta\left(\sum_{j=0}^{n-1} \alpha^j(K)\right)}\right)\right| \leq \|H-K\|,$$

and (iv) follows.

Let Ω be a finite subset of A, and $k \in \mathbb{N}$. Given $n \in \mathbb{N}$ choose $m \in \mathbb{N}$ such that $mk \leq n < (m+1)k$. Set $H_k = \sum_{j=0}^{k-1} \alpha^j(H)$ and $\Omega_k = \cup_{j=0}^{k-1}\alpha^j(\Omega)$. Then, using the Peierls-Bogoliubov inequality once again,

$$\log \mathrm{Tr}_B\left(e^{-\theta\left(\sum_{j=0}^{n-1} \alpha^j(H)\right)}\right) \geq \log \mathrm{Tr}_B\left(e^{-\theta\left(\sum_{j=0}^{m-1} \alpha^{jk}(H_k)\right)}\right) - k\|H\|.$$

Similarly

$$\log \mathrm{Tr}_B\left(e^{-\theta\left(\sum_{j=0}^{n-1} \alpha^j(H)\right)}\right) \leq \log \mathrm{Tr}_B\left(e^{-\theta\left(\sum_{j=0}^{m} \alpha^{jk}(H_k)\right)}\right) + k\|H\|.$$

Since $\cup_{j=0}^{m-1}\alpha^{jk}(\Omega_k) \subset \cup_{j=0}^{n-1}\alpha^j(\Omega) \subset \cup_{j=0}^{m}\alpha^{jk}(\Omega_k)$, it follows that

$$P_\alpha(H,\Omega;\delta) = \frac{1}{k}P_{\alpha^k}\left(\sum_{j=0}^{k-1}\alpha^j(H), \bigcup_{j=0}^{k-1}\alpha^j(\Omega);\delta\right),$$

and hence $P_\alpha(H) = \frac{1}{k}P_{\alpha^k}(\sum_{j=0}^{k-1}\alpha^j(H))$, which proves (v).

Set $H_k = \sum_{j=0}^{k-1} \alpha^j(H)$ and

$$H'_k = \sum_{j=0}^{k-1} \alpha^j(H + \alpha(K) - K) = H_k + \alpha^k(K) - K.$$

Then by (iv) and (v) we have

$$|P_\alpha(H) - P_\alpha(H + \alpha(K) - K)| = \frac{1}{k}|P_{\alpha^k}(H_k) - P_{\alpha^k}(H'_k)| \leq \frac{2\|K\|}{k}.$$

Thus (vi) follows. □

It follows from Proposition 9.1.2 that if φ is an α-invariant state then $-\varphi(H) \leq P_\alpha(H)$ for any $H \in A^{sa}$. It is interesting that the converse is also true.

9.1 Pressure

Proposition 9.1.5. *Suppose $ht(\alpha) < \infty$. Let φ be a self-adjoint linear functional on A. Then φ is an α-invariant state if and only if $-\varphi(H) \le P_\alpha(H)$ for all $H \in A^{sa}$.*

Proof. As we already remarked, in one direction the result follows from the inequality $-\varphi(H) \le P_\alpha(H) - h_\varphi(\alpha)$ for α-invariant states, Proposition 9.1.2.

Conversely, if $-\varphi(H) \le P_\alpha(H)$ for all $H \in A^{sa}$ then by Proposition 9.1.4(vi)

$$-\varphi(\alpha(H) - H) = -\frac{1}{n}\varphi(\alpha(nH) - nH) \le \frac{1}{n}P_\alpha(\alpha(nH) - nH) = \frac{1}{n}P_\alpha(0).$$

Since $P_\alpha(0) = ht(\alpha) < \infty$, letting $n \to \infty$ we get $\varphi(H) \le \varphi(\alpha(H))$. Applying this also to $-H$ we see that φ is α-invariant.

Furthermore, by Proposition 9.1.4(i),(ii)

$$-\varphi(H) = -\frac{1}{n}\varphi(nH) \le \frac{1}{n}P_\alpha(nH) \le \frac{1}{n}ht(\alpha) + \|H\| \xrightarrow[n\to\infty]{} \|H\|,$$

so that $\|\varphi\| \le 1$. For $c \in \mathbb{R}$ we have

$$-c\varphi(1) \le P_\alpha(c1) = ht(\alpha) - c.$$

Hence $\varphi(1) = 1$, and φ is a state. □

Definition 9.1.6. *We say that an α-invariant state φ is an **equilibrium state** at H if*

$$P_\alpha(H) = h_\varphi(\alpha) - \varphi(H).$$

If φ is an equilibrium state then, by Proposition 9.1.2,

$$h_\varphi(\alpha) - \varphi(H) = \sup_\psi (h_\psi(\alpha) - \psi(H)),$$

where the sup is taken over all α-invariant states.

Even though we cannot prove in general that the pressure is a convex function, this is the case in our main examples. The notion of equilibrium state is then closely related to differentiability of the pressure. Recall that if F is a real convex continuous function on a real Banach space X, then a linear functional f on X is called a **subgradient**, or tangent functional, of F at $x \in X$ if

$$F(x + y) - F(x) \ge f(y) \text{ for any } y \in X.$$

We shall usually identify self-adjoint linear functionals on A with real linear functionals on A^{sa}.

Proposition 9.1.7. *Suppose $ht(\alpha) < \infty$ and the pressure is a convex function on A^{sa}. Then*

(i) if φ is an equilibrium state at H then $-\varphi$ is a subgradient for the pressure at H;

(ii) if $-\varphi$ is a subgradient for the pressure at H then φ is an α-invariant state.

Proof. Assume φ is an equilibrium state at H. Let $K \in A^{sa}$. Then by Proposition 9.1.2

$$P_\alpha(H+K) - P_\alpha(H) \geq (h_\varphi(\alpha) - \varphi(H+K)) - (h_\varphi(\alpha) - \varphi(H)) = -\varphi(K),$$

so $-\varphi$ is a subgradient.

Assume now that φ is a subgradient at H. If $K \in A^{sa}$ then by Proposition 9.1.4(vi)

$$-\varphi(\alpha(K) - K) \leq P_\alpha(H + \alpha(K) - K) - P_\alpha(H) = 0.$$

Applying this also to $-K$ we see that φ is α-invariant. Now note that $\|\varphi\| \leq 1$ by Proposition 9.1.4(iv). By Proposition 9.1.4(ii) we have also $-c \geq -c\varphi(1)$ for any $c \in \mathbb{R}$. Hence $\varphi(1) = 1$, and φ is a state. □

9.2 The variational principle

We shall prove the variational principle for the following class of systems.

Definition 9.2.1. *A system (A, α) consisting of a unital C^*-algebra A and an automorphism α of A is called asymptotically abelian with locality if there is a dense α-invariant $*$-subalgebra $\mathcal{A}$ of A such that for each pair $a, b \in \mathcal{A}$ the C^*-algebra generated by a and b is finite dimensional, and for some $p = p(a, b) \in \mathbb{N}$ we have $[\alpha^j(a), b] = 0$ whenever $|j| \geq p$.*

We call elements of $\mathcal{A}$ for **local operators** and finite dimensional C^*-subalgebras of $\mathcal{A}$ for **local algebras**.

We may assume that $1 \in \mathcal{A}$. Since each finite dimensional C^*-algebra is singly generated, an easy induction argument shows that the C^*-algebra generated by a finite set of local operators is finite dimensional. In particular, if A is separable then A is an AF-algebra. Note also that for each local algebra N there is $p \in \mathbb{N}$ such that $[\alpha^j(a), b] = 0$ for all $a, b \in N$ whenever $|j| \geq p$.

Theorem 9.2.2. *Let (A, α) be asymptotically abelian with locality, and $H \in A^{sa}$. Then*

$$P_\alpha(H) = \sup_\varphi \{h_\varphi(\alpha) - \varphi(H)\},$$

where the supremum is taken over all α-invariant states of A. In particular, the topological entropy satisfies

$$ht(\alpha) = \sup_\varphi h_\varphi(\alpha).$$

Consider first the case when there exists a finite dimensional C^*-subalgebra N of A such that $H \in N$, $\alpha^j(N)$ commutes with N for $j \neq 0$, and the C^*-algebra generated by $\alpha^j(N)$, $j \in \mathbb{Z}$, coincides with A.

9.2 The variational principle

Lemma 9.2.3. *Under the above assumptions there exists an α-invariant state φ such that*

$$P_\alpha(H) = h_\varphi(\alpha) - \varphi(H) = \lim_{n\to\infty} \frac{1}{n} \log \operatorname{Tr}_{\vee_{j=0}^{n-1} \alpha^j(N)} \left(e^{-\sum_{j=0}^{n-1} \alpha^j(H)} \right).$$

Proof. Of course, this is essentially an abelian situation, so the result can be deduced from the classical variational principle. But we shall give a self-contained argument.

First note that if A_1 and A_2 are commuting finite dimensional C*-algebras, and $a_i \in A_i$, $a_i \geq 0$, $i = 1, 2$, then

$$\operatorname{Tr}_{A_1 \vee A_2}(a_1 a_2) \leq \operatorname{Tr}_{A_1}(a_1) \operatorname{Tr}_{A_2}(a_2),$$

since if p_i is a minimal projection in A_i, $i = 1, 2$, then $p_1 p_2$ is either zero or minimal in $A_1 \vee A_2$. Hence the sequence

$$\left\{ \log \operatorname{Tr}_{\vee_{j=0}^{n-1} \alpha^j(N)} \left(e^{-\sum_{j=0}^{n-1} \alpha^j(H)} \right) \right\}_n$$

is subadditive, so by Lemma 1.1.2 the limit in the formulation of the lemma exists and coincides with the infimum. We denote it by $\tilde{P}_\alpha(H)$. It is clear that $\tilde{P}_\alpha(H) \geq P_\alpha(H)$.

Consider $M = N^{\otimes \mathbb{Z}}$, the infinite C*-tensor product of N with itself. Denote the n-th factor of M by N_n. Let β be the shift to the right on M, and $\pi \colon M \to A$ the homomorphism which intertwines β with α, and identifies N_0 with N. Set $I = \operatorname{Ker} \pi$. For each $n \in \mathbb{N}$ let

$$M_n = N_0 \otimes \ldots \otimes N_{n-1}, \quad I_n = I \cap M_n, \quad \pi_n = \pi|_{M_n}.$$

Consider the state f_n on $\pi_n(M_n) = \vee_{j=0}^{n-1} \alpha^j(N)$ with density operator

$$\left(\operatorname{Tr}_{\vee_{j=0}^{n-1} \alpha^j(N)} e^{-\sum_{j=0}^{n-1} \alpha^j(H)} \right)^{-1} e^{-\sum_{j=0}^{n-1} \alpha^j(H)}.$$

Then $f_n \circ \pi_n$ is a state on M_n, and identifying M with $M_n^{\otimes \mathbb{Z}}$ we denote the corresponding product-state on M by ψ_n. The state ψ_n is β^n-invariant. Set $\varphi_n = n^{-1} \sum_{j=0}^{n-1} \psi_n \circ \beta^j$. Then φ_n is β-invariant. Using concavity of entropy, Theorem 3.2.5(ii), and that $h_{\psi_n}(\beta^n) = S(f_n \circ \pi_n)$ by Example 3.2.6(i), we obtain

$$h_{\varphi_n}(\beta) = \frac{1}{n} h_{\varphi_n}(\beta^n) \geq \frac{1}{n^2} \sum_{j=0}^{n-1} h_{\psi_n \circ \beta^j}(\beta^n) = \frac{1}{n} h_{\psi_n}(\beta^n) = \frac{1}{n} S(f_n \circ \pi_n)$$

$$= \frac{1}{n} S(f_n) = \frac{1}{n} \log \operatorname{Tr}_{\vee_{j=0}^{n-1} \alpha^j(N)} \left(e^{-\sum_{j=0}^{n-1} \alpha^j(H)} \right) + \frac{1}{n} f_n \left(\sum_{j=0}^{n-1} \alpha^j(H) \right)$$

$$\geq \tilde{P}_\alpha(H) + \frac{1}{n} f_n \left(\sum_{j=0}^{n-1} \alpha^j(H) \right)$$

9 Variational Principle

$$= \tilde{P}_\alpha(H) + \frac{1}{n}\psi_n\left(\sum_{j=0}^{n-1}\beta^j(H)\right) = \tilde{P}_\alpha(H) + \varphi_n(H).$$

Let $\tilde{\varphi}$ be any weak* limit point of the sequence $\{\varphi_n\}_n$. Then $\tilde{\varphi}$ is β-invariant. Let D be a maximal abelian subalgebra of N_0 containing H. Then D is in the centralizer of the state φ_n. Hence, using Proposition 3.1.6, by the same reasoning as in Example 3.2.6(i) we get

$$h_{\varphi_n}(\beta) = h_{\varphi_n}(D;\beta) = \inf_{k\in\mathbb{N}} \frac{1}{k} H_{\varphi_n}(D,\beta(D),\dots,\beta^{k-1}(D)).$$

By Proposition 3.1.12 the map $\psi \mapsto h_\psi(D;\beta)$ is weak* upper semicontinuous. Hence

$$h_{\tilde{\varphi}}(\beta) \geq \tilde{P}_\alpha(H) + \tilde{\varphi}(H).$$

Now observe that $\tilde{\varphi}$ is zero on I. Indeed, if $x \in I_n$ then $\beta^j(x) \in I_m$ for $j = 0,\dots,m-n$ and $m \geq n$, whence

$$|\varphi_m(x)| \leq \frac{1}{m}\sum_{j=m-n+1}^{m-1}|(\psi_m \circ \beta^j)(x)| \leq \frac{n-1}{m}||x||,$$

so $\tilde{\varphi}(x) = 0$. Thus $\tilde{\varphi}$ defines a state φ on A. We have

$$h_\varphi(\alpha) = h_{\tilde{\varphi}}(\beta) \geq \tilde{P}_\alpha(H) + \varphi(H),$$

where the first equality follows from Theorem 3.2.2(ii). Since by Proposition 9.1.2, $h_\varphi(\alpha) - \varphi(H) \leq P_\alpha(H) \leq \tilde{P}_\alpha(H)$, the proof of the lemma is complete. ∎

We shall reduce the general case to the case considered above by replacing α by its powers. For this suppose that N is a local subalgebra of A, and $H \in N$. Choose $p \geq 1$ such that $\alpha^j(N)$ commutes with N whenever $|j| \geq p$. For $k \geq p$ set $M_k = \vee_{j=0}^{k-p}\alpha^j(N)$, $H_k = \sum_{j=0}^{k-p}\alpha^j(H)$. Then $H_k \in M_k$, and $\alpha^{jk}(M_k)$ commutes with M_k for $j \neq 0$.

Lemma 9.2.4. *For every finite subset Ω of N we have*

$$P_\alpha(H,\Omega) \leq \liminf_{k\to\infty}\frac{1}{k}P_{\alpha^k|\vee_{j\in\mathbb{Z}}\alpha^{jk}(M_k)}(H_k).$$

Proof. The idea of the proof is to reduce to the situation of Lemma 9.2.3 by showing that the contribution of the indices in the intervals $[jk-p+1, jk-1]$, $j \in \mathbb{N}$, becomes negligible for large k.

Fix $\delta > 0$. Choose $m_0 \in \mathbb{N}$ such that

$$\frac{2(p-1)||a||}{m_0} < \delta \quad \text{for} \quad a \in \Omega.$$

9.2 The variational principle

Take $k \geq m_0 + p$. Let $n \in \mathbb{N}$. Then $(m-1)k \leq n < mk$ for some $m \in \mathbb{N}$. Set $B_0 = \vee_{j=0}^m \alpha^{jk}(M_k) = \vee_{i \in X} \alpha^i(N)$, where $X = \cup_{j=0}^m [jk, (j+1)k - p]$. Put

$$B = \underbrace{B_0 \oplus \ldots \oplus B_0}_{m_0}.$$

Choose a conditional expectation $E \colon A \to B_0$, and define unital completely positive maps $\psi \colon B \to A$ and $\theta \colon A \to B$ by

$$\psi(b_1, \ldots, b_{m_0}) = \frac{1}{m_0} \sum_{i=1}^{m_0} \alpha^{-i+1}(b_i),$$

$$\theta(a) = (E(a), (E \circ \alpha)(a), \ldots, (E \circ \alpha^{m_0-1})(a)).$$

Let $a \in A$, and put $S = \{i \,|\, 0 \leq i \leq m_0 - 1, \alpha^i(a) \notin B_0\}$. Then

$$(\psi \circ \theta)(a) - a = \frac{1}{m_0} \sum_{i \in S} ((\alpha^{-i} \circ E \circ \alpha^i)(a) - a).$$

Assume now that $a = \alpha^l(b)$ for some $b \in \Omega$ and l, $0 \leq l \leq n-1$. Then $l < mk$, and hence $[l, l + m_0 - 1]$ is contained in $Y = [0, (m+1)k - p]$. The set X is a union of $m+1$ intervals of length $k - p \geq m_0$, and it is obtained from Y by removing m open intervals of length p. Hence $[l, l + m_0 - 1]$ can intersect at most one of the latter intervals. It follows that S contains at most $p - 1$ elements. Hence

$$\|(\psi \circ \theta)(a) - a\| \leq \frac{2(p-1)\|a\|}{m_0} < \delta.$$

Consequently

$$P\left(\sum_{j=0}^{n-1} \alpha^j(H), \bigcup_{j=0}^{n-1} \alpha^j(\Omega); \delta\right) \leq \log \mathrm{Tr}_B \left(e^{-\theta\left(\sum_{j=0}^{n-1} \alpha^j(H)\right)}\right).$$

Denoting by $\#I$ the number of integer points in a set $I \subset \mathbb{R}$, note that for $X_i = [i, i+n-1]$, $0 \leq i \leq m_0 - 1$, we have

$$\#(X_i \triangle X) \leq \#(Y \backslash X_i) + \#(Y \backslash X) = ((m+1)k - p + 1 - n) + m(p-1)$$
$$\leq (m+1)k - p + 1 - (m-1)k + m(p-1) < mp + 2k.$$

Since $\alpha^j(H) \in B_0$ for $j \in X$, we then get

$$\left\|(E \circ \alpha^i)\left(\sum_{j=0}^{n-1} \alpha^j(H)\right) - \sum_{j=0}^{m} \alpha^{jk}(H_k)\right\| = \left\|E\left(\sum_{j \in X_i} \alpha^j(H)\right) - \sum_{j \in X} \alpha^j(H)\right\|$$

9 Variational Principle

$$= \left\| E\left(\sum_{j \in X_i \setminus X} \alpha^j(H)\right) - \sum_{j \in X \setminus X_i} \alpha^j(H) \right\| \leq (mp + 2k)\|H\|.$$

By the Peierls-Bogoliubov inequality, Corollary 2.3.11, we obtain

$$\text{Tr}_{B_0}\left(e^{-(E \circ \alpha^i)(\sum_{j=0}^{n-1} \alpha^j(H))}\right) \leq e^{(mp+2k)\|H\|}\text{Tr}_{B_0}\left(e^{-\sum_{j=0}^{m} \alpha^{jk}(H_k)}\right),$$

so

$$\text{Tr}_B\left(e^{-\theta(\sum_{j=0}^{n-1}\alpha^j(H))}\right) \leq m_0 e^{(mp+2k)\|H\|}\text{Tr}_{B_0}\left(e^{-\sum_{j=0}^{m}\alpha^{jk}(H_k)}\right).$$

Taking the logarithm, dividing by n and letting $n \to \infty$, we get

$$P_\alpha(H,\Omega;\delta) \leq \frac{p\|H\|}{k} + \frac{1}{k}\lim_{m\to\infty}\frac{1}{m}\log\text{Tr}_{\vee_{j=0}^{m-1}\alpha^{jk}(M_k)}\left(e^{-\sum_{j=0}^{m-1}\alpha^{jk}(H_k)}\right)$$

$$= \frac{p\|H\|}{k} + \frac{1}{k}P_{\alpha^k|\vee_{j\in\mathbb{Z}}\alpha^{jk}(M_k)}(H_k),$$

where the last equality follows from Lemma 9.2.3. Since this is true for any $k \geq m_0 + p$, we get the result. □

Proof of Theorem 9.2.2. The inequality $\geq$ has been proved in Proposition 9.1.2. Since the pressure is continuous by Proposition 9.1.4(iv), to prove the converse inequality it suffices to consider local H. Then by Lemma 9.2.4 we have only to show that if H is contained in a local algebra N then with $H_k = \sum_{j=0}^{k-p} \alpha^j(H)$ as in Lemma 9.2.4

$$\sup_\varphi(h_\varphi(\alpha) - \varphi(H)) \geq \liminf_{k\to\infty}\frac{1}{k}P_{\alpha^k|\vee_{j\in\mathbb{Z}}\alpha^{jk}(M_k)}(H_k).$$

By Lemma 9.2.3, for each $k \in \mathbb{N}$ there exists an α^k-invariant state ψ_k on $\vee_{j\in\mathbb{Z}}\alpha^{jk}(M_k)$ such that

$$h_{\psi_k}(\alpha^k|_{\vee_{j\in\mathbb{Z}}\alpha^{jk}(M_k)}) - \psi_k(H_k) = P_{\alpha^k|\vee_{j\in\mathbb{Z}}\alpha^{jk}(M_k)}(H_k).$$

By Proposition 3.2.7 we can extend ψ_k to an α^k-invariant state $\tilde{\varphi}_k$ on A such that

$$h_{\tilde{\varphi}_k}(\alpha^k) \geq h_{\psi_k}(\alpha^k|_{\vee_{j\in\mathbb{Z}}\alpha^{jk}(M_k)}).$$

Set $\varphi_k = \frac{1}{k}\sum_{j=0}^{k-1}\tilde{\varphi}_k \circ \alpha^j$. Then, as in the proof of Lemma 9.2.3,

$$h_{\varphi_k}(\alpha) \geq \frac{1}{k}h_{\tilde{\varphi}_k}(\alpha^k) \geq \frac{1}{k}h_{\psi_k}(\alpha^k|_{\vee_{j\in\mathbb{Z}}\alpha^{jk}(M_k)}).$$

Since

$$\varphi_k(H) = \frac{1}{k}\sum_{j=0}^{k-1}\tilde{\varphi}_k(\alpha^j(H)) \le \frac{1}{k}\tilde{\varphi}_k(H_k) + \frac{p-1}{k}\|H\| = \frac{1}{k}\psi_k(H_k) + \frac{p-1}{k}\|H\|,$$

we get

$$\begin{aligned}h_{\varphi_k}(\alpha) - \varphi_k(H) &\ge \frac{1}{k}h_{\psi_k}(\alpha^k|_{\vee_{j\in\mathbb{Z}}\alpha^{jk}(M_k)}) - \frac{1}{k}\psi_k(H_k) - \frac{(p-1)\|H\|}{k}\\ &= \frac{1}{k}P_{\alpha^k|_{\vee_{j\in\mathbb{Z}}\alpha^{jk}(M_k)}}(H_k) - \frac{(p-1)\|H\|}{k},\end{aligned}$$

and the proof is complete. □

Corollary 9.2.5. *If (A,α) is asymptotically abelian with locality then the pressure $P_\alpha(H)$ is a convex function of H.*

Proof. Indeed, it is the supremum of the affine functions $H \mapsto h_\varphi(\alpha) - \varphi(H)$. □

Corollary 9.2.6. *If (A_1,α_1) and (A_2,α_2) are asymptotically abelian with locality then*

$$ht(\alpha_1 \otimes \alpha_2) = ht(\alpha_1) + ht(\alpha_2).$$

Proof. If φ_i is an α_i-invariant state, $i = 1,2$, then by Theorem 3.2.2(iv), Proposition 6.2.7 and Theorem 6.2.2(iv), used in that order,

$$h_{\varphi_1}(\alpha_1) + h_{\varphi_2}(\alpha_2) \le h_{\varphi_1\otimes\varphi_2}(\alpha_1 \otimes \alpha_2) \le ht(\alpha_1 \otimes \alpha_2) \le ht(\alpha_1) + ht(\alpha_2).$$

Taking the supremum over φ_i we get the conclusion. □

In the next section we shall need an explicit formula for the pressure, which is a consequence of our proof of the variational principle.

Corollary 9.2.7. *Assume A is separable. Let N be a local algebra. Then there exist a sequence $\{A_n\}_n$ of local algebras containing N and three sequences $\{p_n\}_n$, $\{m_n\}_n$, $\{k_n\}_n$ of positive integers such that*
(i) $\alpha^p(A_n)$ commutes with A_n whenever $|p| \ge p_n$;
(ii) $\frac{p_n}{k_n} \to 0$ as $n \to \infty$;
(iii) $P_\alpha(H) = \lim_{n\to\infty} \frac{1}{k_n m_n}\log\mathrm{Tr}_{\vee_{j\in I_n}\alpha^j(A_n)}\left(e^{-\sum_{j\in I_n}\alpha^j(H)}\right)$ *for all $H \in N^{sa}$,*
where $I_n = \cup_{j=0}^{m_n-1}[jk_n,(j+1)k_n - p_n] \cap \mathbb{Z}$.

Proof. Let $\{A_n\}_n$ be an increasing sequence of local algebras containing N such that $\cup_n A_n$ is dense in A, Ω_n a finite subset of A_n such that the linear span of Ω_n is A_n. Let $\{p_n\}_n$ be a sequence satisfying condition (i). By Lemma 9.2.4

$$P_\alpha(H,\Omega_n) \le \liminf_{k\to\infty} \frac{1}{k}P_{\alpha^k|_{\vee_{j\in\mathbb{Z}}\alpha^{jk}(A_{n,k})}}\left(\sum_{j=0}^{k-p_n}\alpha^j(H)\right) \quad \text{for } H \in N^{sa},$$

where $A_{n,k} = \vee_{j=0}^{k-p_n} \alpha^j(A_n)$. On the other hand, by Theorem 9.2.2 and its proof

$$P_\alpha(H) \geq \limsup_{k\to\infty} \frac{1}{k} P_{\alpha^k | \vee_{j\in\mathbb{Z}} \alpha^{jk}(A_{n,k})} \left(\sum_{j=0}^{k-p_n} \alpha^j(H) \right) \quad \text{for } H \in N^{sa}, \ n \in \mathbb{N}.$$

Choose a countable dense subset X of N^{sa}. Since $P_\alpha(H, \Omega_n) \nearrow P_\alpha(H)$ for any $H \in X$, we can find a sequence $\{k_n\}_n$ such that condition (ii) is satisfied and

$$P_\alpha(H) = \lim_{n\to\infty} \frac{1}{k_n} P_{\alpha^{k_n} | \vee_{j\in\mathbb{Z}} \alpha^{jk_n}(A_{n,k_n})} \left(\sum_{j=0}^{k_n-p_n} \alpha^j(H) \right) \quad \text{for } H \in X.$$

Since by Lemma 9.2.3

$$P_{\alpha^{k_n} | \vee_{j\in\mathbb{Z}} \alpha^{jk_n}(A_{n,k_n})} \left(\sum_{j=0}^{k_n-p_n} \alpha^j(H) \right)$$

$$= \lim_{m\to\infty} \frac{1}{m} \log \mathrm{Tr}_{\vee_{j\in I_{n,m}} \alpha^j(A_n)} \left(e^{-\sum_{j\in I_{n,m}} \alpha^j(H)} \right),$$

where $I_{n,m} = \cup_{j=0}^{m-1}[jk_n, (j+1)k_n - p_n]$, we can choose a sequence $\{m_n\}_n$ such that condition (iii) is satisfied for all $H \in X$. But then it is satisfied for all $H \in N^{sa}$ by Proposition 9.1.4(iv) and the Peierls-Bogoliubov inequality, Corollary 2.3.11. □

9.3 KMS-states

Let A be a finite dimensional C*-algebra, $H \in A^{sa}$ a self-adjoint element. By the thermodynamic inequality, Proposition 2.2.5, for any state φ on A we have

$$S(\varphi) - \varphi(H) \leq \log \mathrm{Tr}(e^{-H}),$$

and equality holds if and only if $Q_\varphi = \mathrm{Tr}(e^{-H})^{-1} e^{-H}$, in which case φ is called the **Gibbs state**. Consider the one-parameter automorphism group σ of A defined by $\sigma_t(a) = e^{itH} a e^{-itH}$. Then one can check that if A is a full matrix algebra then the Gibbs state is the unique state with the property $\varphi(ab) = \varphi(b\sigma_i(a))$. In this section we shall show that equilibrium states have a similar property.

Let A be a C*-algebra, σ a strongly continuous one-parameter automorphism group of A. Recall that an element $a \in A$ is called an entire analytic element for σ if the map $\mathbb{R} \ni t \mapsto \sigma_t(a) \in A$ extends to an entire analytic function. Let $\beta \in \mathbb{R}$. Then a state φ on A is called a σ-**KMS**$_\beta$-**state** if

$$\varphi(ab) = \varphi(b\sigma_{i\beta}(a))$$

for any entire analytic element $a \in A$ and any $b \in A$. If $\beta \neq 0$ then φ is automatically σ-invariant by [30, Proposition 5.3.3]. Furthermore, by [30, Corollary 5.3.9 and Theorem 5.3.10] the normal state on $\pi_\varphi(A)''$ defined by φ is faithful, and its modular group is given by $\sigma_t^\varphi(\pi_\varphi(a)) = \pi_\varphi(\sigma_{-\beta t}(a))$. In particular, if φ is faithful and $\beta \neq 0$ then there exists at most one one-parameter group σ for which φ is a KMS$_\beta$-state.

We shall use the following form of the KMS-condition which does not involve analytic continuation of σ and hence is better suited for passing to limits. By [30, Proposition 5.3.12] a state φ is a σ-KMS$_\beta$ if and only if

$$\int_{\mathbb{R}} \hat{f}(t)\varphi(a\sigma_t(b))dt = \int_{\mathbb{R}} \hat{f}(t+i\beta)\varphi(\sigma_t(b)a)dt \tag{9.1}$$

for all $a, b \in A$ and $f \in \mathcal{D}$, where $\mathcal{D}$ is the space of compactly supported C^∞-functions, and $\hat{f}(t) = (2\pi)^{-1/2} \int_{\mathbb{R}} f(x)e^{-ixt}dx$ is the Fourier transform of f.

Return now to our setting of asymptotically abelian systems with locality. Let H be a local self-adjoint element. Our first goal is to show that there is a natural dynamics associated with H. For a subset $I \subset \mathbb{Z}$ and a local operator a put

$$\delta_{H,I}(a) = \sum_{j \in I} [\alpha^j(H), a].$$

Then $\delta_{H,I}$ is a derivation on the algebra of local operators. Put

$$\sigma_t^{H,I} = \exp(it\delta_{H,I}).$$

We shall write δ_H for $\delta_{H,\mathbb{Z}}$, and σ_t^H for $\sigma_t^{H,\mathbb{Z}}$. It will follow from the next lemma that $\sigma^{H,I}$ is a well-defined strongly continuous one-parameter automorphism group of A.

Lemma 9.3.1. *Let $r \in \mathbb{N}$, and let c_n be the number of n-tuples $(k_1, \ldots, k_n) \in \mathbb{Z}^n$ such that*

$$\min_{0 \leq i < j} |k_j - k_i| \leq r \quad \text{for } j = 1, \ldots, n, \tag{9.2}$$

where $k_0 = 0$. Put $c_0 = 1$. Then the series $\sum_{n=0}^\infty \dfrac{c_n}{n!} z^n$ has infinite radius of convergence.

Proof. For $m \in \mathbb{N}$ let $K_n(m)$ denote the set of n-tuples $(k_1, \ldots, k_n)$ such that condition (9.2) is satisfied, and furthermore

$$\max_{0 \leq i \leq n} k_i = \min_{0 \leq i \leq n} k_i + m - 1,$$

146   9 Variational Principle

that is, the minimal interval containing $k_0, \ldots, k_n$ has length $m-1$. Let $c_n(m)$ denote the number of elements in $K_n(m)$ for $n \geq 1$, and put $c_0(1) = 1$ and $c_0(m) = 0$ for $m \geq 2$. Then $c_n = \sum_{m=1}^{\infty} c_n(m)$.

We assert that the following recurrence relation holds:

$$c_{n+1}(m) = mc_n(m) + 2\sum_{k=1}^{r} c_n(m-k),$$

with the convention that $c_n(l) = 0$ for $l \leq 0$. To show this let $(k_1, \ldots, k_n) \in \bigcup_{k=0}^{m-1} K_n(m-k)$. Let $[a, b]$ be the minimal interval containing $k_0, \ldots, k_n$. We want to see how many choices there are for k_{n+1} such that $(k_1, \ldots, k_{n+1}) \in K_{n+1}(m)$. If $(k_1, \ldots, k_n) \in K_n(m)$ then we can take any integer in $[a, b]$ for k_{n+1}, so there are m choices. If $(k_1, \ldots, k_n) \in K_n(m-k)$ with $1 \leq k \leq r$, there are two choices, $k_{n+1} = a - k$ and $k_{n+1} = b + k$. And if $k > r$, no k_{n+1} with the required properties exists. This computation proves the recurrence formula.

By virtue of (9.2) we have $c_n(m) = 0$ for $m > nr + 1$. Thus by the recurrence formula

$$\max_m c_{n+1}(m) \leq (nr + 1 + 2r)\max_m c_n(m),$$

and consequently

$$\max_m c_{n+1}(m) \leq \prod_{k=0}^{n}(1 + r(k+2)).$$

Hence the series

$$f(x, y) = \sum_{n=0}^{\infty}\sum_{m=1}^{\infty} \frac{c_n(m)}{n!} x^n y^m$$

defines an analytic function in a neighborhood of $(0,0) \in \mathbb{C}^2$. We have, using the recurrence formula,

$$\frac{\partial f}{\partial x} = \sum_{n=0}^{\infty}\sum_{m=1}^{\infty} \frac{c_{n+1}(m)}{n!} x^n y^m$$

$$= \sum_{n=0}^{\infty}\sum_{m=1}^{\infty} m\frac{c_n(m)}{n!} x^n y^m + 2\sum_{k=1}^{r}\sum_{n=0}^{\infty}\sum_{m=1}^{\infty} \frac{c_n(m-k)}{n!} x^n y^m$$

$$= \sum_{n=0}^{\infty}\sum_{m=1}^{\infty} m\frac{c_n(m)}{n!} x^n y^m + 2f(x,y)\sum_{k=1}^{r} y^k.$$

On the other hand,

$$y\frac{\partial f}{\partial y} = \sum_{n=0}^{\infty}\sum_{m=1}^{\infty} m\frac{c_n(m)}{n!} x^n y^m.$$

We thus get a partial differential equation

$$\frac{\partial f}{\partial x} - y\frac{\partial f}{\partial y} = 2f(x,y)\sum_{k=1}^{r} y^k.$$

Working in a neighbourhood of a point $(0, y_0)$ with $y_0 \neq 0$ consider the change of variables

$$x = u + v, \quad y = e^{v-u}.$$

Then $\dfrac{\partial x}{\partial u} = 1$ and $\dfrac{\partial y}{\partial u} = -y$, and the above equation can be written as

$$\frac{\partial \log f}{\partial u} = -2\frac{\partial}{\partial u}\sum_{k=1}^{r}\frac{y^k}{k}.$$

Hence

$$\log f = g(v) - 2\sum_{k=1}^{r}\frac{y^k}{k} = g\left(\frac{x + \log y}{2}\right) - 2\sum_{k=1}^{r}\frac{y^k}{k}$$

for a function g. To find g let $x = 0$. Since $f(0, y) = y$, we get

$$g\left(\frac{\log y}{2}\right) = \log y + 2\sum_{k=1}^{r}\frac{y^k}{k},$$

that is,

$$g(v) = 2v + 2\sum_{k=1}^{r}\frac{e^{2kv}}{k}.$$

We thus have

$$\log f(x,y) = x + \log y + 2\sum_{k=1}^{r}\frac{y^k e^{kx}}{k} - 2\sum_{k=1}^{r}\frac{y^k}{k},$$

whence

$$f(x,y) = y\exp\left(x + 2\sum_{k=1}^{r}\frac{y^k}{k}(e^{kx} - 1)\right).$$

We see that f is analytic on $\mathbb{C}^2$. In particular, the series $\sum_{n=0}^{\infty}\dfrac{c_n}{n!}z^n = f(z,1)$ has infinite radius of convergence. $\square$

Proposition 9.3.2. *Let (A, α) be asymptotically abelian with locality, $H \in A^{sa}$ a local self-adjoint element. Then the series*

$$\sigma_z^{H,I}(a) = \sum_{n=0}^{\infty}\frac{(iz)^n}{n!}\delta_{H,I}^n(a)$$

converges absolutely in norm for any $z \in \mathbb{C}$ and any local operator a. Moreover, if N is a local algebra containing H, $\Omega \subset \mathbb{C}$ a compact subset and $\varepsilon > 0$, then

(i) there exists a finite set $I_0 \subset \mathbb{Z}$ such that
$$\|\sigma_z^H(a) - \sigma_z^{H,I}(a)\| \leq \varepsilon \|a\|$$
for any $I \supset I_0$, $z \in \Omega$ and $a \in N$;

(ii) there is $\delta > 0$ such that if $K \in N^{sa}$ with $\|H - K\| < \delta$ then
$$\|\sigma_z^H(a) - \sigma_z^K(a)\| \leq \varepsilon \|a\|$$
for all $z \in \Omega$ and $a \in N$.

Proof. Let N be a local algebra containing H. Let $r \in \mathbb{N}$ be so large that $\alpha^n(N)$ commutes with N for $|n| > r$. For $a \in N$,
$$\delta_H^n(a) = \sum_{k_1,\ldots,k_n \in \mathbb{Z}} [\alpha^{k_n}(H), [\ldots [\alpha^{k_1}(H), a]\ldots]].$$

The commutator in the sum can be nonzero only when the n-tuple $(k_1, \ldots, k_n)$ satisfies condition (9.2), hence with c_n as in Lemma 9.3.1 we have
$$\|\delta_H^n(a)\| \leq c_n 2^n \|H\|^n \|a\|.$$

It follows from that lemma that the series $\sum_{n=0}^\infty (z^n/n!)\delta_H^n(a)$ converges in norm for any $z \in \mathbb{C}$.

If $I \subset \mathbb{Z}$ then in the formula for $\delta_{H,I}^n(a)$ we sum over a smaller set, so convergence holds in this case too. Moreover, for any $m \in \mathbb{N}$, if I contains $\mathbb{Z} \cap [-mr, mr]$ then $\delta_{H,I}^n(a) = \delta_H^n(a)$ for $n \leq m$, whence
$$\|\sigma_z^H(a) - \sigma_z^{H,I}(a)\| \leq 2 \sum_{n > m} \frac{c_n}{n!} 2^n |z|^n \|H\|^n \|a\|.$$

This implies part (i). This also implies that to show (ii) it suffices to prove a similar statement for $\sigma^{H,I}$ and $\sigma^{K,I}$ for some finite I. But then it is immediate, since $\sigma_z^{H,I}(a) = e^{izH_I} a e^{-izH_I}$, where $H_I = \sum_{j \in I} \alpha^j(H)$. □

We can now formulate the main result of the present section.

Theorem 9.3.3. *Let (A, α) be asymptotically abelian with locality. Assume $ht(\alpha) < \infty$. If H is a local self-adjoint operator in A and φ is an equilibrium state at H then φ is a σ^H-KMS_1-state.*

In particular, if $ht(\alpha) = h_\varphi(\alpha)$ then φ is a trace.

By Corollary 9.2.5 the pressure P_α is a convex function of H, so that Proposition 9.1.7 shows that Theorem 9.3.3 will follow from the following more general result.

9.3 KMS-states 149

Theorem 9.3.4. *If $-\varphi$ is a subgradient of P_α at H then φ is a σ^H-KMS_1-state.*

The proof is based on Corollary 9.2.7 which, as we shall see, implies that for generic H a subgradient of the pressure at H can be obtained as a limit of Gibbs states on local algebras.

Note that if B is a finite dimensional C*-algebra, $H_0 \in B^{sa}$, and φ is the state with density operator $Q_\varphi = \text{Tr}_B(e^{-H_0})^{-1}\text{Tr}_B(e^{-H_0})$ then for any $H \in B^{sa}$ we have

$$\frac{d}{dt}\log\text{Tr}_B(e^{-H_0-tH})|_{t=0} = -\text{Tr}_B(e^{-H_0})^{-1}\text{Tr}_B(He^{-H_0}) = -\varphi(H),$$

so that $-\varphi$ is the gradient of the function $H \mapsto \log\text{Tr}_B(e^{-H})$ at H_0. Note also that this function is convex, being the supremum of the affine functions $H \mapsto S(\psi) - \psi(H)$ by the thermodynamic inequality, Proposition 2.2.5.

Lemma 9.3.5. *Let N be a local algebra, $H \in N^{sa}$, $-\varphi \in N^*$ a subgradient of $(P_\alpha)|_{N^{sa}}$ at H. Let $E: A \to N$ be a conditional expectation. Then for every function f in the space $\mathcal{D}$ of C^∞-functions with compact support and $a,b \in N$ we have*

$$\left|\int_\mathbb{R} \hat{f}(t)\varphi(aE(\sigma_t^H(b)))dt - \int_\mathbb{R} \hat{f}(t+i)\varphi(E(\sigma_t^H(b))a)dt\right|$$

$$\leq \|a\|\int_\mathbb{R}(|\hat{f}(t)| + |\hat{f}(t+i)|)\|\sigma_t^H(b) - E(\sigma_t^H(b))\|dt.$$

Proof. We may assume that A is separable. Indeed, for every $K \in A^{sa}$ there exists a sequence $\{\Omega_n\}_{n=1}^\infty$ of finite subsets of A such that $P_\alpha(H, \Omega_n) \nearrow P_\alpha(H)$. Then for any α-invariant C*-subalgebra A_0 containing K and the Ω_n's we have $P_\alpha(K) = P_{\alpha|A_0}(K)$. It follows that there exists a separable α-invariant C*-subalgebra A_0 containing N such that $(A_0, \alpha|_{A_0})$ is asymptotically abelian with locality and $P_{\alpha|A_0}(K) = P_\alpha(K)$ for a dense family of elements K in N^{sa}. But then the equality $P_{\alpha|A_0}(K) = P_\alpha(K)$ holds for all $K \in N^{sa}$ by continuity. Therefore we may replace A by A_0 and thus assume that A is separable.

First consider the case when $(P_\alpha)|_{N^{sa}}$ has a unique subgradient at H. With the notation of Corollary 9.2.7 consider the state f_n on $\vee_{j\in I_n}\alpha^j(A_n)$ with density operator

$$\left(\text{Tr}_{\vee_{j\in I_n}\alpha^j(A_n)}\left(e^{-\sum_{j\in I_n}\alpha^j(H)}\right)\right)^{-1}e^{-\sum_{j\in I_n}\alpha^j(H)}.$$

Then define a positive linear functional φ_n on N by

$$\varphi_n(x) = \frac{1}{k_n m_n}\sum_{j\in I_n}f_n(\alpha^j(x)).$$

150   9 Variational Principle

Note that $\|\varphi_n\| = \varphi_n(1) \le 1$. As we remarked before the formulation of the lemma, the functional $-f_n$ is the gradient of the convex function $x \mapsto \log \operatorname{Tr}_{\vee_{j \in I_n} \alpha^j(A_n)}(e^{-x})$ on $(\vee_{j \in I_n} \alpha^j(A_n))^{sa}$ at the point $\sum_{j \in I_n} \alpha^j(H)$. Hence $-\varphi_n$ is the gradient of the convex function

$$N^{sa} \ni x \mapsto \frac{1}{k_n m_n} \log \operatorname{Tr}_{\vee_{j \in I_n} \alpha^j(A_n)}\left(e^{-\sum_{j \in I_n} \alpha^j(x)}\right)$$

at H. It follows that any limit point of the sequence $\{-\varphi_n\}_n$ is a subgradient of $(P_\alpha)|_{N^{sa}}$ at H. Since the latter is unique by assumption, $\varphi_n \to \varphi$ as $n \to \infty$.

Since f_n is a σ^{H,I_n}-KMS$_1$-state, by (9.1) for every $j \in I_n$ we have

$$\int_{\mathbb{R}} \hat{f}(t) f_n(\alpha^j(a) \sigma_t^{H,I_n}(\alpha^j(b))) dt = \int_{\mathbb{R}} \hat{f}(t+i) f_n(\sigma_t^{H,I_n}(\alpha^j(b)) \alpha^j(a)) dt. \quad (9.3)$$

Note that $\sigma_t^{H,I_n}(\alpha^j(b)) = \alpha^j(\sigma_t^{H,I_n-j}(b))$. Fix $q \in \mathbb{N}$, and put

$$I_{n,q} = \cup_{j=0}^{m_n-1}[jk_n + q, (j+1)k_n - p_n - q] \cap \mathbb{Z},$$

which is nonempty for sufficiently large n. Then $I_n - j$ contains $[-q, q] \cap \mathbb{Z}$ for $j \in I_{n,q}$. By Proposition 9.3.2(i), if q is large enough then $\sigma_t^{H,I_n-j}(b)$ is arbitrarily close to $\sigma_t^H(b)$ for every $j \in I_{n,q}$ and t in a fixed compact subset of $\mathbb{R}$. But then $\sigma_t^{H,I_n}(\alpha^j(b)) - \alpha^j(E(\sigma_t^H(b)))$ is arbitrarily close to $\alpha^j(\sigma_t^H(b) - E(\sigma_t^H(b)))$. It follows that for all sufficiently large $n \in \mathbb{N}$

$$\left| \int_{\mathbb{R}} \hat{f}(t) \frac{1}{k_n m_n} \sum_{j \in I_{n,q}} f_n\left(\alpha^j(a) \sigma_t^{H,I_n}(\alpha^j(b)) - \alpha^j(a) \alpha^j(E(\sigma_t^H(b)))\right) dt \right|$$

$$\le \|a\| \int_{\mathbb{R}} |\hat{f}(t)| \|\sigma_t^H(b) - E(\sigma_t^H(b))\| dt + \varepsilon(q),$$

where $\varepsilon(q) \to 0$ as $q \to \infty$. Since $|I_{n,q}|/|I_n| \to 1$ as $n \to \infty$, letting $n \to \infty$ we may replace averaging over the set $I_{n,q}$ by averaging over I_n, and then obtain

$$\left| \frac{1}{k_n m_n} \sum_{j \in I_n} \int_{\mathbb{R}} \hat{f}(t) f_n\left(\alpha^j(a) \sigma_t^{H,I_n}(\alpha^j(b))\right) dt - \int_{\mathbb{R}} \hat{f}(t) \varphi(a E(\sigma_t^H(b))) dt \right|$$

$$\le \|a\| \int_{\mathbb{R}} |\hat{f}(t)| \|\sigma_t^H(b) - E(\sigma_t^H(b))\| dt + \varepsilon_n,$$

with $\varepsilon_n \to 0$ as $n \to \infty$. Since an analogous estimate holds for the integral $\int_{\mathbb{R}} \hat{f}(t+i) \varphi(E(\sigma_t^H(b)) a) dt$, we obtain the conclusion of the lemma by virtue of (9.3).

To finish the proof we need a few facts about differentiability of convex functions, see e.g. [179, Section 25]. Let F be a convex function on $\mathbb{R}^n$. For any $x \in \mathbb{R}^n$ the function F is differentiable at x if and only if it has a unique subgradient at x, and then this subgradient coincides with the gradient of F at x. The points where F is differentiable form a dense G_δ-set of full measure. Finally, for any $x \in \mathbb{R}^n$ the set of subgradients of F at x coincides with the closed convex hull of subgradients which are limits of sequences of the form $\{f_n(x_n)\}_n$ such that $x_n \to x$, F is differentiable at x_n, and $f_n(x_n)$ is the gradient of F at x_n.

Returning to the proof of the lemma, we conclude that φ lies in the closed convex hull of linear functionals $\tilde{\varphi}$ for which there exist sequences $\{H_n\}_n \subset N^{sa}$ and $\{\varphi_n\}_n \subset N^*$ such that $H_n \to H$, $\varphi_n \to \tilde{\varphi}$, and $-\varphi_n$ is the unique subgradient of $(P_\alpha)|_{N^{sa}}$ at H_n. Since for φ_n the lemma is already proved (for H_n instead of H), using Proposition 9.3.2(ii) we conclude that the conclusion of the lemma is true for $\tilde{\varphi}$. But then it is true for any functional in the closed convex hull of the $\tilde{\varphi}$'s. □

Proof of Theorem 9.3.4. If $-\varphi$ is a subgradient of P_α at H then $-\varphi|_N$ is a subgradient of $(P_\alpha)|_{N^{sa}}$ at H for any local algebra N containing H. Thus applying Lemma 9.3.5 to an increasing sequence of local algebras we conclude that the equality

$$\int_\mathbb{R} \hat{f}(t)\varphi(a\sigma_t^H(b))dt = \int_\mathbb{R} \hat{f}(t+i)\varphi(\sigma_t^H(b)a)dt$$

holds for all $f \in \mathcal{D}$ and all local a, b, hence for all $a, b \in A$. But this is one of the equivalent forms of the KMS-condition. □

Remark 9.3.6. Under the assumptions of Theorem 9.3.3, any weak* limit point of a sequence on which the supremum in the variational principle is attained is a subgradient of the pressure, hence a σ^H-KMS$_1$ state. ♦

In order to define the dynamics σ^H we need some assumptions on α, such as asymptotic abelianness with locality. On the other hand, the question when $h_\varphi(\alpha) = ht(\alpha) < \infty$ forces φ to be tracial makes sense in general. In Chap. 13 we shall give an example of an asymptotically abelian system (A, α) with zero topological entropy and many nontracial invariant states.

In Chaps. 11 and 12 we shall see examples of an automorphism α with a unique invariant state φ such that $0 = h_\varphi(\alpha) < ht(\alpha)$.

9.4 Notes

Let B be a finite dimensional C*-algebra, $A = B^{\otimes \mathbb{Z}}$ the infinite C*-tensor product, α the shift to the right on A. For a subset $\Lambda \subset \mathbb{Z}$ denote by A_Λ the C*-subalgebra of A generated by the factors corresponding to the elements of Λ. Given an α-invariant state φ on A, the quantity

$$s(\varphi) = \lim_{n\to\infty} \frac{1}{n} S(\varphi|_{A_{[0,n-1]}})$$

is called the **mean entropy** of φ. We always have $h_\varphi(\alpha) \leq s(\varphi)$.

By an interaction potential one understands a map Φ which associates a self-adjoint element $\Phi(X) \in A_X$ to every $X \subset \mathbb{Z}$. We assume that Φ is translationally invariant, $\Phi(n + X) = \Phi(X)$, and that the series

$$H = \sum_{X \ni 0} \frac{\Phi(X)}{|X|}$$

is absolutely convergent. For a finite subset $\Lambda \subset \mathbb{Z}$ define the local Hamiltonian by

$$H(\Lambda) = \sum_{X \subset \Lambda} \Phi(X).$$

The pressure (at inverse temperature $\beta = 1$) is defined by

$$P(\Phi) = \lim_{n \to \infty} \frac{1}{n} \log \operatorname{Tr}_{A_{[0,n-1]}}(e^{-H([0,n-1])}).$$

Then the variational principle of Ruelle [185] (classical lattices, B is abelian) and Robinson [180] (quantum lattices, B is a full matrix algebra) states that

$$P(\Phi) = \sup_\varphi \{s(\varphi) - \varphi(H)\}.$$

Moreover, as was proved by Lanford and Robinson [115], see also [30] for a detailed exposition, under suitable conditions on the potential, the limit

$$\sigma_t(x) = \lim_{n \to \infty} (\operatorname{Ad} e^{itH([-n,n])})(x)$$

defines a one-parameter automorphism group of A, and if the supremum above is attained at a state φ, then φ is a σ-KMS$_1$-state. Note that if Φ has finite range in the sense that there exist only finitely many sets $X \ni 0$ such that $\Phi(X) \neq 0$, then the above σ coincides with the dynamics σ^H defined in Sect. 9.3 by a result of Araki [7], who also proved that local operators are σ^H-analytic, Proposition 9.3.2 and Lemma 9.3.1.

In the classical case there is a more general result subsuming also the variational principle for topological entropy. Let (X,T) be a topological dynamical system, and $f \in C(X)$. For an open cover $\mathcal{U}$ of X put

$$q(T, f, \mathcal{U}) = \inf_\mathcal{V} \left\{ \sum_{V \in \mathcal{V}} \inf_{x \in V} e^{f(x)} \right\},$$

where the first infimum is taken over all finite subcovers $\mathcal{V}$ of $\mathcal{U}$, and then denoting $f + f \circ T + \ldots + f \circ T^{n-1}$ by $S_n f$ define

$$P(T, f) = \sup_\mathcal{U} \limsup_{n \to \infty} \frac{1}{n} \log q(T, S_n f, \vee_{k=0}^{n-1} T^{-k} \mathcal{U}).$$

Then $P(T,f) = \sup_\mu \{h_\mu(T) + \int_X f\, d\mu\}$. This was proved by Ruelle [186] under additional assumptions, and by Walters [230] in general, see [231] for details.

It is therefore only natural to try to generalize the variational principle for lattice systems to C*-dynamical systems using dynamical entropy instead of mean entropy. The first attempt in this direction was made by Narnhofer [125]. She studied the question whether a state φ on a lattice system such that $h_\varphi(\alpha) - \varphi(H) = \sup\{h_\psi(\alpha) - \psi(H)\}$ satisfies the KMS-condition. Moriya [123] approached this problem differently by proving that

$$\sup_\psi \{h_\psi(\alpha) - \psi(H)\} = \sup_\psi \{s(\psi) - \psi(H)\}.$$

This implies in particular that the pressure $P(\Phi)$ defined above coincides with the pressure $P_\alpha(H)$.

The pressure $P_\alpha(H)$ for nuclear C*-algebras was introduced by the authors [140], who obtained the main results of this chapter. But of course most of the arguments are already found in the above mentioned papers. The notion of pressure was extended to exact C*-algebras by Kerr and Pinzari [110]. For further results and relevant examples of systems see [105], [110], [213], [77], [3].

There are a number of nonasymptotically abelian systems where the topological entropy is the supremum of the dynamical entropies. E.g. Choda [43] showed that both the topological entropy and the dynamical entropy with respect to a certain state of the canonical endomorphism of the Cuntz algebra $\mathcal{O}_n$ are equal to $\log n$; see [27], [165], [110] for generalizations. Nevertheless in all known examples there always exists an asymptotically abelian system around which completely determines the entropy.

Although, as we mentioned above, for lattice systems we can use both dynamical entropy and mean entropy in the variational principle, it is unknown whether the equality $h_\varphi(\alpha) = s(\varphi)$ always holds. In order to prove it one probably needs a better understanding not only of dynamical entropy, but also of mean entropy. E.g. if φ is pure, we clearly have $h_\varphi(\alpha) = 0$, but it is still an open problem whether $s(\varphi) = 0$. The equality $h_\varphi(\alpha) = s(\varphi)$ holds for states derived from quasi-free states (see Chap. 13), Markov states [160], [154] (which under the assumption of faithfulness can be described as the states such that $\sigma_t^\varphi(A_0) \subset A_{[-1,1]}$ [137], [76]), and of course the states coming from diagonal maximal abelian subalgebras (that is, obtained by composing a state on $D = C^{\otimes \mathbb{Z}}$, where C is a maximal abelian subalgebra of B, with the trace preserving conditional expectation $A \to D$). It is stated in [50] that if Φ is a finite range interaction then $h_\varphi(\alpha) = s(\varphi)$ for the state φ such that $s(\varphi) - \beta\varphi(H) = \sup\{s(\psi) - \beta\psi(H)\}$ for sufficiently small $\beta > 0$ (such φ is known to be unique [8]). However, we were unable to follow the proof. A related open problem is to find conditions on automorphisms of C*-algebras under which the map $\varphi \mapsto h_\varphi(\alpha)$ is weakly* upper semicontinuous.

Part II

Special Topics

10

Relative Entropy and Subfactors

Conditional entropy is, as we saw in Chap. 1, an important concept in the classical theory. In the noncommutative case it was introduced under the name of relative entropy as a tool to take care of approximation of entropy. We shall in the present chapter develop the theory of relative entropy and show its relationship to subfactors of II_1-factors. Then we shall show a formula analogous to the classical formula $h(T) = H(\xi|\xi^-)$. Finally we shall give applications to the canonical shift on the tower of relative commutants defined by an inclusion of II_1-factors and for shifts on the Jones projections.

10.1 Relative Entropy

Let M be a von Neumann algebra, φ a normal state on M, P and Q von Neumann subalgebras of M.

Definition 10.1.1. *The relative entropy of P and Q with respect to φ is*

$$H_\varphi(P|Q) = \sup \sum_i (S(\varphi_i|_P, \varphi|_P) - S(\varphi_i|_Q, \varphi|_Q)),$$

where the supremum is taken over all finite decompositions $\varphi = \sum_i \varphi_i$ of φ into a sum of positive linear functionals.

Note that if $\varphi = \tau$ is a trace, which is the case we are mainly interested in, any $\psi \leq \tau$ has the form $\psi = \tau(\cdot x)$. Hence by Theorem 2.3.1(x) the relative entropy can be written as

$$H_\tau(P|Q) = \sup \sum_i (\tau(\eta(E_Q(x_i))) - \tau(\eta(E_P(x_i)))), \tag{10.1}$$

where the supremum is taken over all finite partitions of unity $1 = \sum_i x_i$ in M, and E_P and E_Q are the τ-preserving conditional expectations on P and Q, respectively. One often suppresses τ in the notation for relative entropy.

The main properties of relative entropy are as follows.

Theorem 10.1.2. *We have:*

(i) $H_\varphi(P|Q) \geq 0$, and if φ is a faithful tracial state then $H_\varphi(P|Q) = 0$ if and only if $P \subset Q$;

(ii) $H_\varphi(P|R) \leq H_\varphi(P|Q) + H_\varphi(Q|R)$;

(iii) $H_\varphi(P|Q)$ is increasing in P and decreasing in Q;

(iv) if $Q \subset P$ and there is a φ-preserving faithful normal conditional expectation $E: P \to Q$, then

$$H_\varphi(P|Q) = \sup \sum_i S(\varphi_i|_P, \varphi_i \circ E),$$

where the supremum is taken over all finite decompositions $\varphi = \sum_i \varphi_i$.

Proof. Taking the trivial decomposition $\varphi = \varphi$ we see that $H_\varphi(P|Q) \geq 0$. If $P \subset Q$ then $H_\varphi(P|Q) = 0$ by monotonicity of relative entropy S, Theorem 2.3.1(vi). Let now $\varphi = \tau$ be a faithful tracial state. If P is not a subset of Q, there exists a projection $e \in P$ such that $e \notin Q$. Then by (10.1)

$$H_\tau(P|Q) \geq \tau(\eta(E_Q(e))) + \tau(\eta(E_Q(1-e))).$$

The element $E_Q(e)$ is not a projection (this follows e.g. by A.13, as if $E_Q(e)$ is a projection then $e \leq s(E_Q(e)) = E_Q(e)$, where $s(a)$ is the support of a self-adjoint element a, and hence $e = E_Q(e)$ by faithfulness of E_Q). Therefore $\tau(\eta(E_Q(e))) > 0$ and similarly $\tau(\eta(E_Q(1-e))) > 0$. This completes the proof of (i).

Part (ii) is immediate from the definition of $H_\varphi(P|Q)$. Part (iii) follows from monotonicity of relative entropy S, while part (iv) follows from Theorem 2.3.1(vii). □

Let us show next that relative entropy coincides with conditional entropy in the abelian case. So let $M = L^\infty(X, \mu)$, ξ and ζ be measurable partitions of X such that ξ is finite, $P = L^\infty(X/\xi)$ and $Q = L^\infty(X/\zeta)$. We want to show that

$$H_\tau(P|Q) = H_\mu(\xi|\zeta).$$

If $p_1, \ldots, p_n$ are the atoms of P, by (10.1) we have

$$H_\tau(P|Q) \geq \sum_i \tau(\eta(E_Q(p_i))).$$

The latter expression is exactly $H_\mu(\xi|\zeta)$. Thus $H_\tau(P|Q) \geq H_\mu(\xi|\zeta)$. It suffices to prove the opposite inequality for finite ζ. Indeed, if $\{\zeta_n\}_n$ is an increasing sequence of finite measurable partitions such that $\vee_n \zeta_n = \zeta$ then

$$H_\tau(P|L^\infty(X/\zeta_n)) \geq H_\tau(P|Q) \geq H_\mu(\xi|\zeta),$$

and $H_\mu(\xi|\zeta_n) \searrow H_\mu(\xi|\zeta)$ by the martingale convergence theorem. Thus if $H_\tau(P|L^\infty(X/\zeta_n)) = H_\mu(\xi|\zeta_n)$, we conclude that $H_\tau(P|Q) = H_\mu(\xi|\zeta)$.

10.1 Relative Entropy   159

So assume ζ is finite. Then

$$H_\tau(P|Q) \le H_\tau(P \vee Q|Q) \text{ and } H_\mu(\xi|\zeta) = H_\mu(\xi \vee \zeta|\zeta).$$

It follows that to prove the inequality $H_\tau(P|Q) \le H_\mu(\xi|\zeta)$ we may assume that $\zeta \prec \xi$, so $Q \subset P$. In this case it is enough to consider partitions of unity in P. Then part (iv) of Theorem 10.1.2 and convexity of relative entropy of positive functionals, Corollary 2.3.2, show that it suffices to consider partitions $1 = \sum_i x_i$ such that each x_i is a scalar multiple of a minimal projection in P. Since $S(\lambda\psi, \lambda\varphi) = \lambda S(\psi, \varphi)$, we then see that this is the same as to consider one partition $1 = \sum_i p_i$, where $p_1, \ldots, p_n$ are the minimal projections in P. But then clearly $H_\tau(P|Q) = H_\mu(\xi|\zeta)$.

Next we want to show that in some cases the computation of relative entropy can be reduced to the finite dimensional case. For this we need the following notion. If P, Q, R are von Neumann subalgebras of a von Neumann algebra M, and $E_P\colon M \to P$, $E_Q\colon M \to Q$, $E_R\colon M \to R$ are conditional expectations, then we say that

$$\begin{array}{ccc} P & \subset & M \\ \cup & & \cup \\ R & \subset & Q \end{array}$$

is a **commuting square** if $E_R = E_Q \circ E_P = E_P \circ E_Q$. Equivalently, $R = P \cap Q$ and $E_Q(P) \subset R$, $E_P(Q) \subset R$.

Remark that if the conditional expectations preserve a faithful normal state φ, then is suffices to check that, say, $E_R = E_Q \circ E_P$. Indeed, if $M \subset B(H)$ and the state φ is defined by a cyclic and separating vector $\xi \in H$, then $E_R(x)\xi = e_R x \xi$ for $x \in M$, where $e_R \in B(H)$ is the projection onto $\overline{R\xi}$. Then $E_R = E_Q \circ E_P$ implies that $e_R = e_Q e_P$. But then $e_R = e_P e_Q$, and so $E_R = E_P \circ E_Q$.

Proposition 10.1.3. *Let M be a von Neumann algebra with a faithful normal state φ, N a von Neumann subalgebra. Suppose $\{M_n\}_n$ and $\{N_n\}_n$ are increasing sequences of von Neumann subalgebras of M such that $N_n \subset M_n$, $M = (\cup_n M_n)''$ and $N = (\cup_n N_n)''$. Suppose there exist φ-preserving conditional expectations $E_{M_n}\colon M \to M_n$ and $E_{N_n}\colon M \to N_n$ such that*

$$\begin{array}{ccc} M_n & \subset & M_{n+1} \\ \cup & & \cup \\ N_n & \subset & N_{n+1} \end{array}$$

is a commuting square for every n. Then $H_\varphi(M|N) = \lim_n H_\varphi(M_n|N_n)$.

Proof. If $\psi \le \varphi$ then $S(\psi|_{M_n}, \varphi|_{M_n}) \nearrow S(\psi, \varphi)$ by Corollary 2.3.5, and a similar convergence holds for N_n and N. This implies that

$$H_\varphi(M|N) \le \liminf_n H_\varphi(M_n|N_n).$$

Since $E_{N_{n+1}} \circ E_{M_n} = E_{N_n}$ for all n, we have

$$E_{N_{n+k+1}} \circ E_{M_n} = E_{N_{n+k+1}} \circ E_{M_{n+k}} \circ E_{M_n} = E_{N_{n+k}} \circ E_{M_n},$$

so by induction $E_{N_{n+k}} \circ E_{M_n} = E_{N_n}$ for all $k \in \mathbb{N}$. Next note that $\{E_{N_n}\}_n$ converges in the pointwise strong operator topology to a φ-preserving conditional expectation $E_N: M \to N$. It follows that for any n

$$\begin{array}{ccc} M_n & \subset & M \\ \cup & & \cup \\ N_n & \subset & N \end{array}$$

is a commuting square. In other words, $E_N|_{M_n} = E_{N_n}|_{M_n}$. Hence for any positive linear functional ψ we have

$$S(\psi, \psi \circ E_N) \geq S(\psi|_{M_n}, \psi \circ E_N|_{M_n}) = S(\psi|_{M_n}, \psi \circ E_{N_n}|_{M_n}).$$

By Theorem 10.1.2(iv) it follows that $H_\varphi(M|N) \geq H_\varphi(M_n|N_n)$, which completes the proof. $\square$

From now onwards we shall only consider tracial states, and our next goal is to compute the relative entropy $H_\tau(M|N)$ when $N \subset M$ are finite dimensional (and τ is a trace on M).

Let $Z(M)$ and $Z(N)$ be the centers of M and N, respectively,

$$M = \bigoplus_{l=1}^{m} M_l, \quad N = \bigoplus_{k=1}^{n} N_k,$$

where $M_l \cong \mathrm{Mat}_{m_l}(\mathbb{C})$ and $N_k \cong \mathrm{Mat}_{n_k}(\mathbb{C})$. Let w_l and z_k be the central projections in M and N such that $M_l = Mw_l$ and $N_k = Nz_k$. Let a_{kl} be the multiplicity of $N_k w_l$ in M_l. Thus if t_l denotes the trace of a minimal projection in M_l and s_k that of a minimal projection in N_k,

$$\tau(w_l) = m_l t_l = \sum_k n_k a_{kl} t_l \quad \text{and} \quad \tau(z_k) = n_k s_k = \sum_l n_k a_{kl} t_l.$$

Theorem 10.1.4. *With the above notation the relative entropy $H_\tau(M|N)$ is*

$$(2H_\tau(M) - H_\tau(Z(M))) - (2H_\tau(N) - H_\tau(Z(N))) + \sum_{k,l} n_k a_{kl} t_l \log c_{kl},$$

where $c_{kl} = \min\left\{\dfrac{n_k}{a_{kl}}, 1\right\}$.

In particular, if M is abelian we have

$$H_\tau(M|N) = H_\tau(M) - H_\tau(N),$$

which we already know, and if $M = \mathrm{Mat}_m(\mathbb{C})$ and $N = \mathrm{Mat}_n(\mathbb{C})$ then

$$H_\tau(M|N) = 2\log m - 2\log n + \log\min\left\{\frac{n^2}{m}, 1\right\} = \min\left\{\log m, 2\log\frac{m}{n}\right\}.$$

A straightforward computation of the formula in the theorem shows that it is equivalent to the formula

$$\begin{aligned}H_\tau(M|N) = &-\sum_l m_l t_l \log t_l + \sum_l m_l t_l \log m_l + \sum_k n_k s_k \log s_k \\ &-\sum_k n_k s_k \log n_k + \sum_{k,l} n_k a_{kl} t_l \log c_{kl}.\end{aligned} \quad (10.2)$$

Note that if $\tau = \sum_i \lambda_i \varphi_i$ is a decomposition of τ into a convex combination of states, then

$$\sum_i S(\lambda_i \varphi_i, \tau) = S(\tau) - \sum_i \lambda_i S(\varphi_i) - \sum_i \eta(\lambda_i),$$

and we have a similar expression for $\sum_i S(\lambda_i \varphi_i|_N, \tau|_N)$. It follows that

$$H_\tau(M|N) = \sup\left\{H_\tau(M) - H_\tau(N) + \sum_i \lambda_i(S(\varphi_i|_N) - S(\varphi_i))\right\}, \quad (10.3)$$

where the supremum is taken over all decompositions $\tau = \sum_i \lambda_i \varphi_i$ of τ into a convex combination of states. This is the expression we shall use in the proof of the theorem. Note also that instead of finite convex decompositions we can use integral decompositions of τ.

Proof of Theorem 10.1.4. We first show that the left side is majorized by the right side in equation (10.2). Consider an integral decomposition $\tau = \int_X \varphi_x d\mu(x)$. By (10.3) we have to estimate

$$H_\tau(M) - H_\tau(N) + \int_X (S(\varphi_x|_N) - S(\varphi_x)) d\mu(x). \quad (10.4)$$

By Theorem 2.2.2(ii) we have

$$S(\varphi_x|_N) = \sum_k S(\varphi_x|_{Nz_k}) \le \sum_{k,l} S(\varphi_x|_{Nz_k w_l}).$$

For fixed k and l consider the state $\psi = \varphi_x(z_k w_l)^{-1}\varphi_x$ on $z_k M w_l z_k$. By Theorem 2.2.2(i) we have

$$S(\psi|_{Nz_k w_l}) - S(\psi) \le S(\psi|_{Nz_k w_l}) \le \log n_k.$$

On the other hand, since $z_k M w_l z_k \cong N z_k w_l \otimes \mathrm{Mat}_{a_{kl}}(\mathbb{C})$, by Theorem 2.2.2(vi) we have

$$S(\psi|_{Nz_k w_l}) - S(\psi) \le \log a_{kl}.$$

Using $S(\lambda\psi) = \lambda S(\psi) + \eta(\lambda)$ with $\lambda = \varphi_x(z_k w_l)$, we therefore get

$$S(\varphi_x|_{Nz_k w_l}) - S(\varphi_x|_{z_k M w_l z_k}) \leq \varphi_x(z_k w_l) \log \min\{n_k, a_{kl}\}.$$

Finally, from Lemma 2.2.4 applied to the state $\varphi_x(w_l)^{-1}\varphi_x$ on Mw_l, we obtain

$$\sum_k S(\varphi_x|_{z_k M w_l z_k}) - S(\varphi_x|_{Mw_l}) \leq \varphi_x(w_l) \sum_k \eta\left(\frac{\varphi_x(z_k w_l)}{\varphi_x(w_l)}\right).$$

Thus, since $\sum_l S(\varphi_x|_{Mw_l}) = S(\varphi_x)$, the integral in (10.4) is estimated from above by

$$\sum_{k,l} \int_X \varphi_x(z_k w_l) \log(\min\{n_k, a_{kl}\}) d\mu(x) + \sum_{k,l} \int_X \varphi_x(w_l) \eta\left(\frac{\varphi_x(z_k w_l)}{\varphi_x(w_l)}\right) d\mu(x).$$

Since $\int_X \varphi_x d\mu(x) = \tau$ and $\tau(z_k w_l) = n_k a_{kl} t_l$, the first summand in the above expression is equal to

$$\sum_{k,l} n_k a_{kl} t_l \log \min\{n_k, a_{kl}\}.$$

On the other hand, using concavity of η and that $\tau(w_l) = m_l t_l$, we can estimate the second summand by

$$m_l t_l \int_X \frac{\varphi_x(w_l)}{m_l t_l} \eta\left(\frac{\varphi_x(z_k w_l)}{\varphi_x(w_l)}\right) d\mu(x) \leq m_l t_l \eta\left(\int_X \frac{\varphi_x(w_l)}{m_l t_l} \frac{\varphi_x(z_k w_l)}{\varphi_x(w_l)} d\mu(x)\right)$$

$$= m_l t_l \eta\left(\frac{n_k a_{kl} t_l}{m_l t_l}\right)$$

$$= -t_l n_k a_{kl} \log \frac{n_k a_{kl}}{m_l}.$$

Thus we estimate (10.4) by

$$H_\tau(M) - H_\tau(N) + \sum_{k,l} n_k a_{kl} t_l \log(\min\{a_{kl}, n_k\}) - \sum_{k,l} n_k a_{kl} t_l \log \frac{n_k a_{kl}}{m_l}$$

$$= H_\tau(M) - H_\tau(N) + \sum_{k,l} n_k a_{kl} t_l \log c_{kl} + \sum_{k,l} n_k a_{kl} t_l \log a_{kl}$$

$$- \sum_{k,l} n_k a_{kl} t_l (\log a_{kl} + \log n_k - \log m_l)$$

$$= H_\tau(M) - H_\tau(N) + \sum_{k,l} n_k a_{kl} t_l \log c_{kl}$$

$$- \sum_k n_k s_k \log n_k + \sum_l m_l t_l \log m_l,$$

where we used that $s_k = \sum_l a_{kl} t_l$ and $m_l = \sum_k n_k a_{kl}$. Since

10.1 Relative Entropy

$$H_\tau(M) = -\sum_l m_l t_l \log t_l \quad \text{and} \quad H_\tau(N) = \sum_k n_k s_k \log s_k,$$

we have shown that the left side of equation (10.2) is majorized by the right side.

In order to prove the opposite inequality choose a pure state φ_l of Mw_l such that

$$\varphi_l(z_k w_l) = \frac{n_k a_{kl}}{m_l},$$

$\varphi_l|_{Nz_k w_l}$ is a trace if $n_k < a_{kl}$,

$\varphi_l|_{(N' \cap M)z_k w_l}$ is a trace if $n_k \geq a_{kl}$.

Explicitly φ_l can be constructed as follows. Identify $\ell_2^{m_l}$ with $\oplus_k(\ell_2^{n_k} \otimes \ell_2^{a_{kl}})$, and let $\{e_i\}_{i \in \mathbb{N}}$ denote the standard basis in ℓ_2. Put

$$\xi_l = \bigoplus_k \left(\frac{n_k a_{kl}}{\min\{n_k, a_{kl}\} m_l}\right)^{1/2} \sum_{i=1}^{\min\{n_k, a_{kl}\}} e_i \otimes e_i.$$

Then let $\varphi_l = \omega_{\xi_l}$ be the vector state defined by ξ_l.

Let K_l denote the subalgebra of $\oplus_k z_k M w_l z_k \subset M w_l$ consisting of elements $\oplus_k x_k$ such that $x_k \in Nz_k w_l$ if $n_k < a_{kl}$, and $x_k \in (N' \cap M)z_k w_l$ if $n_k \geq a_{kl}$. Then φ_l is a pure state on Mw_l such that its restriction to K_l coincides with the restriction of the unique tracial state τ_l on Mw_l to K_l. Furthermore, the map $E_l: Mw_l \to K_l$ defined by

$$E_l(x) = \int_{U(K_l' \cap Mw_l)} (\operatorname{Ad} u)(x) d\mu_l(u),$$

where μ_l is the normalized Haar measure on the unitary group $U(K_l' \cap Mw_l)$ of $K_l' \cap Mw_l$, is a τ_l-preserving conditional expectation. It follows that

$$\tau = \sum_l m_l t_l \int_{U(K_l' \cap Mw_l)} \varphi_l \circ \operatorname{Ad} u \, d\mu_l(u).$$

Thus by (10.3)

$$H_\tau(M|N) \geq H_\tau(M) - H_\tau(N) + \sum_{k,l} m_l t_l S(\varphi_l|_{Nz_k w_l}). \tag{10.5}$$

The restricition of φ_l to $z_k M w_l z_k$ is a scalar multiple of a pure state. Hence by Lemma 2.2.3(i)

$$S\left(\frac{m_l}{n_k a_{kl}} \varphi_l|_{Nz_k w_l}\right) = S\left(\frac{m_l}{n_k a_{kl}} \varphi_l|_{(N' \cap M)z_k w_l}\right) = \log \min\{n_k, a_{kl}\},$$

whence

$$S(\varphi_l|_{N_{z_k w_l}}) = \frac{n_k a_{kl}}{m_l}(\log c_{kl} + \log a_{kl}) - \frac{n_k a_{kl}}{m_l} \log \frac{n_k a_{kl}}{m_l}.$$

Thus the right side of (10.5) equals

$$H_\tau(M) - H_\tau(N) + \sum_{k,l} n_k a_{kl} t_l (\log c_{kl} + \log a_{kl}) - \sum_{k,l} n_k a_{kl} t_l \log \frac{n_k a_{kl}}{m_l},$$

which is exactly what we need. □

10.2 Index of Subfactors

Our goal in this section is to show that relative entropy is related to index of subfactors. We shall briefly recall some basic facts of the latter theory referring the reader to [214, Chapter XIX] for more details.

Let M be a II_1-factor with trace τ. If M acts on a Hilbert space H then H can be embedded into $L^2(M) \otimes K$ for a sufficiently large Hilbert space K. Let $p \in (M \otimes \mathbb{C}1)' = M' \bar{\otimes} B(K)$ be the projection onto H. Then the **dimension of H relative to M** is defined by

$$\dim_M(H) = (\tau' \otimes \mathrm{Tr})(p),$$

where τ' is the unique tracial state on $M' \subset B(L^2(M))$ and Tr is the canonical trace on $B(K)$. The dimension can take any value in $(0, +\infty]$, and if H is countably generated in the sense that it is a direct sum of at most countably many cyclic subspaces for M, then $\dim_M(H)$ is a complete invariant of the unitary equivalence class of the representation $M \to B(H)$.

The dimension $\dim_M(H)$ is finite if and only if the commutant M' of M in $B(H)$ is a II_1-factor. In the latter case

$$\dim_M(H) = \frac{\tau([M'\xi])}{\tau'([M\xi])}, \qquad (10.6)$$

where $\xi \in H$ is any nonzero vector, τ' is the unique tracial state on $M' \subset B(H)$, and $[M\xi]$ denotes the projection onto the subspace $\overline{M\xi}$.

If N is a subfactor of M, the **index** of N in M is

$$[M:N] = \dim_N(L^2(M)).$$

If M acts on H then $\dim_N(H) = [M:N]\dim_M(H)$. It follows that if $N \subset M \subset L$ then $[L:N] = [L:M][M:N]$. It follows also that if $[M:N] < \infty$ and N is represented on H then this representation extends to a representation of M on H. Indeed, decomposing H into a direct sum of cyclic subspaces we may assume that $\dim_N(H) \leq 1$. Choose a representation of M on a Hilbert space H' such that $\dim_M(H') = [M:N]^{-1} \dim_N(H)$. Then $\dim_N(H') = \dim_N(H)$.

Hence the representations of N on H and H' are equivalent. Since the representation of N on H' extends to a representation of M, the same is true for H.

Using Proposition 10.1.3 and Theorem 10.1.4 it is easy to check that if R is the hyperfinite II_1-factor then

$$H_\tau(R \otimes \mathrm{Mat}_n(\mathbb{C}) | R \otimes 1) = 2\log n.$$

On the other hand, $[R \otimes \mathrm{Mat}_n(\mathbb{C}): R \otimes 1] = n^2$. So we see that in this case the relative entropy is the logarithm of the index. This is a particular case of a far more general result. For simplicity of the presentation we stick to the **irreducible** case, that is, when $N' \cap M = \mathbb{C}1$.

Theorem 10.2.1. *Let $N \subset M$ be II_1-factors such that $[M:N] < \infty$ and $N' \cap M = \mathbb{C}1$. Then*

$$H_\tau(M|N) = \log[M:N].$$

To prove this theorem we need two properties of finite index subfactors. The first one is the **Pimsner-Popa inequality** stating that

$$E_N(x) \geq [M:N]^{-1} x \quad \text{for any} \quad x \in M_+,$$

where $E_N: M \to N$ is the trace preserving conditional expectation, see [214, Theorem XIX.4.14].

Proposition 10.2.2. *If $N \subset M$ is a finite index subfactor, then*

$$H_\tau(M|N) \leq \log[M:N].$$

Proof. Let $\lambda = [M:N]^{-1}$. Then by the Pimsner-Popa inequality $\varphi \circ E_N \geq \lambda \varphi$ for any positive linear functional φ on M. Since the relative entropy S is decreasing in the second variable, Theorem 2.3.1(iii), we then have

$$S(\varphi, \varphi \circ E_N) \leq S(\varphi, \lambda \varphi) = -\varphi(1) \log \lambda.$$

By Theorem 10.1.2(iv) it follows that $H_\tau(M|N) \leq -\log \lambda = \log[M:N]$. □

The second property of subfactors which we need, is existence of special projections.

Lemma 10.2.3. *If $N \subset M$ is a finite index subfactor then there exists a projection $q \in M$ such that $E_N(q) = [M:N]^{-1} 1$. Moreover, any other projection with this property is of the form vqv^*, where v is a unitary in N.*

Proof. See [214, Theorem XIX.4.12]. The existence will also essentially be shown in Lemma 10.4.1 below. □

Proof of Theorem 10.2.1. Let $\lambda = [M:N]^{-1}$ and q be the projection from Lemma 10.2.3. The idea of the proof is to consider the state $\varphi = \lambda^{-1} \tau(\cdot q)$

and the decomposition $\tau = \int_{U(N)} \varphi \circ \operatorname{Ad} u\, du$, which formally holds because of $E_{N' \cap M}(q) = \lambda 1$. For II_1-factors the above decomposition does not make sense, but it is still possible to get an approximate version of it.

Let $x = q - \lambda 1$. Then $\tau(x) = 0$. Let

$$K_x = \overline{\operatorname{conv}\{vxv^*\colon v \in U(N)\}},$$

where the closure is in the weak operator topology. Then K_x is a convex w-compact set in M, and $\tau(y) = 0$ for all $y \in K_x$. The weak operator closure of a bounded convex set coincides with the closure with respect to the L^2-norm. Thus there exists a unique $y_0 \in K_x$ such that

$$\|y_0\|_2 = \inf\{\|y\|_2\colon y \in K_x\}.$$

But $vy_0v^* \in K_x$ for all $v \in U(N)$, and $\|vy_0v^*\|_2 = \|y_0\|_2$, so that $vy_0v^* = y_0$, i.e., $y_0 \in N' \cap M$. Since $N' \cap M = \mathbb{C}1$, it follows that y_0 is a scalar, so $y_0 = 0$. Hence for any fixed $\varepsilon > 0$ there are unitaries $v_1, \ldots, v_n \in N$ such that

$$\left\| \frac{1}{n} \sum_{i=1}^n v_i q v_i^* - \lambda 1 \right\|_2 < \varepsilon^2 \lambda.$$

Let $y = (\lambda n)^{-1} \sum_i v_i q v_i^*$. Then $0 \le y \le \lambda^{-1} 1$ and

$$\|y - 1\|_2 \le \varepsilon^2.$$

Let p be the spectral projection of y corresponding to the interval $[0, 1+\varepsilon]$. Set

$$x_i = ((1+\varepsilon)\lambda n)^{-1} v_i q v_i^* \wedge p.$$

Then $\sum_i x_i \le ((1+\varepsilon)\lambda n)^{-1} \sum_i p v_i q v_i^* p = (1+\varepsilon)^{-1} y p \le 1$. Hence

$$H_\tau(M|N) \ge \sum_i \tau(\eta(E_N(x_i)) - \eta(x_i)) = ((1+\varepsilon)\lambda n)^{-1} \sum_i \tau(\eta(E_N(v_i q v_i^* \wedge p))).$$

Using operator monotonicity of log as in the proof of Theorem 2.2.2(ii), we get $\tau(\eta(e+f)) \le \tau(\eta(e)) + \tau(\eta(f))$ for any positive e and f. Hence

$$\tau(\eta(E_N(v_i q v_i^* \wedge p))) \ge \tau(\eta(E_N(v_i q v_i^*))) - \tau(\eta(E_N(v_i q v_i^* - v_i q v_i^* \wedge p)))$$
$$\ge \eta(\lambda) - \eta(\tau(v_i q v_i^* - v_i q v_i^* \wedge p)),$$

where in the second inequality we used that $E_N(q) = \lambda 1$ and that $\eta \circ \tau \ge \tau \circ \eta$ by concavity of η. Thus

$$H_\tau(M|N) \ge -(1+\varepsilon)^{-1} \log \lambda - ((1+\varepsilon)\lambda n)^{-1} \sum_i \eta(\tau(v_i q v_i^* - v_i q v_i^* \wedge p)). \quad (10.7)$$

Now recall that $\tau(e \vee f) - \tau(f) = \tau(e) - \tau(e \wedge f)$ for any projections e and f. Hence

$$\tau(v_i q v_i^* - v_i q v_i^* \wedge p) \leq 1 - \tau(p).$$

Since $y(1-p) \geq (1+\varepsilon)(1-p)$, we have

$$\|y-1\|_2^2 \geq \tau((y-1)^2(1-p)) \geq \varepsilon^2 \tau(1-p),$$

so that $1 - \tau(p) \leq \varepsilon^{-2} \|y-1\|_2^2 \leq \varepsilon^2$. Thus

$$\tau(v_i q v_i^* - v_i q v_i^* \wedge p) \leq \varepsilon^2.$$

Since η is increasing on $[0, e^{-1}]$, for ε small enough we obtain

$$\eta(\tau(v_i q v_i^* - v_i q v_i^* \wedge p)) \leq \eta(\varepsilon^2).$$

From (10.7) we then get

$$H_\tau(M|N) \geq -(1+\varepsilon)^{-1} \log \lambda - ((1+\varepsilon)\lambda)^{-1} \eta(\varepsilon^2).$$

Letting $\varepsilon \to 0$ we conclude that $H_\tau(M|N) \geq -\log \lambda = \log[M:N]$. The opposite inequality follows from Proposition 10.2.2. □

Remark 10.2.4. What we actually used in the proof is not the irreducibility $N' \cap M = \mathbb{C}1$, but that $E_{N' \cap M}(q)$ is a scalar. Subfactors satisfying the latter condition are called **extremal**, and in fact the equality $H_\tau(M|N) = \log[M:N]$ holds if and only if $N \subset M$ is extremal [162, Corollary 4.5].

Note that if $[M:N] < 4$ then automatically $N' \cap M = \mathbb{C}1$, see [214, Corollary XIX.2.10]. Recall also that the possible values of the index is the set $\{4 \cos^2 \pi/n \,|\, n \geq 3\} \cup [4, +\infty]$, see [214, Theorem XIX.2.22].

Finally note that with some extra work the above theorem can be extended to arbitrary subfactors [162, Theorem 4.5]. In that case $H_\tau(M|N) = \infty$ whenever $N' \cap M$ has a diffuse part. If $N' \cap M$ is atomic, and $\{f_k\}_k$ are minimal projections in $N' \cap M$ with sum 1, then

$$H_\tau(M|N) = 2 \sum_k \eta(\tau(f_k)) + \sum_k \tau(f_k) \log[f_k M f_k : f_k N f_k].$$

♦

10.3 Generators and Relative Entropy

In this section we shall prove an analogue of the formula $h(T) = H(\xi|\xi^-)$, see (1.3) and Theorem 1.1.4.

Before we state the main result we introduce some notation. Let (M, τ, α) be a W*-dynamical system, where τ is a faithful normal trace. Let $\{A_n\}_{n=1}^\infty$ be an increasing sequence of finite dimensional C*-subalgebras of M.

We say that $\{A_n\}_{n=1}^\infty$ is a **generating sequence** for α if

168   10 Relative Entropy and Subfactors

(i) $\alpha(A_n) \subset A_{n+1}$, $n \in \mathbb{N}$;

(ii) the von Neumann subalgebra generated by $\alpha^k(A_n)$, $k \in \mathbb{Z}$, $n \in \mathbb{N}$, coincides with M;

(iii) $h_\tau(\alpha) = \lim_{n\to\infty} \dfrac{1}{n} H_\tau(A_n)$.

We say $\{A_n\}_n$ satisfies the **commuting square condition** if

$$\begin{array}{ccc} A_{n+1} & \subset & A_{n+2} \\ \cup & & \cup \\ \alpha(A_n) & \subset & \alpha(A_{n+1}) \end{array}$$

is a commuting square with respect to the τ-preserving conditional expectations for every $n \in \mathbb{N}$. Remark that if M is abelian and $A_n = \vee_{k=0}^{n-1} \alpha^k(A_1)$, then this condition is satisfied if and only if A_1 corresponds to the generating partition of a Markov process.

Write A_n in the form

$$A_n = \bigoplus_{l \in K_n} M_k^n,$$

where $M_k^n \cong \mathrm{Mat}_{m_k^n}(\mathbb{C})$. Let $(a_{kl}^n)_{k \in K_{n-1}, l \in K_n}$ be the inclusion matrix for $\alpha(A_{n-1}) \subset A_n$. Denote by $Z(A_n)$ the center of A_n.

Theorem 10.3.1. *Let (M, τ, α) be a W^*-dynamical system, where τ is a faithful normal trace and M is of type II_1. Suppose $\{A_n\}_{n=1}^\infty$ is a generating sequence for α satisfying the commuting square condition. Assume also that*

$$h_\tau(\alpha) < \infty \quad \text{and} \quad \sup_{n,k,l} \frac{a_{kl}^n}{m_k^{n-1}} < \infty.$$

Then the limit $\lim_{n\to\infty} \dfrac{1}{n} H_\tau(Z(A_n))$ exists, and with $R = (\cup_n A_n)''$ we have

$$h_\tau(\alpha) = \frac{1}{2} H_\tau(R|\alpha(R)) + \frac{1}{2} \lim_{n\to\infty} \frac{1}{n} H_\tau(Z(A_n)).$$

If we combine this theorem with Theorem 10.2.1, we immediately obtain the following corollary.

Corollary 10.3.2. *If we in the above theorem add the assumptions that $\alpha(R)' \cap R = \mathbb{C}1$ and $\lim_n \frac{1}{n} H_\tau(Z(A_n)) = 0$, then $h_\tau(\alpha) = \frac{1}{2} \log[R : \alpha(R)]$.* □

To prove the theorem we need some preparation.

Lemma 10.3.3. *Fix $r > 0$. Let f_n be the central projection in A_n such that $A_n f_n = \oplus_{m_k^n \le r} M_k^n$. Then $\{f_n\}_n$ is a decreasing sequence converging strongly to zero.*

Proof. First note that $R = (\cup_n A_n)''$ is of type II$_1$. Indeed, let z be the central projection in R corresponding to the sum of the type I$_k$ components of R with $k \leq r$, and assume $z \neq 0$. Since $\alpha(R) \subset R$, and type II and type I$_k$ algebras can not be embedded into a type I$_l$ algebra with $l < k$, we have $z \leq \alpha(z)$. As τ is α-invariant and faithful, it follows that $z = \alpha(z)$. The maximal number of mutually equivalent orthogonal projections in Rz is not larger than r. By the definition of a generating sequence, the sequence $\{\alpha^{-n}(R)\}_{n=1}^\infty$ is increasing with union dense in M. Since $z = \alpha^{-n}(z)$ is central in each $\alpha^{-n}(R)$, the projection z is central in M, and the maximal number of mutually equivalent orthogonal projections in Mz is not larger than r. In other words, Mz is a sum of type I$_k$ algebras with $k \leq r$. But this contradicts our assumption on M.

Since a type I$_k$ algebra can not be embedded into a type I$_l$ algebra with $l < k$, it is clear that $\{f_n\}_n$ is a decreasing sequence. Let f be its strong limit. Since f_n is central in A_n, the projection f is central in R. Then $Rf = (\cup_n A_n f)''$, which by the same argument as above contradicts the fact that R is of type II$_1$ unless $f = 0$. □

Denote by t_k^n the trace of a minimal projection in $M_k^n \subset A_n$. Since $H_\tau(A_{n-1}) = H_\tau(\alpha(A_{n-1}))$ and $H_\tau(Z(A_{n-1})) = H_\tau(Z(\alpha(A_{n-1})))$, by Theorem 10.1.4 we have

$$H_\tau(A_n|\alpha(A_{n-1})) = (2H_\tau(A_n) - H_\tau(Z(A_n))) - (2H_\tau(A_{n-1}) - H_\tau(Z(A_{n-1})))$$
$$+ \sum_{k,l} m_k^{n-1} a_{kl}^n t_l^n \log c_{kl}^n, \qquad (10.8)$$

where $c_{kl}^n = \min\{m_k^{n-1}/a_{kl}^n, 1\}$.

Lemma 10.3.4. *Suppose*

$$C = \sup_N \frac{1}{N} \sum_{n=2}^N \sum_{k,l} m_k^{n-1} a_{kl}^n t_l^n \log a_{kl}^n < \infty.$$

Then

$$\lim_{N \to \infty} \frac{1}{N} \sum_{n=2}^N \sum_{k,l} m_k^{n-1} a_{kl}^n t_l^n \log c_{kl}^n = 0.$$

Proof. Since by assumption of the theorem there is $c > 0$ such that $c \leq c_{kl}^n \leq 1$ for all k, l and n, it suffices to prove that

$$\lim_N \frac{1}{N} \sum_{n=2}^N \sum_{k,l: a_{kl}^n > m_k^{n-1}} m_k^{n-1} a_{kl}^n t_l^n = 0.$$

Fix $r > 0$. Let f_n be as in Lemma 10.3.3. Then

$$\frac{1}{N} \sum_{n=2}^N \sum_{k,l: m_k^{n-1} \leq r} m_k^{n-1} a_{kl}^n t_l^n = \frac{1}{N} \sum_{n=2}^N \sum_{k: m_k^{n-1} \leq r} m_k^{n-1} t_k^{n-1} = \frac{1}{N} \sum_{n=2}^N \tau(f_{n-1}),$$

and the latter expression tends to zero by Lemma 10.3.3. On the other hand,

$$\frac{1}{N}\sum_{n=2}^{N}\sum_{k,l:a_{kl}^n>m_k^{n-1}>r} m_k^{n-1}a_{kl}^n t_l^n \leq (\log r)^{-1}\frac{1}{N}\sum_{n=2}^{N}\sum_{k,l} m_k^{n-1}a_{kl}^n t_l^n \log a_{kl}^n$$
$$\leq C(\log r)^{-1}.$$

Since we can take an arbitrary large r, this proves the lemma. □

Proof of Theorem 10.3.1. Let us check that the assumption of the previous lemma is satisfied. Since $\sum_k m_k^{n-1} a_{kl}^n = m_l^n$ and $\sum_l a_{kl}^n t_l^n = t_k^{n-1}$, we have

$$\sum_{k,l} m_k^{n-1} a_{kl}^n t_l^n \log a_{kl}^n \leq \sum_{k,l} m_k^{n-1} a_{kl}^n t_l^n \log \frac{m_l^n}{m_k^{n-1}}$$
$$= \sum_l m_l^n t_l^n \log m_l^n - \sum_k m_k^{n-1} t_k^{n-1} \log m_k^{n-1}$$
$$= H_\tau(A_n) - H_\tau(Z(A_n)) - H_\tau(A_{n-1}) + H_\tau(Z(A_{n-1})).$$

Hence, with $A_0 = \mathbb{C}1$,

$$\frac{1}{N}\sum_{n=1}^{N}\sum_{k,l} m_k^{n-1} a_{kl}^n t_l^n \log a_{kl}^n \leq \frac{1}{N} H_\tau(A_N) - \frac{1}{N} H_\tau(Z(A_N)) \leq \frac{1}{N} H_\tau(A_N).$$

Since the sequence $\{\frac{1}{N} H_\tau(A_N)\}_N$ is bounded by assumption of the theorem, it follows that the assumption of Lemma 10.3.4 is satisfied. Hence if we put $C_n = \sum_{k,l} m_k^{n-1} a_{kl}^n t_l^n \log c_{kl}^n$, we get

$$\frac{1}{N}\sum_{n=1}^{N} C_n \to 0.$$

By equation (10.8) we have

$$\frac{1}{N}\sum_{n=1}^{N} H_\tau(A_n|\alpha(A_{n-1})) = \frac{2}{N} H_\tau(A_N) - \frac{1}{N} H_\tau(Z(A_N)) + \frac{1}{N}\sum_{n=1}^{N} C_n.$$

Since $\{A_n\}_n$ satisfies the commuting square condition, by Proposition 10.1.3

$$H_\tau(A_n|\alpha(A_{n-1})) \to H_\tau(R|\alpha(R)).$$

Since $\frac{1}{N} H_\tau(A_N) \to h_\tau(\alpha)$ by assumption, it follows that $\lim_N \frac{1}{N} H_\tau(Z(A_N))$ exists and
$$H_\tau(R|\alpha(R)) = 2h_\tau(\alpha) - \lim_N \frac{1}{N} H_\tau(Z(A_N)).$$

Thus
$$h_\tau(\alpha) = \frac{1}{2} H_\tau(R|\alpha(R)) + \frac{1}{2}\lim_N \frac{1}{N} H_\tau(Z(A_N)).$$

□

10.4 The Canonical Shift

In the present section we shall apply the results of the preceding sections to an interesting automorphism, called the canonical shift, defined on the tower of relative commutants associated with an inclusion $N \subset M$ of II_1-factors of finite index.

In order to introduce it we need a few more facts from index theory, again see [214, Chapter XIX] for more details. Let τ be the trace on M, $\xi \in L^2(M)$ the cyclic vector defining τ. Denote by e_N the projection onto $\overline{N\xi}$, and by $M_1 = \langle M, e_N \rangle$ the von Neumann subalgebra of $B(L^2(M))$ generated by M and e_N. Since $e_N x e_N = E_N(x) e_N = e_N E_N(x)$ for $x \in M$, we have $\{e_N\}' \cap M = N$. In other words, N' is generated by M' and e_N. So if we denote by J the modular conjugation defined by τ, $Jx\xi = x^*\xi$, so that $JMJ = M'$ and $Je_N J = e_N$, then $JM_1 J = N'$. It follows that M_1 is a II_1-factor and

$$\dim_{M_1}(L^2(M)) = \dim_{N'}(L^2(M)) = \frac{1}{\dim_N(L^2(M))} = \frac{1}{[M:N]},$$

where the second equality follows from (10.6). Hence $[M_1:M] = [M:N]$. Furthermore, by (10.6) the unique tracial state τ' on N' has the property $\tau'(e_N) = [M:N]^{-1}$. The unique tracial state on M_1, which we again denote by τ, is such that $\tau(x) = \tau'(Jx^*J)$. Hence $\tau(e_N) = [M:N]^{-1}$. Since N is a factor, $\tau(ab) = \tau(a)\tau(b)$ for $a \in N$ and $b \in N' \cap M_1$. Hence, for $x \in M$,

$$\tau(xE_M(e_N)) = \tau(xe_N) = \tau(E_N(x)e_N) = \tau(E_N(x))\tau(e_N) = [M:N]^{-1}\tau(x),$$

so we also have $E_M(e_N) = [M:N]^{-1}1$, where $E_M: M_1 \to M$ is the trace preserving conditional expectation.

Thus starting with a finite index inclusion $N \subset M$ we construct a II_1-factor M_1 and a projection $e_N \in M_1$ such that $[M_1:M] = [M:N]$, $M_1 = \langle M, e_N \rangle$, $N = \{e_N\}' \cap M$ and $E_M(e_N) = [M:N]^{-1}1$. This is called the **basic construction**. Iterating it we get the **Jones tower** of II_1-factors

$$M_{-1} - N \subset M_0 = M \subset M_1 \subset \ldots \subset M_k \subset \ldots$$

and projections $e_k \in M_{k+1}$ such that $M_{k+1} = \langle M_k, e_k \rangle$, $k \geq 0$, $e_0 = e_N$.

There is also a downward construction. So there exists a projection e_{-1} and a subfactor $N_{-1} \subset N$ such that

$$N_{-1} \subset N \subset \langle N, e_{-1} \rangle = M$$

is the basic construction. Since $E_N(e_{-1}) = [M:N]^{-1}1$, by Lemma 10.2.3 the projection e_{-1} is defined up to conjugation by a unitary in N. Since $N_{-1} = \{e_{-1}\}' \cap N$, the subfactor N_{-1} is also defined up to conjugation by a unitary in N. Denoting N_{-1} by M_{-2} and iterating the downward construction we get a **tunnel**

$$\ldots \subset M_{-k} \subset \ldots \subset M_{-1} \subset M_0$$

and projections $e_{-k} \in M_{-k+1}$, $k \geq 1$.

Using that $e_k x e_k = e_k E_{M_{k-1}}(x) = E_{M_{k-1}}(x) e_k$ for $x \in M_k$ and $E_{M_k}(e_k) = [M:N]^{-1} 1$ one then checks that the projections e_k, $k \in \mathbb{Z}$, satisfy the relations

$$e_k e_{k\pm 1} e_k = [M:N]^{-1} e_k, \quad e_k e_j = e_j e_k \text{ if } |k - j| \geq 2.$$

Since $[M_{k+1}: M_k] = [M:N] < \infty$, by the discussion at the beginning of Sect. 10.2 we can inductively construct a representation of $\cup_k M_k$ on $L^2(M)$ extending the representation of M_1. This representation is not unique, but as soon as we fix a tunnel there exists a canonical one.

Lemma 10.4.1. *With the above notation there exists a unique extension of the representation of M_1 on $L^2(M)$ to a representation of $\cup_{k=1}^{\infty} M_k$ on $L^2(M)$ such that $e_k = J e_{-k} J$ for $k \geq 1$. In this representation $J M'_k J = M_{-k}$ for every $k \in \mathbb{Z}$.*

Proof. Since $M_{k+1} = \langle M_k, e_k \rangle$, the condition $e_k = J e_{-k} J$ completely determines the representation. We therefore only have to prove its existence. We shall do this by constructing the required representation of M_k by induction on k. For $k = 1$ we have a representation by definition, and $M_1 = J M'_{-1} J$. So assume the representation is well-defined for some k, and $M_j = J M'_{-j} J$ for $j = 1, \ldots, k$, where we identify M_j with its image under the representation.

Since $[M_{k+1}: M_k] < \infty$, the identity map $M_k \to M_k \subset B(L^2(M))$ extends to a representation $\pi: M_{k+1} \to B(L^2(M))$. Let $f = J\pi(e_k)J$. Denote $[M:N]^{-1}$ by λ. We assert that $f \in M_{-k+1}$ and $E_{M_{-k}}(f) = \lambda 1$. Indeed, since $e_k \in M'_{k-1} \cap M_{k+1}$, we have $\pi(e_k) \in M'_{k-1}$, so that $f \in J M'_{k-1} J = M_{-k+1}$. Then

$$E_{M_{-k}}(f) e_{-k+1} = e_{-k+1} f e_{-k+1} = J e_{k-1} J J \pi(e_k) J J e_{k-1} J$$
$$= J \pi(e_{k-1} e_k e_{k-1}) J = \lambda J e_{k-1} J = \lambda e_{-k+1},$$

so that $E_{M_{-k}}(f) = \lambda 1$. But then by Lemma 10.2.3 there exists a unitary $u \in M_{-k}$ such that $e_{-k} = u f u^* = J(J u J \pi(e_k) J u^* J) J$. Since $J u J \in J M_{-k} J = M'_k$, we thus see that $x \mapsto J u J \pi(x) J u^* J$ is a representation of M_{k+1} which coincides with π on M_k and maps e_k onto $J e_{-k} J$. Therefore it is the required representation of M_{k+1}.

Since $M_{-k-1} = \{e_{-k}\}' \cap M_{-k}$, we have

$$J M'_{-k-1} J = \langle J M'_{-k} J, J e_{-k} J \rangle = \langle M_k, e_k \rangle = M_{k+1},$$

which completes the proof of the induction step. □

Since $J M'_{-n} J = M_n$, it follows that $M_{-n} \subset M_0 \subset M_n$ is the basic construction for each $n \in \mathbb{N}$. Hence, more generally, $M_{k-n} \subset M_k \subset M_{k+n}$ is the basic construction for any $k \in \mathbb{Z}$ and $n \in \mathbb{N}$.

Consider now the AF-algebra A_∞ obtained by taking the inductive limit of the algebras $M'_k \cap M_l$ as $k \to -\infty$ and $l \to +\infty$ (note that $M'_k \cap M_l$ is finite dimensional by [214, Corollary XIX.2.9]). By the above lemma we get

10.4 The Canonical Shift

a **mirroring** γ_0, the involutive anti-automorphism of A_∞ defined by $\gamma_0(x) = Jx^*J$ for $x \in \cup_{k,l}(M'_k \cap M_l)$. It has the properties $\gamma_0(M'_k \cap M_l) = M'_{-l} \cap M_{-k}$ and $\gamma_0(e_k) = e_{-k}$.

On the other hand, we can consider

$$\ldots \subset M_k \subset M_{k+1} \subset \ldots$$

as the tower and the tunnel associated with the inclusion $M_0 \subset M_1$, and get an anti-automorphism γ_1 of A_∞ such that $\gamma_1(M'_k \cap M_l) = M'_{-l+2} \cap M_{-k+2}$ and $\gamma_1(e_k) = e_{-k+2}$. The **canonical shift** for the inclusion $N \subset M$ is the automorphism $\Gamma = \gamma_1 \circ \gamma_0$ of A_∞. It has the properties

$$\Gamma(M'_k \cap M_l) = M'_{k+2} \cap M_{l+2} \text{ and } \Gamma(e_k) = e_{k+2}.$$

Since each algebra M_k has a unique trace, the C*-algebra A_∞ has a canonical tracial state, which we continue to denote by τ. Our next goal is to show that Γ is τ-preserving. By [214, Proposition XIX.4.19] in the finite depth case, which we shall consider below, the trace τ is the unique tracial state on A_∞. Hence in that case it is γ_0- and γ_1-invariant. However, this is not true in general. To prove that nevertheless the trace is Γ-invariant we need to compare the constructed representations of $\cup_k M_k$ on $L^2(M_0)$ and $L^2(M_1)$. To simplify the notation we consider $\cup_k M_k$ as a subalgebra of $B(L^2(M_1))$, and denote by π the representation of $\cup_k M_k$ on $L^2(M_0)$.

Lemma 10.4.2. *If we identify $L^2(M_0)$ with $e_1 L^2(M_1)$, then the representation π satisfies*

$$\pi(x) = \lambda^{-k} e_1 \ldots e_k x e_{k+1} e_k \ldots e_1 \text{ for } x \in M_k, \ k \geq 1,$$

where $\lambda = [M:N]^{-1}$.

Proof. Using the relations $e_i e_{i-1} e_i = \lambda e_i$, $i = 1, \ldots, k$, we get

$$e_{k+1} e_k \ldots e_1 e_1 \ldots e_k e_{k+1} = \lambda^k e_{k+1}.$$

Since $e_{k+1} x = x e_{k+1}$ for $x \in M_k$, it follows that the expression in the formulation of the lemma defines a homomorphism. For $x \in M_0$ we have $\pi(x) = x e_1$. Since M_k is generated by M_0 and $e_0, \ldots, e_{k-1}$, all that remains to show is that the equality in the formulation holds for $x = e_0, \ldots, e_{k-1}$. In other words,

$$\pi(e_{k-1}) = e_{k+1} e_1 \text{ for } k \geq 2, \text{ and } \pi(e_0) = \lambda^{-1} e_1 e_0 e_2 e_1.$$

Denote by J_1 and J_0 the modular conjugations on $L^2(M_1)$ and $L^2(M_0)$, respectively. Since $J_0 = J_1|_{e_1 L^2(M_1)}$, for $k \geq 2$ we have

$$\pi(e_{k-1}) = J_0 \pi(e_{-k+1}) J_0 = J_1 e_{-k+1} e_1 J_1 = e_{k+1} e_1,$$

proving the first identity. To prove the second note that if ξ_1 is the cyclic trace vector in $L^2(M_1)$, then

$$(\lambda^{-1} e_1 e_0 e_2 e_1 \xi_1, \xi_1) = \lambda^{-1}(e_2 \xi_1, e_0 \xi_1) = \lambda^{-1}(J_1 e_0 J_1 \xi_1, e_0 \xi_1)$$
$$= \lambda^{-1}(e_0 \xi_1, e_0 \xi_1) = \lambda^{-1} \tau(e_0) = 1.$$

Since $\lambda^{-1} e_1 e_0 e_2 e_1$ is a projection, we conclude that $\xi_1 = \lambda^{-1} e_1 e_0 e_2 e_1 \xi_1$. Since $e_1 e_0 e_2 e_1 \in M'_{-1}$, it follows that the projection $[M_{-1} \xi_1]$ is majorized by $\lambda^{-1} e_1 e_0 e_2 e_1$. On the other hand, since M_1 is generated by M_0 and e_0, we have $e_0 M_1 e_0 = e_0 M_{-1}$. Hence

$$[M_{-1} \xi_1] = [M_{-1} e_1 e_2 e_0 e_1 \xi_1] = [e_1 e_2 e_0 M_{-1} \xi_1] = [e_1 e_2 e_0 M_1 e_0 \xi_1].$$

Since $e_2 \xi_1 = J_1 e_0 J_1 \xi_1 = e_0 \xi_1$ and $e_2 \in M'_1$, the above projection equals

$$[e_1 e_2 e_0 M_1 e_2 \xi_1] = [e_1 e_2 e_0 M_1 \xi_1] = [e_1 e_2 e_0 L^2(M_1)] \geq \lambda^{-1} e_1 e_0 e_2 e_1.$$

By definition of $e_0 = e_N$ we get $\pi(e_0) = [M_{-1} \xi_1] = \lambda^{-1} e_1 e_0 e_2 e_1$, completing the proof of the lemma. $\square$

Proposition 10.4.3. *The canonical shift Γ is trace preserving.*

Proof. We use the same notation as in the proof of the previous lemma. Since $x = \gamma_0(\gamma_1(\Gamma(x)))$, $\pi \circ \gamma_0 = J_0 \pi(\cdot)^* J_0$ and $J_0 = J_1|_{e_1 L^2(M_1)}$, for any $x \in A_\infty$ we have

$$\pi(x) = J_1 \pi(\gamma_1(\Gamma(x^*))) J_1.$$

If $x \in M'_k \cap M_l$ with $k \leq -1$ and $l \geq 1$, then $\gamma_1(\Gamma(x)) = \gamma_0(x) \in M'_{-l} \cap M_{-k}$. Using Lemma 10.4.2 and that $J_1 e_i J_1 = e_{-i+2}$, we conclude that the above expression equals

$$\lambda^k J_1 e_1 \ldots e_{-k} J_1 \Gamma(x) J_1 e_{-k+1} e_{-k} \ldots e_1 J_1 = \lambda^k e_1 e_0 \ldots e_{k+2} \Gamma(x) e_{k+1} \ldots e_0 e_1.$$

Thus

$$\lambda^k e_1 e_0 \ldots e_{k+2} \Gamma(x) e_{k+1} \ldots e_0 e_1 = \pi(x) = \lambda^{-l} e_1 \ldots e_l x e_{l+1} \ldots e_1. \quad (10.9)$$

Applying the trace to the left side and using that $e_i e_{i+1} e_i = \lambda e_i$ and $\Gamma(x) \in M'_{k+2} \cap M_{l+2}$ we get

$$\lambda^k \tau(\Gamma(x) e_{k+1} \ldots e_0 e_1 e_0 \ldots e_{k+1}) = \tau(\Gamma(x) e_{k+1}) = \lambda \tau(\Gamma(x)).$$

Similarly applying τ to the right side of (10.9) we get $\lambda \tau(x)$, so that $\tau(\Gamma(x)) = \tau(x)$. $\square$

Remark 10.4.4. If $x \in M'_k \cap M_l$ then multiplying (10.9) by $e_{k+1} \ldots e_0$ from the left and by $e_0 \ldots e_{k+1}$ from the right, we conclude that $\Gamma(x)$ is an element in $M'_{k+2} \cap M_{l+2}$ such that

$$\Gamma(x) e_{k+1} = \lambda^{k-l} e_{k+1} \ldots e_l x e_{l+1} \ldots e_{k+1}.$$

Since $M'_{k+2} \subset B(L^2(M_1))$ is a factor and $e_{k+1} \in M_{k+2}$, the homomorphism $M'_{k+2} \ni a \mapsto ae_{k+1}$ is faithful. Hence $\Gamma(x)$ is uniquely determined by the above identity. In particular, the restriction of Γ to $\overline{\cup_{n\geq 0}(N' \cap M_n)} \subset A_\infty$ is independent of the choice of the tunnel. Often it is this restriction which is called the canonical shift.

It follows also that if for some $n \in \mathbb{Z}$ we consider

$$\ldots \subset M_k \subset M_{k+1} \subset \ldots$$

as the tower and the tunnel for $M_n \subset M_{n+1}$ and define the corresponding mirrorings γ_n and γ_{n+1}, then $\Gamma = \gamma_{n+1} \circ \gamma_n$. ♦

To compute the entropy of Γ we shall use Theorem 10.3.1. We have to check that its assumptions are satisfied.

To show that $\{M' \cap M_{2n}\}_n$ is a generating sequence for Γ we prove the next general proposition.

Proposition 10.4.5. *Let (P, τ, σ) be a W^*-dynamical system, where τ is a normal tracial state. Let $\{P_n\}_n$ be an increasing sequence of finite dimensional C^*-subalgebras such that $\sigma(P_n) \subset P_{n+1}$ and P is generated by $\sigma^k(P_n)$, $k \in \mathbb{Z}$, $n \in \mathbb{N}$. Assume there exists a sequence $\{k_n\}_n$ of natural numbers such that*

(i) $\dfrac{k_n}{n} \to 1$ *as* $n \to \infty$;

(ii) *the algebras $\sigma^{mk_n}(P_n)$, $m \in \mathbb{N}$, are mutually τ-independent.*

Then $\{P_n\}_n$ is a generating sequence for σ.

Recall that we say that $A_1, \ldots, A_n$ are τ-independent if they mutually commute, and

$$\tau(a_1 \ldots a_n) = \tau(a_1) \ldots \tau(a_n)$$

for $a_i \in A_i$.

Proof of Proposition 10.4.5. Since $\sigma^k(P_l) \subset \sigma^{-[n/2]}(P_n)$ for $-[n/2] \leq k \leq [n/2] - l$, the algebra $\cup_n \sigma^{-[n/2]}(P_n)$ is dense in P. Hence, by the Kolmogorov-Sinai type theorem, Theorem 3.2.3,

$$h_\tau(\sigma) = \lim_{n \to \infty} h_\tau(P_n; \sigma).$$

Since $\sigma^i(P_k) \subset P_{k+i}$, for any $n \in \mathbb{N}$ we have

$$H_\tau(P_k, \sigma(P_k), \ldots, \sigma^n(P_k)) \leq H_\tau(P_{n+k}),$$

whence

$$h_\tau(P_k; \sigma) \leq \liminf_{n \to \infty} \frac{1}{n} H_\tau(P_n).$$

On the other hand, since the algebras $\sigma^{mk_n}(P_n)$, $m \in \mathbb{Z}$, are independent, we have

$$h_\tau(P_n; \sigma^{k_n}) = H_\tau(P_n).$$

Hence
$$h_\tau(\sigma) = \frac{1}{k_n} h_\tau(\sigma^{k_n}) \geq \frac{1}{k_n} h_\tau(P_n; \sigma^{k_n}) = \frac{1}{k_n} H_\tau(P_n).$$

Since $\dfrac{k_n}{n} \to 1$, it follows that

$$h_\tau(\sigma) \geq \limsup_{n \to \infty} \frac{1}{n} H_\tau(P_n).$$

Thus the limit $\lim_n \dfrac{1}{n} H_\tau(P_n)$ exists and equals $h_\tau(\sigma)$, that is, $\{P_n\}_n$ is a generating sequence. □

The proposition implies that $\{M' \cap M_{2n}\}_n$ is a generating sequence for the canonical shift Γ. Indeed, since $\Gamma^{mn}(M' \cap M_{2n}) = M'_{2mn} \cap M_{2(m+1)n}$, we see that $\Gamma^{mn}(M' \cap M_{2n})$ commutes with $M' \cap M_{2n}$ for any $m \geq 1$. Moreover, since M_{2n} is a factor, we have $\tau(xy) = \tau(x)\tau(y)$ for any $x \in M_{2n}$ and $y \in \cup_k(M'_{2n} \cap M_{2n+k})$. Thus the algebras $\Gamma^{mn}(M' \cap M_{2n})$, $m \in \mathbb{N}$, are independent, and we can apply the previous proposition with $k_n = n$.

Next note that the von Neumann algebra generated by $\cup_n(M' \cap M_{2n})$ in the GNS-representation of A_∞ defined by τ has type II_1, since already the von Neumann algebra generated by the projections e_n, $n \geq 1$, is a II_1-factor by [214, Theorem XIX.3.1].

The generating sequence $\{M' \cap M_{2n}\}_n$ satisfies the commuting square condition, that is,
$$\begin{array}{ccc} M' \cap M_{2n} & \subset & M' \cap M_{2n+2} \\ \cup & & \cup \\ M'_2 \cap M_{2n} & \subset & M'_2 \cap M_{2n+2} \end{array}$$
is a commuting square for any $n \in \mathbb{N}$. Indeed, the trace preserving conditional expectation $M' \cap M_{2n+2} \to M' \cap M_{2n}$ is the restriction of the conditional expectation $E_{M_{2n}}: M_{2n+2} \to M_{2n}$. But then we obviously have $E_{M_{2n}}(M'_2 \cap M_{2n+2}) \subset M'_2 \cap M_{2n}$.

Next we check that the entropy is finite.

Proposition 10.4.6. *We have*
$$h_\tau(\Gamma) \leq ht(\Gamma) \leq \log[M:N].$$

Proof. If Ω is a finite subset of $M' \cap M_k$, then $\Gamma^n(\Omega) \subset M' \cap M_{2n+k}$. It follows that
$$ht(\Omega; \Gamma) \leq \limsup_{n \to \infty} \frac{1}{n} \log \operatorname{rank} M' \cap M_{2n}.$$

By [214, Proposition XIX.2.8] if $p_1, \ldots, p_m$ are projections in $M' \cap M_{2n}$ with sum 1, then

$$\sum_i \frac{[p_i M_{2n} p_i : M p_i]}{\tau(p_i)} = [M_{2n} : M].$$

Hence $\sum_i \tau(p_i)^{-1} \leq [M_{2n}: M]$, so by the Cauchy-Schwarz inequality

$$m = \sum_{i=1}^{m} \tau(p_i)^{1/2} \tau(p_i)^{-1/2} \leq [M_{2n}:M]^{1/2}.$$

Therefore rank $M' \cap M_{2n} \leq [M_{2n}:M]^{1/2} = [M:N]^n$, and the proof of the proposition is complete. □

Finally, the multiplicities for the inclusion $M' \cap M_{n-1} \subset M' \cap M_n$ are bounded by $[M:N]$ (in fact, by $[M:N]^{1/2}$). To see this let p be a minimal projection in $M' \cap M_{n-1}$. Then $q = p e_n$ is a minimal projection in $M' \cap M_{n+1}$. Indeed, since $e_n M_{n+1} e_n = M_{n-1} e_n$, we have

$$e_n(M' \cap M_{n+1})e_n = (Me_n)' \cap e_n M_{n+1} e_n = (Me_n)' \cap M_{n-1}e_n = (M' \cap M_{n-1})e_n,$$

so that $q(M' \cap M_{n+1})q = p(M' \cap M_{n-1})pe_n = \mathbb{C}q$. Furthermore, if f is a projection in $M' \cap M_n$ and $f \leq p$, then $qfq = e_n f e_n = E_{M_{n-1}}(f)e_n \neq 0$. Therefore if p majorizes a minimal projection $f \in M' \cap M_n$ then the central support of f in $M' \cap M_{n+1}$ majorizes q. In other words, if $(a_{ij})_{i,j}$ and $(b_{jk})_{j,k}$ are the inclusion matrices for $M' \cap M_{n-1} \subset M' \cap M_n$ and $M' \cap M_n \subset M' \cap M_{n+1}$, respectively, i_0 corresponds p and k_0 corresponds to q, then $b_{jk_0} \neq 0$ as soon as $a_{i_0 j} \neq 0$. If $\{z_k\}_k$ is the set of minimal central projections in $M' \cap M_{n+1}$, then denoting by s_k the trace of a minimal projection majorized by z_k, we get

$$\tau(p) = \sum_{j,k} a_{i_0 j} b_{jk} s_k \geq a_{i_0 j} b_{jk_0} \tau(q) = [M:N]^{-1} a_{i_0 j} b_{jk_0} \tau(p).$$

It follows that $a_{i_0 j} \leq [M:N]$ (in fact, it is known that $a_{i_0 j} = b_{jk_0}$, so that $a_{i_0 j} \leq [M:N]^{1/2}$).

Since $\ldots \subset M_{2k-2} \subset M_{2k} \subset \ldots$ can be considered as the tower and the tunnel associated with $M_{-2n} \subset M_{-2n+2}$, we conclude that the multiplicities for the inclusion $M'_{-2n} \cap M_{-2} \subset M'_{-2n} \cap M$ are bounded by $[M:N]^2$. Since $\gamma_0(M'_2 \cap M_{2n}) = M'_{-2n} \cap M_{-2}$ and $\gamma_0(M' \cap M_{2n}) = M'_{-2n} \cap M$, it follows that the multiplicities for the inclusion $\Gamma(M' \cap M_{2n-2}) \subset M' \cap M_{2n}$ are also bounded by $[M:N]^2$.

We have thus checked that all the assumptions of Theorem 10.3.1 are satisfied, so we get our first and most general result on the entropy of Γ.

Theorem 10.4.7. *Let Γ be the canonical shift for the inclusion $N \subset M$ of II_1-factors of finite index. Let $R = \pi_\tau(\cup_{n=1}^\infty (M' \cap M_{2n}))''$. Then*

$$h_\tau(\Gamma) = \lim_{n \to \infty} \frac{1}{n} H_\tau(M' \cap M_{2n}) = \frac{1}{2} H_\tau(R|\Gamma(R)) + \lim_{n \to \infty} \frac{1}{2n} H_\tau(Z(M' \cap M_{2n})).$$

□

Combining this theorem and Proposition 10.4.6 we obtain the following estimates.

Corollary 10.4.8. *We have*

$$\frac{1}{2}H_\tau(R|\Gamma(R)) \leq h_\tau(\Gamma) \leq ht(\Gamma) \leq \log[M:N].$$

□

In particular, if $H_\tau(R|\Gamma(R)) = 2\log[M:N]$, then both the topological entropy and the dynamical entropy of Γ with respect to τ are equal to $\log[M:N]$. By [169, Theorem 5.3.1] the condition $H_\tau(R|\Gamma(R)) = 2\log[M:N]$ is one of several equivalent characterizations of extremality of $N \subset M$ and strong amenability of the standard invariant of the inclusion. The simplest case when this condition is satisfied, is that of finite depth inclusions.

A finite index inclusion $N \subset M$ of II_1-factors is said to have **finite depth** if

$$\sup_n \dim Z(M' \cap M_n) < \infty.$$

Then, by the discussion prior to [214, Proposition XIX.4.19], there exists n_0 such that if G denotes the matrix of the inclusion $M' \cap M_{2n_0} \subset M' \cap M_{2n_0+1}$, then the inclusions $M' \cap M_{2n} \subset M' \cap M_{2n+1}$ and $M' \cap M_{2n+1} \subset M' \cap M_{2n+2}$ are given by the matrices G and G^t, respectively, for any $n \geq n_0$. Moreover, the matrix GG^t is primitive, and if s denotes the vector formed by the values of the trace on minimal projections in $M' \cap M_{2n_0}$, then $GG^t s = [M:N]s$. We shall now show that this allows us to compute the entropy without relying on the deep results of Popa [169].

Recall that a matrix $B \in \mathrm{Mat}_r(\mathbb{R})$ with nonnegative coefficients is called primitive if there exists $m \in \mathbb{N}$ such that B^m has only positive entries. By the Perron-Frobenius theorem, see e.g. [193, Section 1.1], the spectral radius β of B is an eigenvalue of B, called the Perron-Frobenius eigenvalue, and the following conditions are satisfied:

(i) there is an eigenvector ξ of B with eigenvalue β whose coordinates are all positive; similarly, there is an eigenvector ζ of B^* with eigenvalue β and positive coordinates;

(ii) if ζ is normalized such that $(\xi, \zeta) = 1$ then $\beta^{-n} B^n \to P$ as $n \to +\infty$, where $P = (\cdot, \zeta)\xi$ is the projection onto $\mathbb{R}\xi$ along $\zeta^\perp$.

These properties imply $\cap_{n\in\mathbb{N}} B^n(\mathbb{R}^r_+) = \mathbb{R}_+\xi$. Indeed, assume $\vartheta_n \in \mathbb{R}^r_+$, $n \geq 0$, are such that $\vartheta_0 = \beta^{-n} B^n \vartheta_n$. Since ζ has positive coordinates, there exists $\varepsilon > 0$ such that $\|P\vartheta\| = (\vartheta, \zeta)\|\xi\| \geq \varepsilon \|\vartheta\|$ for $\vartheta \in \mathbb{R}^r_+$. Then if $\varepsilon' < \varepsilon$ and n is such that $\|\beta^{-n} B^n - P\| \leq \varepsilon'$, we get $\|\beta^{-n} B^n \vartheta\| \geq (\varepsilon - \varepsilon')\|\vartheta\|$. It follows that the sequence $\{\vartheta_n\}_n$ is bounded. But then the vectors $\beta^{-n} B^n \vartheta_n$ become arbitrarily close to $P\vartheta_n$ as $n \to \infty$. Hence $\vartheta_0 \in \mathbb{R}_+\xi$.

It follows that ξ is the unique (up to a scalar factor) eigenvector of B with nonnegative coordinates. In particular, if $B = GG^t$ corresponds to a finite depth inclusion $N \subset M$, then the Perron-Frobenius eigenvalue is $[M:N]$.

Proposition 10.4.9. *Let $A = \varinjlim A_n$ be an AF-algebra such that for every n the inclusion $A_n \subset A_{n+1}$ is defined by a fixed primitive matrix B. Then there exists a unique tracial state τ on A, and*

$$\lim_{n\to\infty} \frac{1}{n} H_\tau(A_n) = \log \beta,$$

where β is the Perron-Frobenius eigenvalue of B.

Proof. Let $A_n = \oplus_{k=1}^r \operatorname{Mat}_{m_k^n}(\mathbb{C})$. A tracial state τ on A is defined by a sequence of vectors $s^n \in \mathbb{R}_+^r$, $n \in \mathbb{N}$, such that $\sum_k s_k^1 m_k^1 = 1$ and $s^n = B s^{n+1}$. Let ξ be the Perron-Frobenius eigenvector of B normalized by the condition $\sum_k \xi_k m_k^1 = 1$. Then we can take $s^n = \beta^{-n+1} \xi$ and obtain a tracial state on A. Conversely, given a tracial state τ on A, we have $s^n \in \cap_m B^m(\mathbb{R}_+^r)$. Hence by the discussion before the formulation of the proposition the vector s^n is a scalar multiple of ξ for any n, and the scalar is completely determined by the condition $\tau(1) = 1$.

Then we have

$$H_\tau(A_n) = -\sum_{k=1}^r m_k^n s_k^n \log s_k^n = -\sum_{k=1}^r m_k^n s_k^n \log(\beta^{-n+1} \xi_k)$$

$$= (n-1) \log \beta - \sum_{k=1}^r m_k^n s_k^n \log \xi_k.$$

Since $\sum_k m_k^n s_k^n = 1$, dividing by n and letting $n \to \infty$ we get the result. □

Applying this proposition to $A_n = M' \cap M_{2n}$ for a finite depth inclusion, we then get

$$\lim_{n\to\infty} \frac{1}{n} H_\tau(M' \cap M_{2n}) = \log[M:N].$$

We also obviously have

$$\lim_{n\to\infty} \frac{1}{n} H_\tau(Z(M' \cap M_{2n})) = 0.$$

Thus combining Proposition 10.4.6 and Theorem 10.4.7 we obtain the following result.

Theorem 10.4.10. *Let $N \subset M$ be an inclusion of II_1-factors of finite index and finite depth. Let $R = \pi_\tau(\cup_{n=1}^\infty (M' \cap M_n))''$. Then*

$$\frac{1}{2} H_\tau(R|\Gamma(R)) = h_\tau(\Gamma) = ht(\Gamma) = \log[M:N].$$

□

Remark that by [214, Corollary XIX.4.9] any inclusion of index < 4 has finite depth.

10.5 Shifts on Temperley-Lieb Algebras

Let λ be a real number such that $\lambda^{-1} \in \{4\cos^2 \frac{\pi}{n+1} \mid n \geq 3\} \cup [4, \infty)$. Consider the universal C*-algebra A generated by a sequence of projections $\{e_k\}_{k \in \mathbb{Z}}$ such that
$$e_k e_{k\pm 1} e_k = \lambda e_k, \quad e_k e_j = e_j e_k \text{ if } |k-j| \geq 2.$$
As we saw in the previous section, a representation of this algebra arises naturally from an inclusion of II_1-factors with index λ^{-1}. We also saw that there exists a trace τ on A such that
$$\tau(we_k) = \lambda \tau(w) \tag{10.10}$$
for any $k \in \mathbb{Z}$ and any w in the algebra generated by projections e_j with $j < k$. For $n \geq 2$ denote by A_n the C*-algebra generated by $e_1, \ldots, e_{n-1}$, and set $A_0 = A_1 = \mathbb{C}1$. An easy induction argument shows that
$$A_n = A_{n-1} + A_{n-1} e_{n-1} A_{n-1}, \quad n \geq 2.$$
In particular, the trace is completely determined by (10.10).

The dimension and the trace vectors m^n and s^n of the algebras A_n, $n \in \mathbb{N}$, are described as follows, see §3 of Chapter XIX in [214] for a proof.

If $\lambda^{-1} \geq 4$ then
$$A_n \cong \bigoplus_{k=0}^{[n/2]} \text{Mat}_{m_k^n}(\mathbb{C}), \quad m_k^n = \binom{n}{k} - \binom{n}{k-1}, \quad s_k^n = \lambda^k P_{n-2k}(\lambda),$$
where $\{P_n\}_{n=0}^\infty$ is the sequence of polynomials such that $P_0 = P_1 = 1$,
$$P_{n+1}(t) = P_n(t) - t P_{n-1}(t), \quad n \geq 1.$$

On the other hand, if $\lambda^{-1} = 4\cos^2 \frac{\pi}{n+1}$ with $n \geq 3$ then for any $k \geq 0$ the embeddings $A_{n+2k-1} \hookrightarrow A_{n+2k+1}$ are given by a fixed primitive matrix with the Perron-Frobenius eigenvalue λ^{-1}.

Denote by θ_λ the automorphism of A defined by $\theta_\lambda(e_k) = e_{k+1}$.

Theorem 10.5.1. *We have:*
(i) $h_\tau(\theta_\lambda) = -\frac{1}{2}\log \lambda$, *if* $\lambda^{-1} \leq 4$;

(ii) $h_\tau(\theta_\lambda) = \eta(\alpha) + \eta(1-\alpha)$, *where* $\alpha = \dfrac{1+\sqrt{1-4\lambda}}{2}$, *if* $\lambda^{-1} \geq 4$.

Proof. The algebras $\theta_\lambda^{nm}(A_n)$, $m \in \mathbb{Z}$, are mutually τ-independent. By Proposition 10.4.5 it follows that $\{A_n\}_n$ is a generating sequence, that is,
$$h_\tau(\theta_\lambda) = \lim_{n \to \infty} \frac{1}{n} H_\tau(A_n).$$
If $\lambda^{-1} < 4$, then by Proposition 10.4.9

$$\lim_{n\to\infty} \frac{1}{n} H_\tau(A_{2n}) = \log \lambda.$$

Hence $h_\tau(\theta_\lambda) = -\frac{1}{2}\log \lambda$. It remains to consider the case $\lambda^{-1} \geq 4$.

Assume first that $\lambda^{-1} > 4$. Set

$$\alpha = \frac{1 + \sqrt{1-4\lambda}}{2} \quad \text{and} \quad \beta = 1 - \alpha = \frac{1 - \sqrt{1-4\lambda}}{2}.$$

Then using that $\alpha\beta = \lambda$ we check that

$$P_n(\lambda) = \frac{\alpha^{n+1} - \beta^{n+1}}{\alpha - \beta}, \quad n \geq 0.$$

Since $s_k^{2n} = \lambda^k P_{2n-2k}(\lambda) = \lambda^k \alpha^{2n-2k}(1 - (\beta/\alpha)^{2n-2k+1})(1 - \beta/\alpha)^{-1}$, we see that the difference

$$\log s_k^{2n} - \log(\lambda^k \alpha^{2n-2k}) = \log s_k^{2n} - ((2n-k)\log \alpha + k \log \beta)$$

is uniformly bounded. Since

$$H_\tau(A_{2n}) = -\sum_{k=0}^{n} m_k^{2n} s_k^{2n} \log s_k^{2n} \quad \text{and} \quad \sum_{k=0}^{n} m_k^{2n} s_k^{2n} = 1,$$

the limit of $H_\tau(A_{2n})/2n$ coincides with the limit of

$$-\frac{1}{2n}\sum_{k=0}^{n} m_k^{2n} s_k^{2n}((2n-k)\log \alpha + k\log \beta) = -\log \alpha + \log\left(\frac{\alpha}{\beta}\right) \sum_{k=0}^{n} m_k^{2n} s_k^{2n} \frac{k}{2n}. \tag{10.11}$$

Since $s_k^{2n} = (\alpha\beta)^k(\alpha^{2n-2k+1} - \beta^{2n-2k+1})/(\alpha - \beta)$, we have

$$\sum_{k=0}^{n} m_k^{2n} s_k^{2n} \frac{k}{2n} = \frac{\alpha}{\alpha - \beta} \sum_{k=0}^{n} m_k^{2n} \alpha^{2n-k}\beta^k \frac{k}{2n} - \frac{\beta}{\alpha - \beta} \sum_{k=0}^{n} m_k^{2n} \alpha^k \beta^{2n-k} \frac{k}{2n}.$$

As $\alpha > \beta$, $m_k^{2n} \leq \binom{2n}{k}$ and $4\alpha\beta = 4\lambda < 1$,

$$\sum_{k=0}^{n} m_k^{2n} \alpha^k \beta^{2n-k} \frac{k}{2n} \leq (\alpha\beta)^n \sum_{k=0}^{n} \binom{2n}{k} \leq (4\alpha\beta)^n \to 0.$$

On the other hand,

$$\sum_{k=0}^{n} \binom{2n}{k} \alpha^{2n-k} \beta^k \frac{k}{2n} = \sum_{k=1}^{n} \binom{2n-1}{k-1} \alpha^{2n-k} \beta^k$$

$$= \beta \sum_{k=0}^{n-1} \binom{2n-1}{k} \alpha^{2n-1-k} \beta^k \to \beta,$$

since $\sum_{k=n}^{2n-1} \binom{2n-1}{k} \alpha^{2n-1-k}\beta^k \leq \alpha^{-1}(\alpha\beta)^n \sum_{k=n}^{2n-1}\binom{2n-1}{k} \to 0$. Similarly

$$\lim_{n\to\infty}\sum_{k=0}^{n}\binom{2n}{k-1}\alpha^{2n-k}\beta^k \frac{k}{2n} = \lim_{n\to\infty}\sum_{k=1}^{n}\binom{2n}{k-1}\alpha^{2n-k}\beta^k \frac{k-1}{2n}$$

$$= \lim_{n\to\infty}\frac{\beta}{\alpha}\sum_{k=0}^{n-1}\binom{2n}{k}\alpha^{2n-k}\beta^k \frac{k}{2n} = \frac{\beta^2}{\alpha}.$$

Therefore, since $m_k^{2n} = \binom{2n}{k} - \binom{2n}{k-1}$,

$$\lim_{n\to\infty}\sum_{k=0}^{n} m_k^{2n} s_k^{2n} \frac{k}{2n} = \frac{\alpha}{\alpha-\beta}\lim_{n\to\infty}\sum_{k=0}^{n} m_k^{2n}\alpha^{2n-k}\beta^k \frac{k}{2n}$$

$$= \frac{\alpha}{\alpha-\beta}\left(\beta - \frac{\beta^2}{\alpha}\right) = \beta,$$

so that from (10.11) we get that

$$h_\tau(\theta_\lambda) = -\log\alpha + \log\left(\frac{\alpha}{\beta}\right)\beta = \eta(\alpha) + \eta(\beta).$$

Finally consider the case $\lambda^{-1} = 4$. Then

$$P_n(\lambda) = \frac{n+1}{2^n},$$

whence $s_k^{2n} = 2^{-2n}(2n-2k+1)$, and

$$\frac{1}{2n}H_\tau(A_{2n}) = -\frac{1}{2n}\sum_{k=0}^{n} m_k^{2n} s_k^{2n} \log s_k^{2n} = \log 2 - \sum_{k=0}^{n} m_k^{2n} s_k^{2n} \frac{\log(2n-2k+1)}{2n}.$$

Letting $n \to \infty$ we get $h_\tau(\theta_{1/4}) = \log 2$. $\square$

Denote by M the von Neumann subalgebra of $\pi_\tau(A)''$ generated by e_n, $n \leq -1$, and by R the von Neumann subalgebra generated by e_n, $n \geq 1$. By [214, Theorem XIX.3.1] both M and R are the hyperfinite II$_1$-factors, $N = \theta_\lambda^{-1}(M) \subset M$ is a subfactor of index λ^{-1}, and $N \subset M \subset \theta_\lambda(M) = \langle M, e_0\rangle$ is the basic construction. It follows that the automorphism θ_λ^2 of $\pi_\tau(A)''$ is (the extension to the weak operator closure of) the canonical shift associated with the inclusion $N \subset M$.

It is clear that $A_n \subset M' \cap M_n$ for any $n \geq 0$. It can be shown that if $\lambda^{-1} \leq 4$ then $A_n = M' \cap M_n$. For $\lambda^{-1} < 4$ the inclusion $N \subset M$ has finite depth, and from Theorem 10.4.10 we again obtain

$$h_\tau(\theta_\lambda) = \frac{1}{2}h_\tau(\theta_\lambda^2) = \frac{1}{2}\log[M:N] = -\frac{1}{2}\log\lambda.$$

On the other hand, if $\lambda^{-1} > 4$ then it can be shown that already $M' \cap M_1$ is different from $A_1 = \mathbb{C}1$. It is not difficult to check that the generating sequence $\{A_n\}_n$ satisfies the commuting square condition. Hence by Theorem 10.3.1

$$h_\tau(\theta_\lambda) = \frac{1}{2} H_\tau(R|\theta_\lambda(R)).$$

Therefore $H_\tau(R|\theta_\lambda(R)) = 2(\eta(\alpha) + \eta(\beta))$. This equality, as well as the entropy $h_\tau(\theta_\lambda)$, can also be obtained from the embedding $A \hookrightarrow \mathrm{Mat}_2(\mathbb{C})^{\otimes \mathbb{Z}}$,

$$e_k \mapsto (1-\lambda) e_{11}^k \otimes e_{22}^{k+1} + \lambda e_{22}^k \otimes e_{11}^{k+1} + \sqrt{\lambda(1-\lambda)}(e_{12}^k \otimes e_{21}^{k+1} + e_{21}^k \otimes e_{12}^{k+1}),$$

where $\{e_{ij}^k\}_{i,j=1}^2$ are the matrix units in the k-th factor $\mathrm{Mat}_2(\mathbb{C})$. It can be shown, see [162], that in the GNS-representation of $\mathrm{Mat}_2(\mathbb{C})^{\otimes \mathbb{Z}}$ corresponding to the product-state

$$\psi = \varphi^{\otimes \mathbb{Z}}, \quad Q_\varphi = \begin{pmatrix} \alpha & 0 \\ 0 & \beta \end{pmatrix},$$

the weak operator closure of A coincides with the centralizer of the state. Thus θ_λ is one of the Bernoulli shifts on the hyperfinite factor considered in Example 3.2.6(ii). Moreover, the von Neumann algebra generated by the projections e_n, $n \geq 1$, coincides with the centralizer of ψ on $\pi_\psi(\mathrm{Mat}_2(\mathbb{C})^{\otimes \mathbb{N}})''$. In particular, there exists a projection $p \in R$ which is mapped onto e_{11}^1 under the above embedding, and then

$$\tau(p) = \alpha, \quad p \in \theta_\lambda(R)' \cap R, \quad pRp = \theta_\lambda(R)p, \quad (1-p)R(1-p) = \theta_\lambda(R)(1-p).$$

It should be remarked that the projections e_n, $n \leq -1$, do not generate the centralizer of ψ on $\pi_\psi(\otimes_{n \leq 0} \mathrm{Mat}_2(\mathbb{C}))''$.

10.6 Notes

The notion of relative entropy for subalgebras of a von Neumann algebra with a normal tracial state was introduced by Connes and Størmer [51], who used it as a tool to prove continuity of mutual entropy. The definition was extended to arbitrary states by Connes [49]. To see that relative entropy is relevant for estimates of mutual entropy, observe that for finite dimensional C*-subalgebras $P_1, \ldots, P_n$ and $Q_1, \ldots, Q_n$ of M we have

$$H_\varphi(P_1, \ldots, P_n) \leq H_\varphi(Q_1, \ldots, Q_n) + \sum_{k=1}^n H_\varphi(P_k|Q_k).$$

It was Pimsner and Popa [162] who undertook a serious study of this notion and discovered its connection with Jones' index of II_1-factors. Sects. 10.1 and 10.2 are almost entirely based on their paper. For some examples of computations of relative entropy see [22], [103], [233]. The results of Pimsner and Popa were extended to type III factors by Hiai [86].

Theorem 10.3.1 was proved by Størmer [210] extending results of Choda [40] and Hiai [88]. See [81] for further discussion of generating sequences.

The canonical shift was introduced by Pimsner and Popa [162] and Ocneanu [146]. One usually defines the canonical shift a bit differently. As we remarked after the proof of Lemma 10.4.1, for any $n \in \mathbb{N}$ the algebras $M_{-1} \subset M_n \subset M_{2n+1}$ form a basic construction. Pimsner and Popa [163] found an explicit formula for the corresponding Jones projection in M_{2n+1}. In other words, they found an explicit representation of M_{2n+1} on $L^2(M_n)$ extending the representation of M_{n+1}. Using this representation we can define an antiautomorphism of $M'_{-1} \cap M_{2n+1}$, which for the moment we denote by γ'_n. Then one checks that $\gamma'_{n+2} \circ \gamma'_{n+1}$ and $\gamma'_{n+1} \circ \gamma'_n$ agree on $M_{-1} \subset M_n \subset M_{2n+1}$, so there exists a well-defined endomorphism of $\bigcup_n (M'_{-1} \cap M_{2n+1})$. To see that we get the same endomorphism as the one defined in the present chapter, by Remark 10.4.4 it suffices to check that $\gamma'_n = \gamma_n$ on $M'_{-1} \cap M_{2n+1}$, or even better, the representation of M_{2n+1} on $L^2(M_n)$ alluded to above coincides with the representation defined as in Lemma 10.4.1. Since these representations are determined by the images of $e_{n+1}, \ldots, e_{2n}$, and $\gamma_n(e_{n+k}) = e_{n-k}$ by construction, we just have to check that $\gamma'_n(e_{n+k}) = e_{n-k}$ for $k = 1, \ldots, n$. This is indeed the case, see e.g. [54]. Alternatively one can use the identity in Remark 10.4.4 and an explicit formula for the canonical shift, see e.g. [23].

The entropy of the shifts on the Temperley-Lieb algebras, Theorem 10.5.1, was computed by Pimsner and Popa [162] for all $\lambda \neq 1/4$. For the case $\lambda < 1/4$ instead of the elementary but tedious computations presented above they proved that the shifts are isomorphic to the Bernoulli shifts on the hyperfinite factor considered in Example 3.2.6(ii); see [191] for a more general result. The missing case $\lambda = 1/4$ was treated by Choda [40] and Yin [236]. The latter paper also contains the computation of the entropy for $\lambda < 1/4$, which we used.

For arbitrary subfactors the study of entropic properties of the canonical shift was initiated by Choda [41]. In particular, she obtained the inequalities from Corollary 10.4.8 and computed the entropy in the case of a finite depth subfactor, Theorem 10.4.10. The most general result, Theorem 10.4.7, was proved by Hiai [88]. For similar results in the type III case see [48], [87]. An argument similar to the proof of Proposition 10.4.5 appeared already in the paper of Connes and Størmer [51] and had been used by several authors until it was formalized by Choda [40], [41]. Proposition 10.4.9 was discovered by so many people that it should probably be considered a folklore. It was successfully applied by Choda [40], [41] to a number of models.

11
Systems of Algebraic Origin

Given an automorphism T of a discrete abelian group preserving a 2-cocycle, we define an automorphism α_T of the corresponding twisted group C*-algebra. If the cocycle is trivial, we get an abelian dynamical system defined by the dual automorphism $\hat T$ of $\hat G$. Dynamical properties of such automorphisms are well understood. In this chapter we shall study entropic properties of α_T for general cocycles.

11.1 Twisted Group C*-algebras

For a discrete abelian group G and a 2-cocycle $\omega \in Z^2(G;\mathbb{T})$ consider the twisted group C*-algebra $C^*(G,\omega)$. By definition it is the universal C*-algebra generated by unitaries u_g, $g \in G$, such that

$$u_g u_h = \omega(g,h) u_{g+h}.$$

We have considered a more general class of algebras in Chap. 8. In particular, we know that $C^*(G,\omega)$ is nuclear and the regular representation

$$C^*(G,\omega) \to B(\ell_2(G)), \quad u_g \delta_h = \omega(g,h)\delta_{g+h},$$

is faithful, where $\delta_g \in \ell_2(G)$ is defined by $\delta_g(h) = 0$ for $h \neq g$ and $\delta_g(g) = 1$. Recall also that $\omega(g,0) = \omega(0,g) = \omega(0,0)$ for any $g \in G$.

The vector δ_0 defines the canonical trace τ on $C^*(G,\omega)$ such that $\tau(u_g) = 0$ for $g \neq 0$. Since δ_0 is cyclic and the regular representation is faithful, the trace τ is faithful.

For any 2-cocycle ω the cocycle $\rho_\omega(g,h) = \omega(g,h)\overline{\omega(h,g)}$ is a **skew-symmetric bicharacter**, that is, $\rho_\omega(g,g) = 1$ and

$$\rho_\omega(g_1+g_2, h) = \rho_\omega(g_1,h)\rho_\omega(g_2,h), \quad \rho_\omega(h, g_1+g_2) = \rho_\omega(h,g_1)\rho_\omega(h,g_2).$$

Indeed, using that $u_g u_h = \rho_\omega(g,h) u_h u_g$ in $C^*(G,\omega)$ we get

$$u_{g_1+g_2}u_h = \rho_\omega(g_1+g_2,h)u_h u_{g_1+g_2},$$

and on the other hand

$$u_{g_1+g_2}u_h = \overline{\omega(g_1,g_2)}u_{g_1}u_{g_2}u_h = \overline{\omega(g_1,g_2)}\rho_\omega(g_1,h)\rho_\omega(g_2,h)u_h u_{g_1}u_{g_2}$$
$$= \rho_\omega(g_1,h)\rho_\omega(g_2,h)u_h u_{g_1+g_2},$$

so that $\rho_\omega(g_1+g_2,h) = \rho_\omega(g_1,h)\rho_\omega(g_2,h)$. Note also that any bicharacter is a 2-cocycle.

It is now easy to describe when the twisted group C*-algebra is simple.

Theorem 11.1.1. *For any $\omega \in Z^2(G;\mathbb{T})$ the following conditions are equivalent:*

(i) *the C*-algebra $C^*(G,\omega)$ is simple;*
(ii) *τ is the unique tracial state on $C^*(G,\omega)$;*
(iii) *the skew-symmetric bicharacter ρ_ω is nondegenerate.*

Recall that nondegeneracy of ρ_ω means by definition that for any $h \neq 0$ there exists g such that $\rho_\omega(g,h) \neq 1$. In other words, the subgroup of the dual group $\hat{G}$ formed by the characters $\rho_\omega(g,\cdot)$, $g \in G$, has trivial annihilator in G, so that it is dense in $\hat{G}$.

Proof of Theorem 11.1.1. Since $u_g u_h u_g^* = \rho_\omega(g,h)u_h$, if ρ_ω is nondegenerate then $\varphi(u_h) = 0$ for every tracial state φ. Thus (iii)$\Rightarrow$(ii).

If $h \neq 0$ is in the kernel of ρ_ω, so that $\rho_\omega(g,h) = 1$ for any $g \in G$, then u_h belongs to the center of $C^*(G,\omega)$. Hence $C^*(G,\omega)$ is nonsimple. Moreover, then $\pi_\tau(C^*(G,\omega))''$ has nontrivial center and consequently possesses a normal trace different from τ. Thus (i)$\Rightarrow$(iii) and (ii)$\Rightarrow$(iii).

It remains to prove (iii)$\Rightarrow$(i). Consider the dual action α of $\hat{G}$ on $C^*(G,\omega)$, so

$$\alpha_\chi(u_h) = \langle \chi, h \rangle u_h \quad \text{for} \quad \chi \in \hat{G},\ h \in G.$$

Since ρ_ω is nondegenerate, any character of G can be approximated in the topology of pointwise convergence by characters $\rho_\omega(g,\cdot)$. Since $(\operatorname{Ad} u_g)(u_h) = \rho_\omega(g,h)u_h$, we conclude that the automorphism α_χ is approximately inner for any $\chi \in \hat{G}$ in the sense that it can be approximated by inner automorphisms in the pointwise norm topology. Hence any ideal in $C^*(G,\omega)$ is α_χ-invariant. Since

$$\tau(\cdot)1 = \int_{\hat{G}} \alpha_\chi(\cdot)d\chi,$$

it follows that if $a \geq 0$ lies in an ideal I, then either $1 \in I$ or $\tau(a) = 0$, and hence $a = 0$, as τ is faithful. Therefore $C^*(G,\omega)$ is simple. □

In particular, if G is countable and ρ_ω is nondegenerate, then the weak operator closure $W^*(G,\omega) = \pi_\tau(C^*(G,\omega))''$ of $C^*(G,\omega)$ is a II$_1$-factor. This factor is hyperfinite since $C^*(G,\omega)$ is nuclear.

If ρ_ω is degenerate, the argument in the proof of Theorem 11.1.1 essentially gives a description of the ideal structure of $C^*(G,\omega)$. The following statement will be sufficient for our purposes.

Lemma 11.1.2. *Let H be the kernel of ρ_ω, that is, H consists of the elements $h \in G$ such that $\rho_\omega(g,h) = 1$ for every $g \in G$. Then the center of $C^*(G,\omega)$ is generated by the elements u_h, $h \in H$, so it is isomorphic to $C^*(H, \omega|_H)$.*

Note that it follows from Lemma 11.1.3 below that $C^*(H, \omega|_H)$ is isomorphic to $C(\hat{H})$, but the isomorphism is noncanonical.

Proof of Lemma 11.1.2. It is clear that $C^*(H, \omega|_H)$ is contained in the center. On the other hand, the map E defined by

$$E(x) = \int_{\widehat{G/H}} \alpha_\chi(x) d\chi$$

is a conditional expectation onto $C^*(H, \omega|_H)$. Since $(\operatorname{Ad} u_g)(u_h) = \rho_\omega(g,h)u_h$, and the characters $\rho_\omega(g, \cdot)$, $g \in G$, form a dense subgroup of $\widehat{G/H}$, the automorphism α_χ is approximately inner for any $\chi \in \widehat{G/H}$. Hence $E(x) = x$ for any element x in the center, so the center is contained in $C^*(H, \omega|_H)$. □

Finally remark that the map $\omega \mapsto \rho_\omega$ defines an embedding of $H^2(G; \mathbb{T})$ into the group of skew-symmetric bicharacters of G.

Lemma 11.1.3. *If $\rho_\omega(g,h) = 1$ for every $g,h \in G$, then ω is a coboundary, that is, there exists a function $G \ni g \mapsto z_g \in \mathbb{T}$ such that $\omega(g,h) = z_g z_h \bar{z}_{g+h}$.*

Proof. By assumption the C*-algebra $C^*(G,\omega)$ is abelian. Choose a character f on $C^*(G,\omega)$. Then applying it to the identity $u_g u_h = \omega(g,h) u_{g+h}$ we see that ω is the coboundary of $G \ni g \mapsto f(u_g)$. □

Notice that in the notation of the lemma the elements $v_g = \bar{z}_g u_g$ satisfy the relations $v_g v_h = v_{g+h}$, so that $C^*(G,\omega) \cong C^*(G) \cong C(\hat{G})$.

11.2 Estimates of Topological Entropy

Let T be an automorphism of a discrete countable abelian group G. If a cocycle $\omega \in Z^2(G; \mathbb{T})$ is T-invariant, then T defines an automorphism α_T of $C^*(G,\omega)$ such that $\alpha_T(u_g) = u_{Tg}$. We shall write $\alpha_{T,\omega}$ for α_T and $u_{g,\omega}$ for u_g when we want to stress that we consider an automorphism of $C^*(G,\omega)$.

Consider the dual automorphism $\hat{T}$ of $\hat{G}$, so that $\langle \hat{T}\chi, g \rangle = \langle \chi, T^{-1}g \rangle$. Then $\hat{T}$ is a homeomorphism of the compact space $\hat{G}$.

Theorem 11.2.1. *For the automorphism α_T of $C^*(G,\omega)$ we have*

$$\frac{1}{2} h_{top}(\hat{T}) \leq ht(\alpha_T) \leq h_{top}(\hat{T}).$$

If $\omega \equiv 1$ then α_T is the automorphism of $C(\hat{G})$ defined by $\hat{T}$, so that $ht(\alpha_T) = h_{top}(\hat{T})$. Thus the upper bound in the theorem is optimal. In Example 11.2.7 below we shall see that the lower bound is also optimal.

Proof of the lower bound in Theorem 11.2.1. Let $\bar{\omega}$ be the cocycle defined by $\bar{\omega}(g,h) = \overline{\omega(g,h)}$. The elements $u_{g,\omega} \otimes u_{g,\bar{\omega}}$, $g \in G$, generate a C*-subalgebra D of $C^*(G,\omega) \otimes C^*(G,\bar{\omega})$, which is isomorphic to $C^*(G) \cong C(\hat{G})$. Under this isomorphism the automorphism $\alpha_{T,\omega} \otimes \alpha_{T,\bar{\omega}}$ becomes the automorphism of $C(\hat{G})$ defined by $\hat{T}$. Using that the topological entropy is subadditive and monotone we then get

$$ht(\alpha_{T,\omega}) + ht(\alpha_{T,\bar{\omega}}) \geq ht(\alpha_{T,\omega} \otimes \alpha_{T,\bar{\omega}}) \geq ht((\alpha_{T,\omega} \otimes \alpha_{T,\bar{\omega}})|_D) = h_{top}(\hat{T}).$$

Thus we just have to prove that $ht(\alpha_{T,\omega}) = ht(\alpha_{T,\bar{\omega}})$.

Assume first that $\omega(g,h) = \overline{\omega(h,g)}$ for any $g, h \in G$. Then $C^*(G,\bar{\omega})$ can be identified with the opposite algebra $C^*(G,\omega)^{op}$, since for any $g, h \in G$ we have $u_{h,\bar{\omega}} u_{g,\bar{\omega}} = \bar{\omega}(h,g) u_{g+h,\bar{\omega}} = \omega(g,h) u_{g+h,\bar{\omega}}$. Hence it suffices to prove that if α is an automorphism of a C*-algebra and we consider it as an automorphism α^{op} of the opposite algebra, then $ht(\alpha) = ht(\alpha^{op})$. By A.1 a map $\gamma: A \to B$ is completely positive if and only if $\sum_{i,j}^n b_i^* \gamma(a_i^* a_j) b_j \geq 0$ for any n and $a_1, \ldots, a_n \in A$, $b_1, \ldots, b_n \in B$. It follows that any completely positive map $A \to B$ remains completely positive being considered as a map $A^{op} \to B^{op}$. Consequently, if Ω is a finite subset of a C*-algebra A, then $rcp_A(\Omega, \delta) = rcp_{A^{op}}(\Omega, \delta)$. Hence $ht(\alpha) = ht(\alpha^{op})$.

In the general case consider the cocycle $\omega^{op}(g,h) = \omega(h,g)$, so that $C^*(G,\omega^{op})$ can be identified with $C^*(G,\omega)^{op}$. Then it is enough to show that $ht(\alpha_{T,\omega^{op}}) = ht(\alpha_{T,\bar{\omega}})$. Since $\rho_{\omega^{op}} = \rho_{\bar{\omega}}$, by Lemma 11.1.3 the cocycles ω^{op} and $\bar{\omega}$ are cohomologous, so that there exists a function $G \ni g \mapsto z_g \in \mathbb{T}$ such that

$$\bar{\omega}(g,h) = \omega^{op}(g,h) z_g z_h \bar{z}_{g+h}.$$

We thus get an isomorphism $\gamma: C^*(G,\bar{\omega}) \to C^*(G,\omega^{op})$ defined by $u_{g,\bar{\omega}} \mapsto z_g u_{g,\omega^{op}}$. Even though this isomorphism is not equivariant, the elements $\alpha_{T,\bar{\omega}}(u_{g,\bar{\omega}})$ and $(\gamma^{-1} \circ \alpha_{T,\omega^{op}} \circ \gamma)(u_{g,\bar{\omega}})$ differ only by a scalar factor of modulus one. It follows that for any finite set Ω consisting of elements $u_{g,\bar{\omega}}$ we have

$$ht(\Omega, \delta; \alpha_{T,\bar{\omega}}) = ht(\Omega, \delta; \gamma^{-1} \circ \alpha_{T,\omega^{op}} \circ \gamma).$$

Hence $ht(\alpha_{T,\bar{\omega}}) = ht(\gamma^{-1} \circ \alpha_{T,\omega^{op}} \circ \gamma) = ht(\alpha_{T,\omega^{op}})$, and the proof of the lower bound for $ht(\alpha_T)$ is complete. □

To obtain the upper bound we need a way to compute $h_{top}(\hat{T})$ by looking at the action of T on G.

For a finite set $F \subset G$ let

$$h(F;T) = \lim_{n \to \infty} \frac{1}{n} \log |F + T(F) + \ldots + T^{n-1}(F)|.$$

Then define $h(T) = \sup_{F \subset G} h(F;T)$.

11.2 Estimates of Topological Entropy

Lemma 11.2.2. *We have $ht(\alpha_T) \leq h(T)$.*

Proof. To estimate $ht(\alpha_T)$ we shall use the maps Φ_f and Ψ_f from Chap. 8. Let X be a finite subset of G. Fix $\delta > 0$. Since G is amenable, there exists a finitely supported positive function f on G such that $\|_g f - f\|_1 < \delta^2$ for $g \in X$, where $_g f(h) = f(h-g)$, and $\|f\|_1 = 1$. Let F be the support of f. For $n \in \mathbb{N}$ consider the function

$$f^{(n)} = f * (f \circ T^{-1}) * \ldots * (f \circ T^{-n+1}).$$

Notice that for any $h \in G$ and $0 \leq k \leq n-1$ we can write

$$_h(f^{(n)}) = f * \ldots * (f \circ T^{-k+1}) *\, _h(f \circ T^{-k}) * (f \circ T^{-k-1}) * \ldots * (f \circ T^{-n+1}).$$

Then for $g \in X$ and $0 \leq k \leq n-1$ we have

$$\|_{T^k g}(f^{(n)}) - f^{(n)}\|_1 \leq \|f\|_1 \ldots \|_{T^k g}(f \circ T^{-k}) - f \circ T^{-k}\|_1 \ldots \|f \circ T^{-n+1}\|_1$$
$$= \|(_g f) \circ T^{-k} - f \circ T^{-k}\|_1 = \|_g f - f\|_1 < \delta^2.$$

Let $\Omega = \{u_g \mid g \in X\}$. Using the maps $\Phi_{f^{(n)}}$ and $\Psi_{f^{(n)}}$ from Chap. 8 we then get by Proposition 8.1.3(iii) that

$$\mathrm{rcp}(\Omega \cup \alpha_T(\Omega) \cup \ldots \cup \alpha_T^{n-1}(\Omega), \delta) \leq |\mathrm{supp}\, f^{(n)}|.$$

Since the support of $f^{(n)}$ is contained in $F + T(F) + \ldots + T^{n-1}(F)$, we thus obtain $ht(\Omega, \delta; \alpha_T) \leq h(F;T)$. Hence $ht(\alpha_T) \leq h(T)$. □

To finish the proof of Theorem 11.2.1 it is now sufficient to prove the following.

Theorem 11.2.3. *We have $h(T) = h_{top}(\hat{T})$.*

For this we need yet another formula for the entropy of automorphisms of compact groups.

Theorem 11.2.4. *Let S be a continuous automorphism of a compact separable group Γ. Then*

$$h_{top}(S) = h_\mu(S) = \sup_V \limsup_{n \to \infty} -\frac{1}{n} \log \mu(V \cap S^{-1}(V) \cap \ldots \cap S^{-n+1}(V)),$$

where μ is the Haar measure on Γ and the supremum is taken over all neighbourhoods V of the unit element $e \in \Gamma$.

Proof. Since $h_\mu(S) \leq h_{top}(S)$, it suffices to prove that

$$h_{top}(S) \leq \sup_V \limsup_{n \to \infty} -\frac{1}{n} \log \mu(V \cap S^{-1}(V) \cap \ldots \cap S^{-n+1}(V)) \leq h_\mu(S). \quad (11.1)$$

Choose a left Γ-invariant metric d on Γ. For a fixed $\varepsilon > 0$ let $V = B_{\varepsilon/2}(e)$, the open ball of radius $\varepsilon/2$ with center at e. If x and y are (n, ε)-separated with respect to S, then the sets $x(V \cap S^{-1}(V) \cap \ldots \cap S^{-n+1}(V))$ and $y(V \cap S^{-1}(V) \cap \ldots \cap S^{-n+1}(V))$ are disjoint. Indeed, otherwise for each $k = 0, \ldots, n-1$ we could find $a, b \in V$ such that $xS^{-k}(a) = yS^{-k}(b)$, or equivalently, $S^k(x)a = S^k(y)b$. But then $d(S^k(x), S^k(y)) < \varepsilon$, so that x and y are not (n, ε)-separated. Hence if $sr_n(\varepsilon)$ is the maximal number of (n, ε)-separated points, then

$$sr_n(\varepsilon) \mu(V \cap S^{-1}(V) \cap \ldots \cap S^{-n+1}(V)) \leq 1,$$

which implies the first inequality in (11.1).

To prove the second inequality fix a neighbourhood V of e. Let $\varepsilon > 0$ be such that $B_\varepsilon(e) \subset V$. Choose a measurable partition $\xi = \{X_1, \ldots, X_m\}$ such that the diameter of each atom is smaller than ε. If $x \in X_{i_0} \cap S^{-1}(X_{i_1}) \cap \ldots \cap S^{-n+1}(X_{i_{n-1}})$, then $X_{i_k} \subset B_\varepsilon(S^k(x)) \subset S^k(x)V$ for $k = 0, \ldots, n-1$, so that $X_{i_0} \cap S^{-1}(X_{i_1}) \cap \ldots \cap S^{-n+1}(X_{i_{n-1}}) \subset x(V \cap S^{-1}(V) \cap \ldots \cap S^{-n+1}(V))$ and hence

$$\mu(X_{i_0} \cap S^{-1}(X_{i_1}) \cap \ldots \cap S^{-n+1}(X_{i_{n-1}})) \leq \mu(V \cap S^{-1}(V) \cap \ldots \cap S^{-n+1}(V)).$$

Taking the logarithm, multiplying by $\mu(X_{i_0} \cap S^{-1}(X_{i_1}) \cap \ldots \cap S^{-n+1}(X_{i_{n-1}}))$ and summing over all multi-indices $(i_0, \ldots, i_{n-1})$ we get

$$H_\mu(\xi \vee S^{-1}\xi \vee \ldots \vee S^{-n+1}\xi) \geq -\log \mu(V \cap S^{-1}(V) \cap \ldots \cap S^{-n+1}(V)).$$

Dividing by n and letting $n \to \infty$ we get the second inequality in (11.1). $\square$

Note that instead of the supremum over all neighbourhoods we can take the limit along a basis of neighbourhoods.

Before we return to the computation of the entropy of α_T, we shall illustrate this theorem by a computation of the entropy of toral automorphisms.

Example 11.2.5. Let $T \in \mathrm{GL}_n(\mathbb{Z})$. Then T defines an automorphism S_T of the n-dimensional torus $\mathbb{T}^n \cong \mathbb{R}^n/\mathbb{Z}^n$. Then

$$h_{top}(S_T) = \sum_{i:|\lambda_i|>1} \log |\lambda_i|,$$

where $\mathrm{Sp}\, T = \{\lambda_1, \ldots, \lambda_n\}$ (counting with multiplicities).

To see this consider the quotient map $\pi \colon \mathbb{R}^n \to \mathbb{T}^n$. Let U be a relatively compact neighbourhood of $0 \in \mathbb{R}^n$ such that π is injective on $U \cup T(U)$. Let $V = \pi(U)$. If $x \in U$ is such that $\pi(T(x)) = S_T(\pi(x)) \in V$, then $T(x) \in U$. It follows that

$$V \cap S_T^{-1}(V) \cap \ldots \cap S_T^{-n+1}(V) = \pi(U \cap T^{-1}(U) \cap \ldots \cap T^{-n+1}(U)).$$

Hence

$$h_{top}(S_T) = \lim_U \limsup_{n\to\infty} -\frac{1}{n}\log\mu(U \cap T^{-1}(U) \cap \ldots \cap T^{-n+1}(U)),$$

where μ is now the Lebesgue measure on $\mathbb{R}^n$. Since for any two relatively compact neighbourhoods U_1, U_2 of $0 \in \mathbb{R}^n$ we have $c^{-1}U_1 \subset U_2 \subset cU_1$ for some $c > 0$, the lim sup above is actually the same for all U's. Replacing $\mathbb{R}^n$ by $\mathbb{C}^n$, which simply doubles the lim sup, we may assume that T is an operator on a complex vector space. By considering the normal Jordan form of T, we can reduce the computation to the case when $T \in \mathrm{GL}_m(\mathbb{C})$ has only one eigenvalue λ. If we write T in the normal Jordan form then the matrix STS^{-1} can be made arbitrarily close to the matrix $\lambda 1$ by a suitable choice of a diagonal matrix S. In other words, for any $\varepsilon > 0$, we can introduce a norm on $\mathbb{C}^m$ such that $\|T^{\pm 1} - \lambda^{\pm 1}1\| < \varepsilon$. Take the unit ball in this norm for U. Then $T^{-k}(U)$ is contained in the ball of radius $(|\lambda|^{-1} + \varepsilon)^k$ and contains the ball of radius $(|\lambda| + \varepsilon)^{-k}$. So if $|\lambda| + \varepsilon \leq 1$ then $U \subset T^{-1}(U)$ and the lim sup is zero, and if $|\lambda| + \varepsilon \geq 1$ then it lies between $-2m\log(|\lambda|^{-1} + \varepsilon)$ and $2m\log(|\lambda| + \varepsilon)$. It follows that the lim sup is zero when $|\lambda| < 1$, and it is $2m\log|\lambda|$ when $|\lambda| \geq 1$. ♦

We can now prove one half of Theorem 11.2.3.

Lemma 11.2.6. *We have $h(T) \leq h_{top}(\hat{T})$.*

Proof. It is clear that $h(F; T)$ is nondecreasing in F. Thus in proving that $h(F; T) \leq h_{top}(\hat{T})$ we may assume that $F = -F$ and $0 \in F$. For $m \in \mathbb{N}$ set

$$F^{(m)} = \underbrace{F + \ldots + F}_{m}.$$

By replacing, if necessary, F by a larger set we may also assume that

$$\frac{|F^{(m)}|}{|F^{(m+1)}|} \to 1 \quad \text{as} \quad m \to \infty. \tag{11.2}$$

Indeed, consider the group generated by F. It is a finite product $G_1 \times \ldots \times G_n$ of cyclic groups. We may then assume that F is of the form $F_1 \times \ldots \times F_n$. Thus it suffices to prove the claim for cyclic groups. But then it is obvious: if the group is finite, replace F by the group itself, and if the group is isomorphic to $\mathbb{Z}$, take a sufficiently large symmetric interval.

We assert now that for any $n, m \in \mathbb{N}$ we have

$$|F + T(F) + \ldots + T^{n-1}(F)|$$
$$\leq \frac{|F^{(m+1)}|^{2n}}{(\mathbb{1}_{F^{(m)}} * \mathbb{1}_{F^{(m)}} * \mathbb{1}_{T(F^{(m)})} * \mathbb{1}_{T(F^{(m)})} * \ldots * \mathbb{1}_{T^{n-1}(F^{(m)})} * \mathbb{1}_{T^{n-1}(F^{(m)})})(0)}, \tag{11.3}$$

where $\mathbb{1}_X$ denotes the characteristic function of a set $X \subset G$. Indeed, denote by N the denominator in the above expression. It is exactly the number of elements $(g_1, \ldots, g_{2n}) \in F^{(m)} \times F^{(m)} \times \ldots \times T^{n-1}(F^{(m)}) \times T^{n-1}(F^{(m)})$ such that $g_1 + \ldots + g_{2n} = 0$. For every such $(g_1, \ldots, g_{2n})$ and an element $h = h_1 + \ldots + h_n \in F + T(F) + \ldots + T^{n-1}(F)$ consider the element

$$(h_1 + g_1, g_2, \ldots, h_n + g_{2n-1}, g_{2n})$$

of $F^{(m+1)} \times F^{(m+1)} \times \ldots \times T^{n-1}(F^{(m+1)}) \times T^{n-1}(F^{(m+1)})$. This way we get N different elements representing h, which proves (11.3).

Fix m and consider the function $f = |F^{(m)}|^{-2} \mathbf{1}_{F^{(m)}} * \mathbf{1}_{F^{(m)}}$ on G. Then taking the logarithm of the both sides of (11.3) and dividing by n we get

$$\frac{1}{n} \log |F + T(F) + \ldots + T^{n-1}(F)|$$

$$\leq -\frac{1}{n} \log(f * (f \circ T^{-1}) * \ldots * (f \circ T^{-n+1}))(0) + 2 \log \frac{|F^{(m+1)}|}{|F^{(m)}|}. \quad (11.4)$$

Denote by φ the Fourier transform of f, so $\varphi(\chi) = \sum_g \langle \chi, g \rangle f(g)$. Then, since $F^{(m)} = -F^{(m)}$, φ is a positive continuous function on $\hat{G}$ such that $\varphi(e) = \sum_g f(g) = 1$. Fix $\varepsilon > 0$ and put

$$V = \{\chi \in \hat{G} \,|\, \varphi(\chi) > 1 - \varepsilon\}.$$

Then V is an open neighbourhood of the unit $e \in \hat{G}$, and

$$(f * (f \circ T^{-1}) * \ldots * (f \circ T^{-n+1}))(0)$$

$$= \int_{\hat{G}} \varphi(\chi)(\varphi \circ \hat{T}^{-1})(\chi) \ldots (\varphi \circ \hat{T}^{-n+1})(\chi) d\mu(\chi) \geq (1 - \varepsilon)^n \mu \left(\bigcap_{k=0}^{n-1} \hat{T}^k(V) \right).$$

This and (11.4) then imply that

$$\frac{1}{n} \log |F + T(F) + \ldots + T^{n-1}(F)|$$

$$\leq -\frac{1}{n} \log \mu \left(\bigcap_{k=0}^{n-1} \hat{T}^k(V) \right) - \log(1 - \varepsilon) + 2 \log \frac{|F^{(m+1)}|}{|F^{(m)}|},$$

whence by Theorem 11.2.4

$$h(F; T) \leq h_{top}(\hat{T}) - \log(1 - \varepsilon) + 2 \log \frac{|F^{(m+1)}|}{|F^{(m)}|}.$$

In view of (11.2) and since ε could be taken arbitrarily small, we get $h(F; T) \leq h_{top}(\hat{T})$. □

By virtue of Lemma 11.2.2 the previous lemma completes the proof of Theorem 11.2.1.

11.2 Estimates of Topological Entropy

Proof of Theorem 11.2.3. By Lemmas 11.2.2 and 11.2.6 we have

$$ht(\alpha_T) \leq h(T) \leq h_{top}(\hat{T}).$$

But if $\omega \equiv 1$ then α_T is the automorphism of $C(\hat{G})$ defined by $\hat{T}$, so that $ht(\alpha_T) = h_{top}(\hat{T})$. Hence $h(T) = h_{top}(\hat{T})$. □

Examples of nontrivial invariant cocycles can be obtained as follows.

Let $R = \mathbb{Z}[t, t^{-1}]$ be the ring of Laurent polynomials over $\mathbb{Z}$. For $g \in R$ define $\tilde{g}(t) = g(t^{-1})$. Let $f \in \mathbb{Z}[t]$, $f \neq 1$, be a polynomial with nonzero constant term and no cyclotomic factors, that is, its roots are not roots of unity. Consider the abelian group

$$G = R/(\tilde{f}) \oplus R/(f).$$

Let T be the automorphism of G of multiplication by $t \in R$. Then for any character χ of the abelian group $R/(f)$ we have a pairing

$$(g_1, g_2) \mapsto \chi(\tilde{g}_1 g_2) \qquad (11.5)$$

of $R/(\tilde{f})$ with $R/(f)$, and a T-invariant bicharacter ω on G defined by

$$\omega((g_1, g_2), (h_1, h_2)) = \chi(\tilde{h}_1 g_2).$$

The corresponding bicharacter ρ_ω is then

$$\rho_\omega((g_1, g_2), (h_1, h_2)) = \chi(\tilde{h}_1 g_2 - \tilde{g}_1 h_2).$$

This bicharacter is nondegenerate if and only if the pairing (11.5) is nondegenerate. Since $\tilde{g}_1 g_2 = \tilde{\tilde{g}_2 g_1}$, it suffices to check nondegeneracy in the second variable. Let S be the automorphism of $\widehat{R/(f)}$, which is dual to the automorphism of multiplication by t. Then the right hand side of (11.5) can be written as $\chi(g_1(T^{-1})g_2) = \langle g_1(S)\chi, g_2 \rangle$. Hence (11.5) is nondegenerate if and only if the group generated by the S-orbit of χ is dense in $\widehat{R/(f)}$. By Lemma 11.3.2 below and our assumption on f, the automorphism S is ergodic, so the orbit of χ is dense for almost every χ.

Note that if the leading coefficient and the constant term of f are equal to 1, then $G \cong \mathbb{Z}^{2n}$, where n is the degree of f. Then by Example 11.2.5, if $\lambda_1, \ldots, \lambda_n$ are the roots of f, we have

$$h_{top}(\hat{T}) = 2 \sum_{j:|\lambda_j|>1} \log|\lambda_j| = 2 \int_0^1 \log|f(e^{2\pi is})|ds,$$

where the second equality follows from the fact that $\int_0^1 \log|\lambda - e^{2\pi is}|ds$ is zero if $|\lambda| \leq 1$, and it is $\log|\lambda|$ if $|\lambda| \geq 1$. The number $m(f) = \int_0^1 \log|f(e^{2\pi is})|ds$ is called the logarithmic Mahler measure of f, and the formula $h_{top}(S) = m(f)$ holds in fact for arbitrary f. It follows from Theorem 11.2.1 that

$$m(f) \leq ht(\alpha_T) \leq 2m(f).$$

Note also that if we denote $R/(\tilde{f})$ by G_1 and $R/(f)$ by G_2, then the pairing (11.5) defines a homomorphism $G_2 \to \hat{G}_1$. Using this homomorphism we can define the action by right translations of G_2 on $\hat{G}_1$. Then $C^*(G, \omega) \cong C(\hat{G}_1) \rtimes G_2$. In particular $C(\hat{G}_1)$ is a subalgebra of $C^*(G, \omega)$, which gives an alternative way of obtaining the lower bound for $ht(\alpha_T)$.

Example 11.2.7. Let $f = 2$. Then

$$G = R/(2) \oplus R/(2) \cong \bigoplus_{n \in \mathbb{Z}} (\mathbb{Z}/2\mathbb{Z} \oplus \mathbb{Z}/2\mathbb{Z}),$$

T is the shift to the right. We have $h_{top}(\hat{T}) = 2 \log 2$, so that

$$\log 2 \leq ht(\alpha_T) \leq 2 \log 2.$$

A character on $\oplus_{n \in \mathbb{Z}} \mathbb{Z}/2\mathbb{Z}$ is just a sequence $\chi = (\chi_n)_n$ of numbers $\chi_n = \pm 1$. Consider the simplest nontrivial character, namely, $\chi_0 = -1$ and $\chi_n = 1$ for $n \neq 0$. In other words, if we consider χ as a character on $R = \mathbb{Z}[t, t^{-1}]$ then $\chi(1) = -1$ and $\chi(t^n) = 1$ for $n \in \mathbb{Z} \backslash \{0\}$. Then the pairing (11.5) is

$$\bigoplus_{n \in \mathbb{Z}} (\mathbb{Z}/2\mathbb{Z} \oplus \mathbb{Z}/2\mathbb{Z}) \ni (g_1, g_2) \mapsto \prod_{n \in \mathbb{Z}} (-1)^{g_1(n) g_2(n)}.$$

It follows that not only the bicharacter ρ_ω is nondegenerate, but its restriction to the subgroup

$$G^{(n)} = \bigoplus_{i=1}^{n} (\mathbb{Z}/2\mathbb{Z} \oplus \mathbb{Z}/2\mathbb{Z}) \subset G$$

is nondegenerate for any $n \in \mathbb{N}$. It follows by Theorem 11.1.1 that the C*-subalgebra $C^*(G^{(n)}, \omega|_{G^{(n)}})$ of $C^*(G, \omega)$ is a simple C*-algebra of dimension 4^n, hence it is isomorphic to $\mathrm{Mat}_{2^n}(\mathbb{C})$. But then

$$ht(\alpha_T) \leq \limsup_{n \to \infty} \frac{1}{n} \log \mathrm{rank}\, C^*(G^{(n)}, \omega|_{G^{(n)}}) = \log 2.$$

Thus $ht(\alpha_T) = \log 2$. This example shows that the lower bound in Theorem 11.2.1 is optimal. ◆

11.3 K-systems

Keeping the notation of the previous section, it is known that if the dual automorphism $\hat{T}$ is ergodic then $\hat{T}$, considered as a transformation preserving the Haar measure, is a Bernoulli shift, see Notes at the end of the chapter. Although we cannot prove such a result for the automorphism α_T of $C^*(G, \omega)$, we shall show that if the cocycle ω is in a sense asymptotically trivial then the powers of α_T can be approximated by shifts on UHF-algebras. As we shall see, this implies that we get an entropic K-system, in particular $h_\tau(\alpha_T) > 0$. Notice that since by Theorem 11.2.4 the topological entropy of $\hat{T}$ coincides with the measure entropy with respect to the Haar measure μ on $\hat{G}$, Theorem 11.2.1 implies that

$$h_\tau(\alpha_T) \le h_\mu(\hat{T}) = h_{top}(\hat{T}).$$

However, since the dynamical entropy is superadditive we can not obtain a lower bound for $h_\tau(\alpha_T)$ analogous to that in Theorem 11.2.1.

We start with the following classical characterization of ergodicity. Extend α_T to $W^*(G,\omega) = \pi_\tau(C^*(G,\omega))''$ and continue to denote the extended automorphism by α_T.

Lemma 11.3.1. *The following conditions are equivalent:*

(i) *the automorphism α_T of $W^*(G,\omega)$ is ergodic;*

(ii) *the automorphism T of G is aperiodic, that is, the orbit of any nonzero element is infinite.*

Proof. If T is not aperiodic, then there exist $n \in \mathbb{N}$ and $g \ne 0$ such that $T^n g = g$. Then the element $x = \sum_{k=0}^{n-1} \alpha_T^k(u_g)$ of $W^*(G,\omega)$ is α_T-invariant and not a scalar, since the vector

$$x\delta_0 = \sum_{k=0}^{n-1} \omega(0,0)\delta_{T^k g} \in \ell_2(G)$$

is nonzero and orthogonal to δ_0. Thus α_T is not ergodic.

Conversely, assume that T is aperiodic. The automorphism α_T is implemented by the unitary U on $\ell_2(G)$ defined by $U\delta_g = \delta_{Tg}$. To prove ergodicity of α_T it suffices to show that if $U\xi = \xi$ then ξ is a scalar multiple of the trace vector δ_0. This is indeed the case, since if $\xi = \sum_g a_g \delta_g$ then $a_{T^n g} = a_g$ for every $n \in \mathbb{Z}$ and $g \in G$, which is possible only when $a_g = 0$ for every $g \ne 0$. □

In particular, $\hat{T}$ is ergodic if and only if T is aperiodic. As we already mentioned, then $\hat{T}$ is a Bernoulli shift. At the level of G the Bernoullicity manifests itself in the following property of T.

Proposition 11.3.2. *Let G be a discrete abelian group, T an aperiodic endomorphism of G, Y a finite subset of G containing $0 \in G$. Then there exists $m \in \mathbb{N}$ such that whenever $y_1,\ldots,y_n \in Y$ and $m_1,\ldots,m_n \in \mathbb{N}$ satisfy $m_{k+1} - m_k \ge m$, $k = 1,\ldots,n-1$, then*

$$\sum_{k=1}^{n} T^{m_k} y_k = 0 \qquad (11.6)$$

only if $y_1 = \ldots = y_n = 0$.

Proof. First consider the case when G is finitely generated. Then the periodic part of G is finite. Since T acts on it aperiodically, the periodic part is trivial, so $G \cong \mathbb{Z}^N$ for some $N \in \mathbb{N}$. Then T is defined by a nondegenerate matrix with integral coefficients, which we denote by the same letter T. It is clear that the aperiodicity is equivalent to the fact that T has no roots of unity as eigenvalues.

Let P be the spectral projection of T corresponding to the part of the spectrum lying outside the closed unit disc. Thus $V_0 = \operatorname{Ker} P \subset \mathbb{C}^N$ is the sum of the root spaces of T corresponding to the eigenvalues lying in the closed unit disc, and $V_1 = P(\mathbb{C}^N)$ is the sum of the root spaces corresponding to the eigenvalues lying outside the closed unit disc. We claim that $\mathbb{Z}^N \cap V_0 = 0$. Indeed, if W is the linear span of $\mathbb{Z}^N \cap V_0 \cong \mathbb{Z}^l$, then $T|_W$ is represented by an l by l matrix with integral coefficients, all of whose eigenvalues lie in the unit disc, hence (if $l \neq 0$) on the unit circle, since the determinant is a nonzero integer. But then the eigenvalues are roots of unity by the Dirichlet theorem, see e.g. [178, 10.1.B], which contradicts our assumptions. It follows that $W = 0$.

It follows that (11.6) is equivalent to

$$\sum_{k=1}^{n} T^{m_k} P y_k = 0. \tag{11.7}$$

Since the eigenvalues of $T|_{V_1}$ have absolute values greater than 1, writing $T|_{V_1}$ in the normal Jordan form we conclude that there exist $\delta > 0$ and $C > 0$ such that

$$\|T^{-k}|_{V_1}\| \leq C(1-\delta)^k \quad \text{for all } k \in \mathbb{N}.$$

Choose $C_1 > 0$ such that $C_1^{-1} \leq \|Py\| \leq C_1$ for $y \in Y \setminus \{0\}$. Now choose $m \in \mathbb{N}$ such that

$$\sum_{k=m}^{\infty} CC_1(1-\delta)^k < C_1^{-1}.$$

Then if $m_{k+1} - m_k \geq m$, we have

$$\left\| \sum_{k=1}^{n-1} T^{m_k - m_n} P y_k \right\| \leq \sum_{k=1}^{n-1} CC_1(1-\delta)^{m_n - m_k} < C_1^{-1}.$$

So if (11.7) holds then $\|Py_n\| < C_1^{-1}$, and consequently $y_n = 0$ and $\sum_{k=1}^{n-1} T^{m_k} P y_k = 0$. Thus we recursively get $y_n = y_{n-1} = \ldots = y_1 = 0$.

Consider now the general case and prove the proposition by induction on $|Y|$. Let H_0 be the group generated by $T^k(Y)$, $k = 0, 1, \ldots, \infty$. Set $H_k = T^k(H_0)$, $H_\infty = \cap_k H_k$, $Y' = Y \cap H_\infty$.

Suppose $Y' \neq Y$. Since Y is finite, there exists $l \in \mathbb{N}$ such that $Y' = Y \cap H_l$. We claim that if $\sum_{k=1}^{n} T^{m_k} y_k \in H_\infty$ and $m_{k+1} - m_k \geq l$, then $y_1, \ldots, y_n \in Y'$.

If $g \in H_\infty \subset T(H_0)$, then $g = Th$ for a unique element h, since $\operatorname{Ker} T = 0$. As $g \in T^k(H_0)$, we conclude that $h \in T^{k-1}(H_0)$ for any $k \in \mathbb{N}$, so that

$h \in H_\infty$. In other words, T defines an automorphism of H_∞. It follows that $\sum_{k=1}^n T^{m_k - m_1} y_k \in H_\infty$. Hence $y_1 \in H_\infty + \sum_{k=2}^n T^{m_k - m_1} y_k$, so $y_1 \in Y \cap H_l = Y' \subset H_\infty$ and therefore $\sum_{k=2}^n T^{m_k} y_k \in H_\infty$. Continuing this process we get that $y_1, \ldots, y_n \in Y'$.

In particular, if (11.6) holds with $m_{k+1} - m_k \geq l$, then $y_1, \ldots, y_n \in Y'$. Since $|Y'| < |Y|$, we may then apply the inductive assumption.

On the other hand, if $Y' = Y$, then $Y \subset H_1$. Hence there exists $l \in \mathbb{N}$ such that if H is the group generated by $Y, T(Y), \ldots, T^l(Y)$, then $Y \subset T(H)$, so that $H \subset T(H)$. Then H is a finitely generated group, T^{-1} is a well-defined aperiodic endomorphism of H, and (11.6) can be written as $\sum_{k=1}^n (T^{-1})^{m_n - m_k} y_k = 0$. Thus we can apply the case of a finitely generated group to the endomorphism T^{-1} of H, the elements $y_n, \ldots, y_1 \in Y$ and the numbers $0, m_n - m_{n-1}, \ldots, m_n - m_1$. □

Using this proposition we shall show that under an additional assumption on the cocycle the powers of α_T can be approximated by shifts on UHF-algebras.

Theorem 11.3.3. *Let T be an aperiodic automorphism of a discrete abelian group G, $\omega \in Z^2(G; \mathbb{T})$ a T-invariant cocycle such that*

$$\sum_{n=1}^\infty |1 - \rho_\omega(T^n g, h)| < \infty \text{ for any } g, h \in G.$$

Then for any finite subset $\Omega \subset C^*(G, \omega)$ and $\varepsilon > 0$ there exist a full matrix algebra B and $n_0 \in \mathbb{N}$ such that for any $n \geq n_0$ there exist trace preserving unital completely positive maps $\Phi \colon C^*(G, \omega) \to A = B^{\otimes \mathbb{Z}}$ and $\Psi \colon A \to C^*(G, \omega)$ such that

(i) $\alpha \circ \Phi = \Phi \circ \alpha_T^n$ *and* $\alpha_T^n \circ \Psi = \Psi \circ \alpha$, *where α is the shift to the right on A;*

(ii) $\Phi(\Omega) \subset_\varepsilon B_0$ *and* $\|(\Psi \circ \Phi)(a) - a\| < \varepsilon$ *for any $a \in \Omega$, where $B_0 \cong B$ is the 0-th factor of A.*

Before we turn to the proof, let us show that this theorem implies a strong form of positivity of entropy. Recall that according to Definition 4.3.1 we say that a W*-dynamical system (M, φ, α) with $M \neq \mathbb{C}1$ and faithful φ is an entropic K-system if $h_\varphi(\gamma; \alpha^n) \to H_\varphi(\gamma)$ as $n \to \infty$ for any channel γ.

Corollary 11.3.4. *Under the assumptions of Theorem 11.3.3 the system $(W^*(G, \omega), \tau, \alpha_T)$ is an entropic K-system (assuming $G \neq 0$). In particular, $h_\tau(\alpha_T) > 0$.*

Proof. Let $\gamma \colon C \to W^*(G, \omega)$ be a channel. By definition $H_\tau(\gamma)$ is the supremum of $H_\tau(\gamma; \{x_i\})$ over all finite partitions of unity $1 = \sum_i x_i$ in $W^*(G, \omega)$. Any partition of unity in $W^*(G, \omega)$ can be approximated in the strong operator topology by a partition of unity in $C^*(G, \omega)$ (this follows from A.10, since a partition of unity can be considered as a unital completely positive map from an abelian algebra into $W^*(G,\omega)$). Thus it suffices to consider partitions of unity in $C^*(G,\omega)$.

Fix a partition $\Omega = \{x_i\}_i$ such that $H_\tau(\gamma;\{x_i\})$ is δ-close to $H_\tau(\gamma)$ for some $\delta > 0$. Then fix $\varepsilon > 0$ and choose n_0 and B according to Theorem 11.3.3. Let $n \geq n_0$ and let Φ and Ψ be the corresponding maps. Let $E \colon A \to B_0$ be a conditional expectation. Put $b_i = E(\Phi(x_i))$, and then for $k \in \mathbb{N}$ let

$$y_{i_1\ldots i_k} = \Psi(b_{i_1}\alpha(b_{i_2})\ldots \alpha^{k-1}(b_{i_k})).$$

Thus for each k we get a partition of unity in $C^*(G,\omega)$. Since Ψ is assumed to be trace preserving, and the unique trace on A is the product trace, we have

$$\tau(y_{i_1\ldots i_k}) = \tau(y_{i_1})\ldots\tau(y_{i_k}).$$

Since also $y_{i_l}^{(l)} = \Psi(\alpha^{l-1}(b_{i_l})) = \alpha_T^{(l-1)n}(y_{i_l})$ for $l = 1,\ldots,k$, by definition, see (3.3), we obtain

$$H_\tau(\gamma, \alpha_T^n \circ \gamma, \ldots, \alpha_T^{(k-1)n} \circ \gamma; \{y_{i_1\ldots i_k}\}) = kH_\tau(\gamma;\{y_i\}).$$

Hence $h_\tau(\gamma;\alpha_T^n) \geq H_\tau(\gamma;\{y_i\})$. Since $\|b_i - \Phi(x_i)\| \leq \varepsilon$ by assumption on Φ, we also have $\|y_i - x_i\| \leq 2\varepsilon$. Thus we could choose ε such that $H_\tau(\gamma;\{y_i\})$ is δ-close to $H_\tau(\gamma;\{x_i\})$. Then $h_\tau(\gamma;\alpha_T^n)$ is (2δ)-close to $H_\tau(\gamma)$ for any $n \geq n_0$. Hence $h_\tau(\gamma;\alpha_T^n) \to H_\tau(\gamma)$ as $n \to \infty$. □

Remark that by Theorem 4.3.5 the convergence $\rho_\omega(T^n g, h) \to 1$ as $n \to \infty$ is a necessary condition for $(W^*(G,\omega), \tau, \alpha_T)$ to be a K-system.

Proof of Theorem 11.3.3. Without loss of generality we may assume that Ω consists of the elements u_g for g in a finite subset of G. By Proposition 8.1.3(iii) we can find a finite subset $X \subset G$ such that the maps $\Phi_X \colon\colon C^*(G,\omega) \to B(\ell_2(X))$ and $\Psi_X \colon B(\ell_2(X)) \to C^*(G,\omega)$ from Chap. 8 corresponding to the function $f = 1_X$ satisfy

$$\|(\Psi_X \circ \Phi_X)(a) - a\| < \frac{\varepsilon}{3} \quad \text{for } a \in \Omega. \tag{11.8}$$

Set $X' = X \cup \{0\}$ and $Y = X' + X' - X' - X'$, and let m be chosen in accordance with Proposition 11.3.2. Fix $n \geq m$. For $k,l \in \mathbb{Z}$, $k \leq l$, let $X_{k,l} = \sum_{i=k}^{l} T^{in}(X)$. Denote by $\Phi_{k,l}$ the restriction of the compression map $B(\ell_2(G)) \to B(\ell_2(X_{k,l}))$ to $C^*(G,\omega)$. So for $k = l = 0$ we get the map Φ_X. By our choice of m we have a unitary isomorphism

$$U_{k,l} \colon \overset{l}{\underset{i=k}{\otimes}} \ell_2(X) \to \ell_2(X_{k,l}), \quad \delta_{x_k} \otimes \ldots \otimes \delta_{x_l} \mapsto \alpha_T^{kn}(u_{x_k})\ldots\alpha_T^{ln}(u_{x_l})\delta_0.$$

(Note that for $k = l = 0$ this is not the identity map, but the multiplication by $\omega(0,0)$.) Letting $B = B(\ell_2(X))$ we get an isomorphism $\text{Ad}\, U_{k,l}$ of $\otimes_{i=k}^{l} B$ onto $B(\ell_2(X_{k,l}))$, so we can consider $\Phi_{k,l}$ as a map

$$C^*(G,\omega) \to \overset{l}{\underset{i=k}{\otimes}} B.$$

More explicitly, let $g \in G$ and $x_k,\ldots,x_l \in X$. Then if $g + \sum_{i=k}^{l} T^{in}x_i$ does not belong to $X_{k,l}$, we have

$$\Phi_{k,l}(u_g)(\delta_{x_k} \otimes \ldots \otimes \delta_{x_l}) = 0.$$

On the other hand, if $g = \sum_{i=k}^{l} T^{in}(y_i - x_i)$ for some $y_k,\ldots,y_l \in X$, which are uniquely determined by our choice of m, then

$$u_g u_{T^{kn}x_k} \ldots u_{T^{ln}x_l} = z u_{T^{kn}y_k} \ldots u_{T^{ln}y_l} \tag{11.9}$$

for some $z \in \mathbb{T}$, so that

$$U_{k,l}\Phi_{k,l}(u_g)(\delta_{x_k} \otimes \ldots \otimes \delta_{x_l}) = z U_{k,l}(\delta_{y_k} \otimes \ldots \otimes \delta_{y_l}).$$

For $x \in X$, multiplying (11.9) by $u_{T^{(l+1)n}x}$ from the right we get

$$\Phi_{k,l+1}(u_g)(\delta_{x_k} \otimes \ldots \otimes \delta_{x_l} \otimes \delta_x) = z(\delta_{y_k} \otimes \ldots \otimes \delta_{y_l} \otimes \delta_x),$$

and multiplying (11.9) by $u_{T^{(k-1)n}x}$ from the left we get

$$\Phi_{k-1,l}(u_g)(\delta_x \otimes \delta_{x_k} \otimes \ldots \otimes \delta_{x_l}) = z\rho_\omega(g, T^{(k-1)n}x)(\delta_x \otimes \delta_{y_k} \otimes \ldots \otimes \delta_{y_l}).$$

Hence $\Phi_{k,l+1}(u_g) = \Phi_{k,l}(u_g) \otimes 1$ and

$$\Phi_{k-1,l}(u_g) = \left(\sum_{x \in X} \rho_\omega(g, T^{(k-1)n}x) e_{xx}\right) \otimes \Phi_{k,l}(u_g), \tag{11.10}$$

where $\{e_{xy}\}_{x,y \in X}$ are the matrix units in $B(\ell_2(X))$. Therefore

$$\|\Phi_{k-1,l}(u_g) - 1 \otimes \Phi_{k,l}(u_g)\| \leq \max_{x \in X}|1 - \rho_\omega(g, T^{(k-1)n}x)|$$

and $\Phi_{k,l+1}(u_g) = \Phi_{k,l}(u_g) \otimes 1$ if g belongs to $X_{k,l} - X_{k,l}$, and $\Phi_{k,l}(u_g) = 0$ if $g \notin X_{k,l} - X_{k,l}$. Thus if we consider $\otimes_{i=k}^{l} B$ as a subalgebra of $A = B^{\otimes \mathbb{Z}}$, we see that for every k the sequence $\{\Phi_{k,l}(u_g)\}_{l=k}^{+\infty}$ of elements in A stabilizes. So letting $l \to +\infty$ we get unital completely positive maps $\Phi_k \colon C^*(G,\omega) \to A$. Moreover, $\Phi_k(u_g) = 0$ if $g \notin \sum_{i=k}^{\infty} T^{in}(X-X)$, and otherwise

$$\|\Phi_{k-1}(u_g) - \Phi_k(u_g)\| \leq \max_{x \in X}|1 - \rho_\omega(g, T^{(k-1)n}x)|.$$

By the assumption of the theorem we conclude that the elements $\Phi_k(u_g)$ converge as $k \to -\infty$. Hence as $k \to -\infty$ we get a unital completely positive map $\Phi \colon C^*(G,\omega) \to A$. Furthermore, since $\Phi_X = \Phi_{0,0}$, if we identify B with the 0-th factor B_0 of A then for any $g \in X - X$ we have

200   11 Systems of Algebraic Origin

$$\|\Phi(u_g) - \Phi_X(u_g)\| \le \sum_{k=1}^{\infty} \max_{x \in X} |1 - \rho_\omega(g, T^{-kn}x)| = \sum_{k=1}^{\infty} \max_{x \in X} |1 - \rho_\omega(T^{kn}g, x)|.$$

In particular, choosing n to be sufficiently large we may assume that

$$\|\Phi(a) - \Phi_X(a)\| < \frac{\varepsilon}{3} \quad \text{for } a \in \Omega. \tag{11.11}$$

Applying α_T^n to (11.9) we see that $\Phi_{k+1,l+1} \circ \alpha_T^n = \alpha \circ \Phi_{k,l}$. Hence $\Phi \circ \alpha_T^n = \alpha \circ \Phi$. As $\text{Tr}(\Phi_{k,l}(u_g)) = 0$ for any $g \ne 0$, we also conclude that Φ is trace preserving.

Consider now the map $\Psi_{k,l} \colon \otimes_{i=k}^{l} B \to C^*(G, \omega)$,

$$e_{x_k y_k} \otimes \ldots \otimes e_{x_l y_l} \mapsto |X|^{-l+k-1} \alpha_T^{kn}(u_{x_k}) \ldots \alpha_T^{ln}(u_{x_l}) \alpha_T^{ln}(u_{y_l})^* \ldots \alpha_T^{kn}(u_{y_k})^*,$$

which is a unital completely positive map by the same reasoning as in A.10. The matrix unit $e_{st} \in B_0$, considered as an element of $\otimes_{i=k}^{l} B$ for $k \le 0$ and $l \ge 0$, can be written as

$$\sum_{x_k, \ldots, x_{-1}, x_1, \ldots, x_l} e_{x_k x_k} \otimes \ldots \otimes e_{x_{-1} x_{-1}} \otimes e_{st} \otimes e_{x_1 x_1} \otimes \ldots \otimes e_{x_l x_l},$$

so that

$$\Psi_{k,l}(e_{st}) = \frac{1}{|X|^{-k+1}} \sum_{x_k, \ldots, x_{-1}} u_{T^{kn} x_k} \ldots u_{T^{-n} x_{-1}} u_s u_t^* u_{T^{-n} x_{-1}}^* \ldots u_{T^{kn} x_k}^*$$

$$= \prod_{i=k}^{-1} \left(\frac{1}{|X|} \sum_{x \in X} \rho_\omega(T^{in} x, s - t) \right) \frac{u_s u_t^*}{|X|}.$$

We see that as $k \to -\infty$ and $l \to +\infty$ the elements $\Psi_{k,l}(e_{st})$ converge in norm. A similar computation shows that the elements $\Psi_{k,l}(b)$ converge for any $b \in \otimes_{i=k'}^{l'} B \subset A$. Thus we get a unital completely positive map $\Psi \colon A \to C^*(G, \omega)$ such that for $e_{st} \in B_0$

$$\Psi(e_{st}) = \prod_{k=1}^{\infty} \left(\frac{1}{|X|} \sum_{x \in X} \rho_\omega(T^{kn}(t-s), x) \right) \Psi_{0,0}(e_{st}).$$

Since $\Psi_X = \Psi_{0,0}$, for n sufficiently large we have

$$\|\Psi|_{B_0} - \Psi_X\| < \frac{\varepsilon}{3}. \tag{11.12}$$

Since $\Psi_{k+1,l+1} \circ \alpha = \alpha_T^n \circ \Psi_{k,l}$, we have $\Psi \circ \alpha = \alpha_T^n \circ \Psi$. As each map $\Psi_{k,l}$ is trace preserving, we also conclude that Ψ is trace preserving.

Inequalities (11.8), (11.11) and (11.12) show then that $\Phi(\Omega) \subset_{\varepsilon/3} B_0$ and

$$\|(\Psi \circ \Phi)(a) - a\| < \varepsilon \quad \text{for } a \in \Omega.$$

Thus the proof of the theorem is complete. $\qquad\square$

11.4 Zero Entropy Systems

In the previous section we saw that if the cocycle ω is asymptotically trivial, then similarly to the classical case the automorphism α_T is a K-automorphism, in particular it has positive entropy. In the present section we shall first show that if ω is sufficiently chaotic on orbits of T, then the entropy is zero.

Recall that by Definition 5.2.3 a system (A, α) is called asymptotically highly anticommutative if there exists a set $\{x_i\}_{i \in I}$ of elements of A which together with the unit spans a dense subspace of A and satisfies the following property. For any $i \in I$, $\varepsilon > 0$ and $n \in \mathbb{N}$ there exist $m_1, \ldots, m_n \in \mathbb{N}$ such that $\|\alpha^{m_k}(x_i^*)\alpha^{m_l}(x_i) + \alpha^{m_l}(x_i)\alpha^{m_k}(x_i^*)\| < \varepsilon$ for all $k, l \leq n$ with $k \neq l$.

Theorem 11.4.1. *Let T be an aperiodic automorphism of G, $\omega \in Z^2(G; \mathbb{T})$ a T-invariant cocycle. Assume that for any $g \neq 0$ there exists a closed subgroup $\Gamma_g \subset \mathbb{T}$ containing -1 such that for any $h \neq 0$ in the subgroup generated by the T-orbit of g the sequence $\{\rho_\omega(T^n g, h)\}_{n=1}^\infty$ is uniformly distributed on Γ_g. Then the system $(C^*(G, \omega), \alpha_T)$ is asymptotically highly anticommutative. In particular, $h_\tau(\alpha_T) = 0$.*

Recall that a sequence $\{g_n\}_n$ in a compact group Γ is called uniformly distributed if
$$\frac{1}{n} \sum_{k=0}^n f(g_k) \to \int_\Gamma f(g) d\mu(g)$$
for any continuous function f on Γ, where μ is the Haar measure on Γ. If Γ is abelian then it is enough to check the convergence on characters of Γ.

The closed subgroups of $\mathbb{T}$ are either finite cyclic or the whole group $\mathbb{T}$. In our examples the group Γ_g will be either $\{\pm 1\}$ or $\mathbb{T}$.

Proof of Theorem 11.4.1. Since $u_g u_h = \rho_\omega(g, h) u_h u_g$, it suffices to prove that for any $g \in G \backslash \{0\}$, $\varepsilon > 0$, $M \in \mathbb{N}$ and $N \in \mathbb{N}$ there exist $m_1, \ldots, m_N \in \mathbb{N}$ such that $m_{i+1} - m_i \geq M$ and
$$|\rho_\omega(T^{m_i} g, T^{m_j} g) + 1| < \varepsilon \tag{11.13}$$
for all $i, j \leq N$, $i \neq j$. The proof is by induction on N. If $N = 1$ there is nothing to prove. So assume the statement is true for $N - 1$. For the induction step consider two cases.

Assume first $\Gamma_g \subset \mathbb{T}$ is finite, so it is the group of roots of unity of order m, where m is an even number since $-1 \in \Gamma_g$. Since by assumption any root of unity of order m appears in the sequence $\{\rho_\omega(T^n g, g)\}_n$, we have $lg \neq 0$ for $l \neq 0$ with $|l| < m$, since otherwise $\rho_\omega(T^n g, g)^l = 1$. By Proposition 11.3.2 there exists M_0 such that $\sum_{i=1}^{N-1} l_i T^{k_i} g \neq 0$ for any $k_1, \ldots, k_{N-1}$ such that $k_{i+1} - k_i \geq M_0$ and any $(l_1, \ldots, l_{N-1}) \neq (0, \ldots, 0)$ such that $|l_i| < m$. We may assume that $M \geq M_0$ and choose $m_1, \ldots, m_{N-1}$ satisfying (11.13). Consider the function

$$f(z) = \sum_{l=0}^{m-1} (-1)^l z^l,$$

which is zero on $\Gamma_g \setminus \{-1\}$. Then for any $k \in \mathbb{N}$

$$\prod_{i=1}^{N-1} f(\rho_\omega(T^k g, T^{m_i} g)) = \sum_{l_1,\ldots,l_{N-1}=0}^{m-1} (-1)^{l_1+\ldots+l_{N-1}} \rho_\omega \left(T^k g, \sum_{i=1}^{N-1} l_i T^{m_i} g \right).$$

Since by assumption $n^{-1} \sum_{k=0}^{n-1} \rho_\omega(T^k g, h) \to 0$ for any nonzero element h in the subgroup generated by the orbit of g, we conclude that the only term in the above expression which contributes to the limit

$$\lim_{n \to \infty} \frac{1}{n} \sum_{k=0}^{n-1} \prod_{i=1}^{N-1} f(\rho_\omega(T^k g, T^{m_i} g))$$

is the one corresponding to $l_1 = \ldots = l_{N-1} = 0$. Hence the limit is 1. In particular, there exist infinitely many k's such that $f(\rho_\omega(T^k g, T^{m_i} g)) \neq 0$ for $i = 1, \ldots, N-1$, so that $\rho_\omega(T^k g, T^{m_i} g) = -1$. We can take any such $k \geq m_{N-1} + M$ for m_N.

Assume now that $\Gamma_g = \mathbb{T}$. The proof in this case is similar but the details become a bit more involved. Choose sufficiently small $\delta > 0$ such that there exists a trigonometric polynomial f satisfying

(i) $|f(z)| \leq 1$ for any $z \in \mathbb{T}$;
(ii) $|f(z)| \leq \delta$ for any $z \in \mathbb{T}$ such that $|z+1| \geq \varepsilon$;
(iii) $|\int_\mathbb{T} f(z) d\mu(z)| > \delta^{1/(N-1)}$.

Let $f(z) = \sum_{|l|<m} a_l z^l$. Our assumption on g implies that g has infinite order. Hence by Proposition 11.3.2 there exists M_0 such that $\sum_{i=1}^{N-1} l_i T^{k_i} g \neq 0$ for any $k_1, \ldots, k_{N-1}$ such that $k_{i+1} - k_i \geq M_0$ and $(l_1, \ldots, l_{N-1}) \neq (0, \ldots, 0)$ such that $|l_i| < m$. Assuming that $M \geq M_0$ choose $m_1, \ldots, m_{N-1}$ satisfying (11.13). Then by the same argument as before we get

$$\frac{1}{n} \sum_{k=0}^{n-1} \prod_{i=1}^{N-1} f(\rho_\omega(T^k g, T^{m_i} g)) \to a_0^{N-1} = \left(\int_\mathbb{T} f(z) d\mu(z) \right)^{N-1}.$$

Hence for infinitely many k's

$$\prod_{i=1}^{N-1} |f(\rho_\omega(T^k g, T^{m_i} g))| > \delta,$$

so that $|1 + \rho_\omega(T^k g, T^{m_i} g)| < \varepsilon$ for $i = 1, \ldots, N-1$. Any such $k \geq m_{N-1} + M$ can be taken for m_N. □

Finishing our discussion of entropic properties of the automorphisms α_T, we shall show that the ergodic zero entropy systems are exactly those which are uniquely ergodic.

11.4 Zero Entropy Systems

Theorem 11.4.2. *Let T be an aperiodic automorphism of G, $\omega \in Z^2(G;\mathbb{T})$ a T-invariant cocycle. Then the following conditions are equivalent:*

(i) $h_\tau(\alpha_T) > 0$;

(ii) *there exists an α_T-invariant state different from τ.*

Proof. Let φ be an α_T-invariant state, $\varphi \neq \tau$. Using φ we shall construct a nontrivial stationary coupling with a Bernoulli shift. Consider the positive unital map $P \colon C^*(G,\omega) \to C(\hat{G})$ defined by

$$P(x)(\chi) = \varphi(\alpha_\chi(x)) \quad \text{for } x \in C^*(G,\omega), \ \chi \in \hat{G},$$

where α_χ is the dual automorphism, so $\alpha_\chi(u_g) = \langle \chi, g \rangle u_g$. Since $\tau(\cdot)1 = \int_{\hat{G}} \alpha_\chi(\cdot) d\mu(\chi)$, where μ is the Haar measure on $\hat{G}$, we have $\mu \circ P = \tau$.

Consider the automorphism $\alpha_{\hat{T}}$ of $C(\hat{G})$ defined by the dual automorphism $\hat{T}$. Then

$$\alpha_{\hat{T}}(P(a))(\chi) = P(a)(\hat{T}^{-1}\chi) = \varphi(\alpha_{\hat{T}^{-1}\chi}(a)) = \varphi((\alpha_T^{-1} \circ \alpha_\chi \circ \alpha_T)(a))$$
$$= \varphi((\alpha_\chi \circ \alpha_T)(a)) = P(\alpha_T(a))(\chi),$$

so that $\alpha_{\hat{T}} \circ P = P \circ \alpha_T$.

Since $\varphi \neq \tau$, there exists $g \neq 0$ such that $\varphi(u_g) \neq 0$. Then $P(u_g) \neq 0$, so that $P \neq \tau(\cdot)1$. Fix a self-adjoint element $a \in C^*(G,\omega)$ such that $P(a) \neq \tau(a)1$. Let $\varepsilon = \|P(a) - \tau(a)1\|$. By Theorem 11.3.3 applied to the trivial cocycle there exist $n \in \mathbb{N}$, a full matrix algebra B and a unital completely positive map $\Phi \colon C(\hat{G}) \to A = B^{\otimes \mathbb{Z}}$ such that $\alpha \circ \Phi = \Phi \circ \alpha_{\hat{T}}^n$, where α is the shift to the right on A, $\psi \circ \Phi = \mu$, where ψ is the unique tracial state on A, $\|\Phi(P(a)) - a_0\| < \varepsilon/4$ for some self-adjoint element a_0 in the 0-th factor B_0 of A, and $\|\Phi(P(a)) - \tau(a)1\| \geq \varepsilon/2$. Let C be an abelian C*-subalgebra of B_0 containing a_0. Put $D = C^{\otimes \mathbb{Z}}$, an let $E \colon A \to D$ be the ψ-preserving conditional expectation. Consider the map $\Theta = E \circ \Phi \circ P \colon C^*(G,\omega) \to D$. Then $\psi \circ \Theta = \tau$ and $\alpha \circ \Theta = \Theta \circ \alpha_T^n$. Moreover, $\Theta(a) \neq \tau(a)1$, since otherwise in view of $\|\Phi(P(a)) - a_0\| < \varepsilon/4$ we get $\|\tau(a)1 - a_0\| < \varepsilon/4$, so that $\|\Phi(P(a)) - \tau(a)1\| < \varepsilon/2$, which is a contradiction. It follows that the formula

$$\lambda(x \otimes y) = \psi(\Theta(x)y)$$

defines a nontrivial stationary coupling of $(C^*(G,\omega), \tau, \alpha_T^n)$ with the classical Bernoulli shift $(D, \psi|_D, \alpha|_D)$. By Proposition 5.1.2 we get $h_\tau(\alpha_T^n) > 0$, and hence $h_\tau(\alpha_T) > 0$.

Note that we could make the proof a bit shorter if we used the fact that the system $(\hat{G}, \mu, \hat{T})$ is already a Bernoulli shift.

Conversely, let $h_\tau(\alpha_T) > 0$. Then there exists a nontrivial coupling with an abelian system. Thus there exists an abelian C*-algebra B, an automorphism β of B, an $(\alpha_T \otimes \beta)$-invariant state λ on $C^*(G,\omega) \otimes B$ and a nonzero element $g \in G$ such that $\lambda(u_g \otimes \cdot) \neq 0$. Then there exists a unitary $v \in B$ such that $\lambda(u_g \otimes v) \neq 0$. Multiplying v by a scalar we may also assume that $\lambda(u_g \otimes (v + v^*)) \neq 0$. By Proposition 11.3.2 applied to the set $\{\pm 2g, \pm g, 0\}$ there exists $n \in \mathbb{N}$ such that if

$$\sum_{i=1}^{N} a_i T^{in} g = 0$$

for some $N \in \mathbb{N}$ and $a_1, \ldots, a_N \in \{-2, -1, 0, 1, 2\}$, then $a_i g = 0$ for all i. We claim that there exists a unital completely positive map $\Phi \colon C^*(G, \omega) \to C^*(G, \omega) \otimes B$ such that $\Phi \circ \alpha_T^n = (\alpha_T^n \otimes \beta^n) \circ \Phi$ and

$$\Phi(u_g) = \begin{cases} \dfrac{1}{2} u_g \otimes v, & \text{if } 2g \neq 0, \\ \dfrac{1}{2} u_g \otimes (v + v^*), & \text{if } 2g = 0. \end{cases}$$

Assuming that the claim is proved we then get an α_T^n-invariant state $\psi = \lambda \circ \Phi$ such that $\psi(u_g) \neq 0$, so $\psi \neq \tau$. Consider

$$\varphi = \frac{1}{n} \sum_{k=0}^{n-1} \psi \circ \alpha_T^k.$$

Then φ is an α_T-invariant state. If $\varphi = \tau$, then $\psi \leq n\tau$ and hence ψ extends to a normal α_T^n-invariant state on $W^*(G, \omega)$. Since T^n is aperiodic, the automorphism α_T^n of $W^*(G, \omega)$ is ergodic by Lemma 11.3.1. Hence $\psi = \tau$, which is a contradiction. Thus $\varphi \neq \tau$.

It remains to construct the map Φ. For $k, l \in \mathbb{Z}$ such that $k < l$ and $\varepsilon = (\varepsilon_k, \ldots, \varepsilon_l) \in \{0, 1\}^{l-k+1}$ set

$$u(\varepsilon) = u_{T^{kn}g}^{\varepsilon_k} \ldots u_{T^{ln}g}^{\varepsilon_l} \quad \text{and} \quad v(\varepsilon) = \beta^{kn}(v^{\varepsilon_k}) \ldots \beta^{ln}(v^{\varepsilon_l}).$$

Consider the element $T_{k,l} \in C^*(G, \omega) \otimes C^*(G, \omega) \otimes B$ defined by

$$T_{k,l} = 2^{-(l-k+1)/2} \sum_{\varepsilon_k, \ldots, \varepsilon_l = 0, 1} u(\varepsilon)^* \otimes u(\varepsilon) \otimes v(\varepsilon),$$

and define a completely positive map $\Phi_{k,l} \colon C^*(G, \omega) \to C^*(G, \omega) \otimes B$ by

$$\Phi_{k,l}(a) = (\tau \otimes \mathrm{id} \otimes \mathrm{id})(T_{k,l}(a \otimes 1 \otimes 1) T_{k,l}^*)$$
$$= 2^{-(l-k+1)} \sum_{\varepsilon, \varepsilon'} \tau(a u(\varepsilon') u(\varepsilon)^*) u(\varepsilon) u(\varepsilon')^* \otimes v(\varepsilon) v(\varepsilon')^*.$$

If $a = u_h$ then $\tau(a u(\varepsilon') u(\varepsilon)^*) = 0$ unless $h = \sum_{i=k}^{l} (\varepsilon_i - \varepsilon_i') T^{in} g$. In the latter case $u(\varepsilon) u(\varepsilon')^* = z u_h$ for a scalar $z \in \mathbb{T}$, so that $u(\varepsilon') u(\varepsilon)^* = \bar{z} u_h^*$ and thus

$$\tau(u_h u(\varepsilon') u(\varepsilon)^*) u(\varepsilon) u(\varepsilon')^* = u_h \quad \text{and} \quad v(\varepsilon) v(\varepsilon')^* = \prod_{i=k}^{l} \beta^{in}(v^{\varepsilon_i - \varepsilon_i'}).$$

Therefore
$$\Phi_{k,l}(u_h) = 2^{-(l-k+1)} u_h \otimes \left(\sum_{\varepsilon,\varepsilon'} \prod_{i=k}^{l} \beta^{in}(v^{\varepsilon_i - \varepsilon'_i}) \right),$$
where the summation is over all ε and ε' such that $h = \sum_{i=k}^{l}(\varepsilon_i - \varepsilon'_i) T^{in} g$.

By our choice of n, if $2g \neq 0$, a representation of an element h in the form $h = \sum_{i=k}^{l} a_i T^{in} g$, $a_i \in \{-1, 0, 1\}$, is unique if it exists. In this case the numbers ε_i and ε'_i in $\{0,1\}$ such that $\varepsilon_i - \varepsilon'_i = a_i$ are uniquely determined if $a_i \neq 0$, and we have exactly two choices $\varepsilon_i = \varepsilon'_i = 0$ and $\varepsilon_i = \varepsilon'_i = 1$ for them if $a_i = 0$. Hence
$$\Phi_{k,l}(u_h) = 2^{-N} u_h \otimes \left(\prod_{i=k}^{l} \beta^{in}(v^{a_i}) \right),$$
where N is the number of i's such that $a_i \neq 0$. If $2g = 0$, then the numbers a_i are determined only up to a sign, and we similarly get
$$\Phi_{k,l}(u_h) = 2^{-N} u_h \otimes \left(\prod_{i: a_i \neq 0} \beta^{in}(v + v^{-1}) \right).$$
If no representation of h in the form $\sum_{i=k}^{l} a_i T^{in} g$ exists, then $\Phi_{k,l}(u_h) = 0$. In particular, we see that $\Phi_{k,l}$ is a unital map, and for any fixed $h \in G$ the image $\Phi_{k,l}(u_h)$ of u_h is the same for all sufficiently small $k \in \mathbb{Z}$ and sufficiently large $l \in \mathbb{Z}$. Since also $(\alpha_T^n \otimes \beta^n) \circ \Phi_{k,l} = \Phi_{k+1, l+1} \circ \alpha_T^n$, letting $k \to -\infty$ and $l \to +\infty$ we get the required map Φ. $\square$

Remark 11.4.3. If the assumptions of Theorem 11.3.3 are satisfied, invariant states different from the trace can be obtained from product-states on UHF-algebras. Such states can be thought of as noncommutative analogues of Riesz products.

Explicitly, let $g \in G$, and assume $2g \neq 0$. Consider the map Φ from Theorem 11.3.3 corresponding to the set $X = \{0, g\}$. On $B(\ell_2(X))^{\otimes \mathbb{Z}}$ consider the product state defined by the density matrix
$$\begin{pmatrix} 1/2 & \overline{\omega(0,0)}/2 \\ \omega(0,0)/2 & 1/2 \end{pmatrix}.$$

Let ψ be the composition of this state with the map Φ. Using (11.10) and that $\Phi_X(u_g) = \omega(0,0) e_{g0}$ we then get
$$\psi(u_g) = \frac{1}{2} \prod_{k=1}^{\infty} \left(\frac{1}{2} + \frac{\rho_\omega(T^{kn} g, g)}{2} \right).$$

So if n is sufficiently large, ψ is an α_T^n-invariant state different from the trace.

Using the method of the proof of the previous theorem the same state can be obtained as the limit of the states

$$\psi_{k,l}(a) = 2^{-(l-k+1)} \sum_{\substack{\varepsilon_k,\ldots,\varepsilon_l=0,1 \\ \varepsilon'_k,\ldots,\varepsilon'_l=0,1}} \tau(u(\varepsilon)au(\varepsilon')^*).$$

In fact, the map Φ from the previous theorem can be obtained as the composition of the map Φ and a modification of the map Ψ from Theorem 11.3.3. Although those maps were constructed under additional assumptions on the cocycle, their composition makes sense in general. ◆

11.5 Automorphisms of Noncommutative Tori

Let $G = \mathbb{Z}^2$. Consider the $\mathbb{Z}$-valued skew-symmetric form σ defined by

$$\sigma(g,h) = g_1 h_2 - g_2 h_1.$$

For $\theta \in \mathbb{R}$ set $\omega_\theta(g,h) = e^{\pi i \theta \sigma(g,h)}$. The algebra $A_\theta = C^*(\mathbb{Z}^2, \omega_\theta)$ is called the **irrational rotation algebra**. Since σ is nondegenerate, by Theorem 11.1.1 the algebra A_θ is simple if θ is irrational. Denote by u_θ and v_θ the unitaries in $C^*(\mathbb{Z}^2, \omega_\theta)$ corresponding to the elements $(1,0)$ and $(0,1)$, respectively. Then

$$u_\theta v_\theta = e^{2\pi i \theta} v_\theta u_\theta.$$

One can alternatively define A_θ as the universal C*-algebra generated by unitaries u_θ and v_θ satisfying the above relation.

Any matrix $T \in \mathrm{SL}_2(\mathbb{Z})$ preserves the form σ and hence defines an automorphism of A_θ, which we denote by $\alpha_{T,\theta}$. We also denote by ρ_θ the bicharacter defined by ω_θ, and by τ_θ the canonical trace on A_θ.

Let $\{\lambda, \lambda^{-1}\}$ be the spectrum of T. Assume $|\lambda| > 1$. Then it is known that λ is irrational (otherwise λ, being a root of a monic polynomial with integral coefficients, must be an integer, which is impossible as $\lambda + \lambda^{-1} = \mathrm{Tr}(T)$ is an integer). For any $n \in \mathbb{Z}$ we have

$$T^n = \frac{1}{\lambda^{-1} - \lambda}(\lambda^n(\lambda^{-1} - T) + \lambda^{-n}(T - \lambda)) = \frac{\lambda^n}{1 - \lambda^2}(1 - \lambda T) + \frac{\lambda^{-n}}{\lambda^{-1} - \lambda}(T - \lambda),$$

as can be checked on the eigenvectors of T. Hence for any $g, h \in \mathbb{Z}^2$ there exists $C > 0$ such that for any $n \in \mathbb{N}$

$$\left| \sigma(T^n g, h) - \frac{\lambda^n}{1 - \lambda^2}(\sigma(g,h) - \lambda \sigma(Tg,h)) \right| \le C|\lambda|^{-n}. \tag{11.14}$$

We need the following classical result of Weyl.

Lemma 11.5.1. *For any real number λ, $|\lambda| > 1$, the sequence $\{e^{i\theta \lambda^n}\}_{n=1}^\infty$ is uniformly distributed on $\mathbb{T}$ for almost all $\theta \in \mathbb{R}$.*

11.5 Automorphisms of Noncommutative Tori

Proof. Fix $m \in \mathbb{Z}$, $m \neq 0$. For $n \in \mathbb{N}$ put

$$V_n(\theta) = \frac{1}{n} \sum_{k=1}^{n} e^{im\theta \lambda^k}.$$

We have to prove that $V_n(\theta) \to 0$ as $n \to \infty$ for almost all $\theta \in \mathbb{R}$.

Let $a, b \in \mathbb{R}$, $a < b$. We have

$$\int_a^b |V_n(\theta)|^2 d\theta = \frac{1}{n^2} \int_a^b \sum_{k,j=1}^{n} e^{im\theta(\lambda^k - \lambda^j)} d\theta$$

$$= \frac{b-a}{n} + \frac{1}{n^2} \sum_{k \neq j} \frac{e^{imb(\lambda^k - \lambda^j)} - e^{ima(\lambda^k - \lambda^j)}}{im(\lambda^k - \lambda^j)}.$$

Since $|\lambda^k - \lambda^j| \geq |\lambda|^{k+1} - |\lambda|^k$ for $j > k$,

$$\sum_{k \neq j} \frac{1}{|\lambda^k - \lambda^j|} \leq 2 \sum_{k=1}^{n-1} \sum_{j=k+1}^{n} \frac{1}{|\lambda|^{k+1} - |\lambda|^k} \leq 2n \sum_{k=1}^{n-1} \frac{1}{|\lambda|^{k+1} - |\lambda|^k} \leq \frac{2n}{(|\lambda| - 1)^2}.$$

Thus $\int_a^b |V_n(\theta)|^2 d\theta \leq C/n$ for a constant C depending on λ, m, a and b. In particular,

$$\int_a^b \sum_{k=1}^{\infty} |V_{k^2}(\theta)|^2 d\theta < \infty.$$

It follows that $V_{k^2}(\theta) \to 0$ as $k \to \infty$ for almost all $\theta \in [a, b]$. If this convergence holds for some θ, then for $k^2 \leq n < (k+1)^2$ we have

$$|V_n(\theta)| = \frac{1}{n} \left| \sum_{l=1}^{(k+1)^2} e^{im\theta \lambda^l} - \sum_{l=n+1}^{(k+1)^2} e^{im\theta \lambda^l} \right|$$

$$\leq \frac{(k+1)^2}{n} |V_{(k+1)^2}(\theta)| + \frac{(k+1)^2 - n}{n}.$$

Hence $V_n(\theta) \to 0$ as $n \to \infty$. $\square$

We are now in position to apply the results of the previous sections to the automorphisms of noncommutative tori.

Theorem 11.5.2. *For any $T \in \mathrm{SL}_2(\mathbb{Z})$ with $\mathrm{Sp}\, T = \{\lambda, \lambda^{-1}\}$ and $|\lambda| > 1$ we have:*

(i) $\frac{1}{2} \log |\lambda| \leq ht(\alpha_{T,\theta}) \leq \log |\lambda|$ *for any $\theta \in \mathbb{R}$;*

(ii) *if $\theta \in \mathbb{Q}$ then $h_{\tau_\theta}(\alpha_{T,\theta}) = ht(\alpha_{T,\theta}) = \log |\lambda|$;*

(iii) *if θ lies in the additive group generated by 1 and $\lambda^m(1 - \lambda^2)$, $m \in \mathbb{Z}$, then $(\pi_{\tau_\theta}(A_\theta)'', \tau_\theta, \alpha_{T,\theta})$ is an entropic K-system; in particular, $h_{\tau_\theta}(\alpha_{T,\theta}) > 0$ for any $\theta \in \mathbb{Q}(\lambda) = \mathbb{Q}\lambda + \mathbb{Q}$;*

(iv) $h_{\tau_\theta}(\alpha_{T,\theta}) = 0$ *for almost all $\theta \in \mathbb{R}$, and for any such θ the trace τ_θ is the unique $\alpha_{T,\theta}$-invariant state on A_θ.*

Proof. Part (i) follows from Theorem 11.2.1 and Example 11.2.5.

For any $n \in \mathbb{N}$ we have an embedding $A_{n^2\theta} \hookrightarrow A_\theta$ given by $u_{n^2\theta} \mapsto u_\theta^n$, $v_{n^2\theta} \mapsto v_\theta^n$. In particular, if θ is rational then we have an embedding of $C(\mathbb{T}^2)$ into A_θ, so that the entropy $h_{\tau_\theta}(\alpha_{T,\theta})$ is at least as large as in the classical case, where it is $\log|\lambda|$ by Example 11.2.5. This proves (ii).

If $\theta = \lambda^m(1-\lambda^2)$, then by virtue of (11.14) for any g and h we have

$$|\theta\sigma(T^n g, h) - \lambda^{n+m}\sigma(g,h) + \lambda^{n+m+1}\sigma(Tg,h)| \leq C|\lambda|^{m-n}.$$

Since $\lambda^{n+m} = \mathrm{Tr}(T^{n+m}) - \lambda^{-n-m}$ and $\lambda^{n+m+1} = \mathrm{Tr}(T^{n+m+1}) - \lambda^{-n-m-1}$, and the numbers $\mathrm{Tr}(T^{n+m})$ and $\mathrm{Tr}(T^{n+m+1})$ are integers, we see that there exist integers m_n such that

$$|\theta\sigma(T^n g, h) - m_n| \leq C_0|\lambda|^{-n}$$

for some new constant C_0. Hence

$$|1 - \rho_\theta(T^n g, h)| = |1 - e^{2\pi i \theta \sigma(T^n g, h)}| \leq 2\pi C_0 |\lambda|^{-n}.$$

Therefore $\sum_{n=1}^\infty |1 - \rho_\theta(T^n g, h)| < \infty$. The same is true if we assume that θ lies in the group H generated by 1 and $\lambda^m(1-\lambda^2)$, $m \in \mathbb{Z}$. By Corollary 11.3.4 we then conclude that $(\pi_{\tau_\theta}(A_\theta)'', \tau_\theta, \alpha_{T,\theta})$ is an entropic K-system.

Since $\lambda \in \mathbb{Q}\lambda^2 + \mathbb{Q}$ and $1, \lambda^2 \in H$, for any $\theta \in \mathbb{Q}(\lambda)$ we can find $n \in \mathbb{N}$ such that $n^2\theta \in H$. Then, since $A_{n^2\theta}$ embeds into A_θ,

$$h_{\tau_\theta}(\alpha_{T,\theta}) \geq h_{\tau_{n^2\theta}}(\alpha_{T,n^2\theta}) > 0,$$

which completes the proof of (iii). Remark that in fact by Proposition 5.1.8 we have $h_{\tau_\theta}(\alpha_{T,\theta}) = h_{\tau_{n^2\theta}}(\alpha_{T,n^2\theta})$.

To prove the first statement in (iv) note that if $\{a_n\}_n$ and $\{b_n\}_n$ are two sequences of real numbers such that $b_n \to 0$, then $\{e^{i(a_n+b_n)}\}_n$ is uniformly distributed on $\mathbb{T}$ if and only if $\{e^{ia_n}\}_n$ is uniformly distributed. So by (11.14) and Theorem 11.4.1 it suffices to check that the sequence

$$\{e^{2\pi i \theta \lambda^n (1-\lambda^2)^{-1}(\sigma(g,h) - \lambda\sigma(Tg,h))}\}_{n=1}^\infty$$

is uniformly distributed for all $g, h \neq 0$ and almost all θ. But this is true by Lemma 11.5.1, since $\sigma(g,h) - \lambda\sigma(Tg,h) \neq 0$. Indeed, since λ is irrational the latter expression is zero only if $\sigma(g,h) = \sigma(Tg,h) = 0$. But then both g and Tg are scalar multiples of h, which is impossible as T has no rational eigenvalues.

The last statement in (iv) follows from Theorem 11.4.2. □

Finally we give an example of systems for which the tensor product formula for the entropy fails.

Choose any θ such that $h_{\tau_\theta}(\alpha_{T,\theta}) = 0$. Then the same argument as in the proof of the lower bound in Theorem 11.2.1 shows that also $h_{\tau_{-\theta}}(\alpha_{T,-\theta}) = 0$, while

$$h_{\tau_\theta \otimes \tau_{-\theta}}(\alpha_{T,\theta} \otimes \alpha_{T,-\theta}) \geq \log|\lambda|.$$

11.6 Notes

Theorem 11.1.1 is a result of Slawny [202].

The idea of using the diagonal subalgebra of $C^*(G,\omega) \otimes C^*(G,\bar\omega)$ to estimate the entropy of the tensor product automorphism is due to Narnhofer; it was first employed in [128] for binary shifts, which corresponds to the group $G = \bigoplus_\mathbb{Z} \mathbb{Z}/2\mathbb{Z}$, see Chap. 12. As was observed by Kerr and Li [108] in the context of higher dimensional noncommutative tori, it allows one to estimate not only the topological entropy of α_T, but also the topological entropy of $\alpha_\chi \circ \alpha_T$ for $\chi \in \hat G$. We use a similar argument in the proof of the lower bound in Theorem 11.2.1 to show that $ht(\alpha_{T,\omega})$ depends only on the cohomology class of ω. The upper bound in Theorem 11.2.1 was obtained by Voiculescu [227] for $G = \mathbb{Z}^n$, and by Brown and Germain [37] in general. Theorem 11.2.3 is due to Peters [157]. Theorem 11.2.4 was proved by Bowen [29]. The fact that the Haar measure is the measure of maximal entropy for automorphisms of compact groups was established earlier by Berg [20]. The entropy of automorphisms of the two-dimensional torus was computed already by Sinai [199]. For general compact groups it was done by Juzvinskii [97]. For more information on entropic properties of automorphisms (and, more generally, automorphic actions of $\mathbb{Z}^n$) of compact groups we refer the reader to the book by Schmidt [192].

The examples of cocycles at the end of Sect. 11.2 are due to Golodets and Neshveyev [79]. They were used to construct automorphisms α of the hyperfinite II$_1$-factor with a fixed Cartan subalgebra A such that $h_\tau(\alpha) = t$ and $h_{\tau|A}(\alpha|_A) = s$ for any given t and s, $0 \le s \le t \le +\infty$. In this respect an interesting open problem is to find an automorphism of a II$_1$-factor with entropy larger than the supremum of the entropies of the restrictions of the automorphism to invariant abelian subalgebras. Existence of such an automorphism was announced in [74], but the details have never appeared.

The discussion before Example 11.2.7 shows that the example can be interpreted as follows. There exist an action of $G = \bigoplus_\mathbb{Z} \mathbb{Z}/2\mathbb{Z}$ on $X = \{0,1\}^\mathbb{Z}$ and an automorphism α of $C(X) \rtimes G$ such that

$$ht(\alpha) = ht(\alpha|_{C(X)}) = ht(\alpha|_{C^*(G)}) = \log 2.$$

In particular, $ht(\alpha) < ht(\alpha|_{C(X)}) + ht(\alpha|_{C^*(G)})$. Such an example was promised in Notes to Chap. 8.

It was proved by Benatti, Narnhofer and Sewell [17] that for a fixed $T \in \mathrm{SL}_2(\mathbb{Z})$ the automorphism $\alpha_{T,\theta}$ of A_θ can be asymptotically abelian only for a countable number of θ's. Later Narnhofer observed that the automorphism has indeed a strong form of asymptotic abelianness for certain irrational θ's in the quadratic field generated by the eigenvalues of T, and suggested that we then get an entropic K-system [127]. The proof of the more general result in Sect. 11.3 is an elaboration on the work of Neshveyev [138]. Note that Lemma 11.3.1 is a classical result of Halmos [84], while the argument in the proof of Proposition 11.3.2 essentially goes back to Rohlin [182], who proved that ergodic automorphisms of compact abelian group have the K-property. Later, using powerful Ornstein's machinery, it was proved that any ergodic automorphism of a compact separable group is Bernoullian (here one should allow the Bernoulli shift with infinite entropy). The result was obtained by Katznelson [101] for tori, and by Lind [118] and Miles and Thomas [122] in general.

It was Narnhofer who suggested that a sufficiently chaotic behavior of the cocycle should lead to zero dynamical entropy. Such results were independently obtained by Narnhofer and Thirring [134] for binary shifts and noncommutative tori and by Sauvageot [189] for noncommutative tori. Our exposition in Sect. 11.4 is based on the unpublished work of Sauvageot [189].

Lemma 11.5.1 is a classical result of Weyl [234].

Finally remark that the results of this chapter are valid with minor changes for the automorphisms of the form $\alpha_\chi \circ \alpha_T$, $\chi \in \hat{G}$. Some of these automorphisms may look even more natural than α_T. E.g. if $T = \begin{pmatrix} a & b \\ c & d \end{pmatrix} \in \mathrm{SL}_2(\mathbb{Z})$ then with an appropriate choice of χ we get the automorphism of A_θ such that

$$u_\theta \mapsto u_\theta^a v_\theta^c, \quad v_\theta \mapsto u_\theta^b v_\theta^d.$$

12
Binary Shifts

A rich source of C*-dynamical systems is obtained from bitstreams, i.e. sequences of 0's and 1's. Given such a sequence we shall construct a C*-algebra, which in the interesting cases is the UHF-algebra of type 2^∞, and an automorphism, called a binary shift. Such a system is an example of systems considered in Chap. 11. Different bitstreams can give rise to quite different C*-dynamical systems, and in this chapter we shall study their entropic properties.

12.1 The C*-algebra of a Bitstream

With every bitstream $\{\varepsilon_n\}_{n=1}^\infty$, where $\varepsilon_n \in \{0,1\}$, we associate a subset X of $\mathbb{N}$ by
$$X = \{n \in \mathbb{N} \mid \varepsilon_n = 1\}.$$
Conversely each subset $X \subset \mathbb{N}$ gives rise to a bitstream in this way. Let $G = \bigoplus_\mathbb{Z} \mathbb{Z}/2\mathbb{Z}$. Denote by g_i the element 1 in the i-th summand $(\mathbb{Z}/2\mathbb{Z})_i$ of G. Define a bicharacter ω on G by

$$\omega(g_i, g_j) = \begin{cases} -1, & \text{if } j - i \in X, \\ +1, & \text{otherwise,} \end{cases}$$

so that
$$\omega\left(\sum_{i \in I} g_i, \sum_{j \in J} g_j\right) = \prod_{i \in I, j \in J} \omega(g_i, g_j).$$

As in Chap. 11 consider the algebra $A(X) = C^*(G, \omega)$, which is the universal C*-algebra generated by unitaries $u_g, g \in G$, such that $u_g u_h = \omega(g,h) u_{g+h}$. Recall that the canonical trace τ is defined by $\tau(u_g) = 0$ for $g \neq 0$.

Let $s_i = u_{g_i}$. Then s_i is a symmetry, that is, a self-adjoint unitary. We have
$$s_i s_j = \omega(g_i, g_j) u_{g_i + g_j} = \begin{cases} -u_{g_i + g_j}, & \text{if } j - i \in X, \\ u_{g_i + g_j}, & \text{otherwise.} \end{cases}$$

Thus, if χ denotes the characteristic function of X, we have

$$s_i s_j = (-1)^{\chi(|i-j|)} s_j s_i.$$

We could alternatively define $A(X) = C^*(G,\omega)$ as the universal C*-algebra generated by symmetries s_n, $n \in \mathbb{Z}$, such that the above relation is satisfied.

Let $G_{[n,m]} = \oplus_{i=n}^{m}(\mathbb{Z}/2\mathbb{Z})_i \subset G$. We shall write G_n instead of $G_{[1,n]}$. Let $A_n = C^*(G_n, \omega|_{G_n}) \subset C^*(G,\omega)$. So A_n is the C*-algebra generated by $s_1, \ldots, s_n$. Our goal in this section is to determine the structure of the C*-algebras $A(X)$ and $A_+(X) = \overline{\cup_{n=1}^{\infty} A_n}$. From Chap. 11 we know that for this we have to study the bicharacter $\omega(g,h)\omega(h,g)$, which can be written as $(-1)^{B(g,h)}$, where $B: G \times G \to \mathbb{Z}/2\mathbb{Z}$ is the skew-symmetric bilinear form given by

$$B(g_i, g_j) = \begin{cases} 1, & \text{if } |j-i| \in X, \\ 0, & \text{otherwise.} \end{cases}$$

We consider G as a vector space over the field $\mathbb{F}_2 = \mathbb{Z}/2\mathbb{Z}$. Then g_i, $i \in \mathbb{Z}$, form a basis. Let H_n denote the kernel of the restriction of B to G_n, so

$$H_n = \{h \in G_n \mid B(h,g) = 0 \text{ for every } g \in G_n\}.$$

By Lemma 11.1.2 the center $Z(A_n)$ of A_n coincides with $C^*(H_n, \omega|_{H_n})$. Set $c_n = \dim_{\mathbb{F}_2} H_n$. Thus $Z(A_n)$ is an abelian algebra of dimension 2^{c_n}. Note that $c_1 = 1$.

Denote by T the shift to the right on G, so $T(G_{[n,m]}) = G_{[n+1,m+1]}$. The form B is clearly T-invariant.

Lemma 12.1.1. *For each $n \geq 1$ we have one of the following possibilities:*
(i) $c_{n+1} = c_n + 1$ and $H_n = G_n \cap H_{n+1} = T^{-1}(H_{n+1}) \cap H_{n+1}$, and if in addition $H_n \neq 0$ then $H_{n+1} = H_n + T(H_n)$;
(ii) $c_{n+1} = c_n - 1$ and $H_{n+1} = H_n \cap T(H_n)$, and if in addition $H_{n+1} \neq 0$ then $H_n = H_{n+1} + T^{-1}(H_{n+1})$.

Proof. Let us first show that if V is a finite dimensional vector space, W a subspace of V of codimension one, and F a skew-symmetric bilinear form on V, then $\dim \operatorname{Ker} F = \dim \operatorname{Ker}(F|_W) \pm 1$.

Fix a vector $e \in V \setminus W$. Consider the functional $F(\cdot, e)$ on $\operatorname{Ker}(F|_W)$. Assume first that this functional is nonzero, so there exists $h \in \operatorname{Ker}(F|_W)$ such that $F(h, e) \neq 0$. Then $F(h, g+e) = F(h,e) \neq 0$ for every $g \in W$, which implies that $\operatorname{Ker} F$ contains no elements in $V \setminus W$, so that $\operatorname{Ker} F \subset W$. But then $\operatorname{Ker} F$ is exactly the kernel of the nonzero functional $F(\cdot, e)$ on $\operatorname{Ker}(F|_W)$. So $\operatorname{Ker} F \subset \operatorname{Ker}(F|_W)$ and $\dim \operatorname{Ker} F = \dim \operatorname{Ker}(F|_W) - 1$.

Assume now that $F(h,e) = 0$ for every $h \in \operatorname{Ker}(F|_W)$. Then $\operatorname{Ker}(F|_W) \subset \operatorname{Ker} F$. Since F defines a nondegenerate form on $W/\operatorname{Ker}(F|_W)$, there exists $f \in W$ such that $F(g,e) = F(g,f)$ for every $g \in W$. It follows that $f - e \in \operatorname{Ker} F$ (note that $F(e, f-e) = F(e,f) = F(f,f) = 0$). Thus the inclusion $\mathrm{Ker}(F|_W) \subset \mathrm{Ker}\, F$ is proper. In particular, $\dim \mathrm{Ker}\, F \geq \dim \mathrm{Ker}(F|_W) + 1$. On the other hand, we always have $W \cap \mathrm{Ker}\, F \subset \mathrm{Ker}(F|_W)$, so that $\dim \mathrm{Ker}\, F - 1 \leq \dim \mathrm{Ker}(F|_W)$. Hence $\dim \mathrm{Ker}\, F = \dim \mathrm{Ker}(F|_W) + 1$ and $W \cap \mathrm{Ker}\, F = \mathrm{Ker}(F|_W)$.

We have thus proved that $c_{n+1} = c_n \pm 1$.

Assume $c_{n+1} = c_n + 1$. It follows from our considerations that this implies $H_n = G_n \cap H_{n+1}$. Applying this to the inclusion $T(G_n) \subset G_{n+1}$ instead of $G_n \subset G_{n+1}$ we can also conclude that $T(H_n) = T(G_n) \cap H_{n+1}$. Hence $H_n \subset T^{-1}(H_{n+1}) \cap H_{n+1}$. Since $T^{-1}(H_{n+1}) \cap H_{n+1} \subset G_{[1,n]}$, the opposite inclusion is obvious.

In particular, $H_n + T(H_n) \subset H_{n+1}$. If $H_n \neq 0$ then H_n is a proper subspace of $H_n + T(H_n)$, whence $H_{n+1} = H_n + T(H_n)$.

Assume now $c_{n+1} = c_n - 1$. As we have seen, this implies $H_{n+1} \subset H_n$. Applying this to $T(G_n) \subset G_{n+1}$ we also conclude $H_{n+1} \subset T(H_n)$. Hence $H_{n+1} \subset H_n \cap T(H_n)$. The opposite inclusion is obvious.

In particular, $H_{n+1} + T^{-1}(H_{n+1}) \subset H_n$. If $H_{n+1} \neq 0$ then H_{n+1} is a proper subspace of $H_{n+1} + T^{-1}(H_{n+1})$. Hence $H_n = H_{n+1} + T^{-1}(H_{n+1})$. □

From the previous lemma we get further restrictions on the c_n's.

Lemma 12.1.2. *If $c_{n-1} = c_{n+1} = c_n + 1$ for some $n \geq 2$ then $c_n = 0$.*

Proof. Since $c_n = c_{n-1} - 1$, by Lemma 12.1.1(ii) we have

$$H_n = H_{n-1} \cap T(H_{n-1}) \subset G_{[2,n-1]}.$$

On the other hand, since $c_n = c_{n+1} - 1$, by Lemma 12.1.1(i) we have $H_n \subset H_{n+1}$. Since we always have $T^{-1}(H_{n+1}) \cap G_n \subset H_n$, we get $T^{-1}(H_n) \cap G_n \subset H_n$. Since $T^{-1}(H_n) \subset G_{[1,n-2]} \subset G_n$, we see that $T^{-1}(H_n) \subset H_n$. This is possible only when $H_n = 0$. □

It follows that the sequence $\{c_n\}_{n=1}^{\infty}$ increases by one to some point, then it decreases by one to zero, and so on. We shall next study the case when the sequence only increases after some n.

We say a subset X of $\mathbb{N}$ is **periodic** if the subset $-X \cup X$ of $\mathbb{Z}$ is periodic. In terms of the form B this means that there exists $m \geq 1$ such that

$$B(T^m g, h) = B(g, h) \quad \text{for} \quad g, h \in G.$$

Otherwise we say X is **nonperiodic**.

Lemma 12.1.3. *The following conditions are equivalent:*

(i) *X is periodic;*

(ii) *there exists $n \geq 0$ such that $c_{n+k} = k$ for any $k \geq 0$.*

Here we put for convenience $G_0 = 0$ and $c_0 = 0$.

Proof of Lemma 12.1.3. If $B(T^m g, h) = B(g, h)$ for $g, h \in G$, then $g_1 + g_{m+1}$ is in the kernel of B, so $c_n \geq 1$ for all $n \geq m + 1$. Hence there exists a largest number $n \leq m$ such that $c_n = 0$, and then $c_{n+k} = k$ for any $k \geq 1$ by Lemmas 12.1.1 and 12.1.2.

Conversely, assume (ii) is satisfied. Then H_{n+1} is one-dimensional, and H_{n+k} is spanned by $H_{n+1}, T(H_{n+1}), \ldots, T^{k-1}(H_{n+1})$ by Lemma 12.1.1(i). In particular, the kernel H_∞ of the restriction of B to $G_\infty = \cup_n G_n$ is spanned by $T^k(H_{n+1})$, $k \geq 0$, so that $T(H_\infty) \subset H_\infty$. Hence T induces a linear operator $\bar{T}$ on G_∞/H_∞.

Since H_{n+k} is k-dimensional and has trivial intersection with G_n (otherwise $c_n \neq 0$), we have $G_{n+k} = G_n \oplus H_{n+k}$ for any $k \in \mathbb{N}$. Hence $G_\infty = G_n \oplus H_\infty$. Thus G_∞/H_∞ is a finite set. It follows that there exist $k \geq 0$ and $m \geq 1$ such that $\bar{T}^{k+m} = \bar{T}^k$. Hence $B(T^{k+m}g, h) = B(T^k g, h)$ for any $g, h \in G_\infty$. Since B is T-invariant, we conclude that $B(T^m g, h) = B(g, h)$ for all $g, h \in G$, so that X is periodic. □

Let us now summarize properties of the sequence $\{A_n\}_n$.

Theorem 12.1.4. *There exist sequences $\{c_n\}_{n=0}^\infty$ and $\{d_n\}_{n=1}^\infty$ of nonnegative integers and a strictly increasing sequence $\{n_l\}_{l=1}^N$ of even integers, where $N \in \mathbb{N} \cup \{\infty\}$, such that*

(i) $A_n \cong \mathrm{Mat}_{2^{d_n}}(\mathbb{C}) \otimes \mathbb{C}^{2^{c_n}}$, *and the trace τ takes the value $2^{-c_n - d_n}$ on every minimal projection in A_n;*

(ii) $n = c_n + 2d_n$ *for any $n \geq 1$;*

(iii) $n_1 = 0$, *and if $m_l = (n_{l+1} - n_l)/2$ then*

$$c_{n_l + k} = \begin{cases} k, & \text{if } 0 \leq k < m_l, \\ 2m_l - k, & \text{if } m_l \leq k \leq 2m_l, \end{cases}$$

$Z(A_{n_l+k}) \subset Z(A_{n_l+k+1})$ *for $0 \leq k \leq m_l - 1$ and $Z(A_{n_l+k}) \supset Z(A_{n_l+k+1})$ for $m_l \leq k \leq 2m_l - 1$.*

Furthermore, $N < \infty$ if and only if X is periodic.

Proof. Let G'_n be a subspace of G_n such that $G_n = G'_n \oplus H_n$. Since $u_g u_h = u_h u_g$ for $g \in G'_n$ and $h \in H_n$, we have

$$A_n \cong C^*(G'_n, \omega|_{G'_n}) \otimes C^*(H_n, \omega|_{H_n}),$$

and the canonical trace on A_n is the tensor product of the canonical traces.

By Lemma 12.1.1 the numbers n and c_n have the same parity, so G'_n has even dimension $2d_n = n - c_n$, which also follows from nondegeneracy of the skew-symmetric form B on G'_n. By Lemma 11.1.2 the nondegeneracy implies also that $C^*(G'_n, \omega|_{G'_n})$ is simple, so it is a full matrix algebra of dimension 2^{2d_n}. Thus $C^*(G'_n, \omega|_{G'_n}) \cong \mathrm{Mat}_{2^{d_n}}(\mathbb{C})$.

The algebra $Z(A_n) = C^*(H_n, \omega|_{H_n})$ is an abelian C*-algebra of dimension 2^{c_n}. Moreover, it is a tensor product of c_n twisted group C*-algebras of the group $\mathbb{Z}/2\mathbb{Z}$. Since the latter algebras have minimal projections of trace $1/2$, we conclude that every minimal projection in $Z(A_n)$ has trace 2^{-c_n}.

We have thus proved (i) and (ii). The rest of the theorem follows from Lemmas 12.1.1–12.1.3. □

Corollary 12.1.5. *Let $X \subset \mathbb{N}$. Then*

$$A_+(X) \cong \begin{cases} \mathrm{Mat}_{2^m}(\mathbb{C}) \otimes C(\{0,1\}^{\mathbb{N}}), & \text{if } X \text{ is periodic,} \\ \otimes_{n=1}^{\infty} \mathrm{Mat}_2(\mathbb{C}), & \text{if } X \text{ in nonperiodic.} \end{cases}$$

The same is true for $A(X)$.

Proof. In the notation of Theorem 12.1.4 if X is nonperiodic then $N = \infty$, so that $A_{2n} \cong \mathrm{Mat}_{2^n}(\mathbb{C})$ for an infinite number of n's. Hence $A_+(X)$ is the UHF-algebra of type 2^∞.

If X is periodic, then there exists $n = 2m$ such that $c_{n+k} = k$ for all $k \geq 0$. Then $A_n \cong \mathrm{Mat}_{2^m}(\mathbb{Z})$ and $A_{n+k} \cong A_n \otimes Z(A_{n+k})$. Since $Z(A_{n+k}) \subset Z(A_{n+k+1})$, we get $A_+(X) \cong A_n \otimes (\cup_k Z(A_{n+k})) \cong \mathrm{Mat}_{2^m}(\mathbb{C}) \otimes C(\{0,1\}^{\mathbb{N}})$.

To see that the same is true for $A(X)$, denote by $\tilde{A}_{2n}$ the algebra generated by $s_{-n+1}, \ldots, s_n$, and by $\tilde{A}_{2n+1}$ the algebra generated by $s_{-n}, \ldots, s_n$. Then $\{\tilde{A}_n\}_{n=1}^{\infty}$ is an increasing sequence, and $A(X) = \overline{\cup_n \tilde{A}_n}$. It is then not difficult to see that all the properties of the sequence $\{A_n\}_n$ stated in Theorem 12.1.4 remain true for the sequence $\{\tilde{A}_n\}_n$. So the same proof as the one for $A_+(X)$ gives the result for $A(X)$. □

For the rest of the section we discuss further properties of the sequence $\{A_n\}_n$, which are of independent interest, but are not used later.

Note first that the proof of Lemma 12.1.3 gives a more precise information about the center of $A_+(X)$ for periodic X than that formulated in the above corollary. Namely, let $n = 2m$ be such that $c_{n+k} = k$ for $k \geq 0$. Then in the notation of Lemma 12.1.3 the space H_∞ is spanned by $T^k h$, $k \geq 0$, where h is the unique nonzero element in H_{n+1}. Hence the center $Z = C^*(H_\infty, \omega|_{H_\infty})$ of $A_+(X)$ is isomorphic to $C(\{0,1\}^{\mathbb{N}})$ in such a way that the endomorphism $\alpha|_Z$ becomes the usual shift endomorphism of $C(\{0,1\}^{\mathbb{N}})$. It follows also that the kernel H of B is spanned by $T^k h$, $k \in \mathbb{Z}$. Thus the center $Z(X)$ of $A(X)$ is isomorphic to $C(\{0,1\}^{\mathbb{Z}})$ in such a way that the automorphism $\alpha|_{Z(X)}$ becomes the shift automorphism of $C(\{0,1\}^{\mathbb{Z}})$.

Let us call a word in A every nonempty product $w = s_{i_1} \ldots s_{i_k}$ with $i_1 < \ldots < i_k$. Then we can reformulate the above remark as follows. If X is periodic then there exists a unique word $w \in A_+(X)$ such that the center $Z_+(X)$ of $A_+(X)$ is generated by $\alpha^k(w)$, $k \geq 0$, and then the center $Z(X)$ of $A(X)$ is generated by $\alpha^k(w)$, $k \in \mathbb{Z}$. If $w = s_{i_1} \ldots s_{i_k}$, then $i_1 = 1$ (otherwise $\alpha^{-1}(w)$ would be in the center), $i_k = 2m+1$ for some $m \geq 0$, and

$$A_+(X) \cong \mathrm{Mat}_{2^m}(\mathbb{C}) \otimes Z_+(X) \quad \text{and} \quad A(X) \cong \mathrm{Mat}_{2^m}(\mathbb{C}) \otimes Z(X).$$

The sequences $\{c_n\}_n$ and $\{d_n\}_n$ in Theorem 12.1.4 are determined by $\{n_l\}_l$. Any sequence $\{n_l\}_l$ can arise this way (assuming that it is strictly increasing, consists of even numbers, and $n_1 = 0$). In other words, the restrictions on $\{c_n\}_n$ given by Lemmas 12.1.1 and 12.1.2 are the only ones. To see this note that $H_1, \ldots, H_n$ depend only on $X \cap [1, n-1]$. Hence we can construct X inductively. So let $\{c_n\}_n$ be the sequence defined by $\{n_l\}_{l=1}^N$ according to part (iii) of Theorem 12.1.4, and assume we have constructed $X_n \subset \{1, \ldots, n-1\}$ such that for the corresponding form B_n we have $\dim H_k = c_k$ for $k \leq n$. We have to set either $X_{n+1} = X_n$ or $X_{n+1} = X_n \cup \{n\}$ such that for the corresponding form B_{n+1} we get $\dim H_{n+1} = c_{n+1}$. If $c_n = 0$, then $c_{n+1} = 1$ and $\dim H_{n+1} = 1$ independently of how we define X_{n+1}. Similarly, if $c_n > 0$ and $c_n = c_{n-1} - 1$, then $c_{n+1} = c_n - 1$ and $\dim H_{n+1} = c_n - 1$ independently of the definition. So we may assume that for some $m < n$ we have $c_k = k - m$ for $m \leq k \leq n$. Let $h \in H_{m+1}$ be the unique nonzero element. By Lemma 12.1.1 the space H_n is spanned by $h, Th, \ldots, T^{n-m-1}h$. Since $H_m = 0$, we have $h = g_1 + h'$ for some $h' \in G_{[2,m+1]}$. Since

$$B_{n+1}(g, g_{n+1}) = B_{n+1}(T^{-1}g, g_n) = B_n(T^{-1}g, g_n)$$

for $g \in G_{[2,n+1]}$, we have $Th, \ldots, T^{n-m-1}h \in H_{n+1}$ independently of how we define X_{n+1}. On the other hand, since

$$B_{n+1}(h, g_{n+1}) = B_{n+1}(g_1, g_{n+1}) + B_n(T^{-1}h', g_n),$$

we see that we can arrange so that we have both $h \in H_{n+1}$ or $h \notin H_{n+1}$. In the first case $H_n \subset H_{n+1}$, so that $\dim H_{n+1} = c_n + 1$ by Lemma 12.1.1. In the second case H_n is not contained in H_{n+1}, so that $\dim H_{n+1} = c_n - 1$. So whatever c_{n+1} is, we can define X_{n+1} such that $\dim H_{n+1} = c_{n+1}$.

The above argument shows also that if $N = \infty$ then for infinitely many n's it does not matter how we define X_{n+1} from X_n. Thus there exist uncountably many sets X giving the same sequence $\{n_l\}_{l=1}^N$.

In the next section we shall see that an interesting class of systems is obtained when X is finite. We finish this section by showing that in this case the sequence $\{c_n\}_n$ is periodic.

Proposition 12.1.6. *Let $X \subset \mathbb{N}$ be finite and nonempty. Then there exists $m \in \mathbb{N}$ such that $c_{n+m} = c_n$ for any $n \geq 0$.*

Proof. Let d be the largest number in X. If $g = \sum_{j=1}^n \xi_j g_j$, $\xi_j \in \mathbb{F}_2$, then $g \in H_n$ if and only if $B(g, g_i) = 0$ for $i = 1, \ldots, n$, that is, $\sum_{j=1}^n \chi(|j-i|)\xi_j = 0$, where χ is the characteristic function of X. Letting $\xi_{-d+1} = \ldots = \xi_0 = \xi_{n+1} = \ldots = \xi_{n+d} = 0$ we can write this condition as

$$\xi_{i-d} + \sum_{k=-d+1}^{d-1} \chi(|k|)\xi_{i+k} + \xi_{i+d} = 0 \tag{12.1}$$

for $i = 1, \ldots, n$. Consider (12.1) for all $i \in \mathbb{Z}$ as a system of equations. If $\xi_{-d+1}, \ldots, \xi_{n+d}$ satisfy (12.1) for $i = 1, \ldots, n$ then all other ξ_j's are uniquely determined. Thus there is a bijection between H_n and the set of solutions $\xi = (\xi_i)_{i \in \mathbb{Z}}$ of (12.1) such that $\xi_{-d+1} = \ldots = \xi_0 = \xi_{n+1} = \ldots = \xi_{n+d} = 0$.

We consider sequences $\xi = (\xi_i)_{i \in \mathbb{Z}}$ as elements of $\bar{G} = \prod_{-\infty}^{\infty} \mathbb{Z}/2\mathbb{Z}$. Let $\bar{T}$ be the shift to the right on $\bar{G}$, so $(\bar{T}\xi)_i = \xi_{i-1}$. Then equations (12.1) can be written as $p(\bar{T})\xi = 0$, where $p(t) \in \mathbb{F}_2[t]$ is the polynomial defined by

$$p(t) = t^{2d} + \sum_{k=-d+1}^{d-1} \chi(|k|)t^{d-k} + 1.$$

To summarize, there is an isomorphism between H_n and the space of $\xi \in \bar{G}$ such that $p(\bar{T})\xi = 0$ and $\xi_{-d+1} = \ldots = \xi_0 = \xi_{n+1} = \ldots = \xi_{n+d} = 0$.

The ring $\mathbb{F}_2[t]/(p(t))$ is finite. The element t is invertible in this ring, so it is of finite order. In other words, there exists $m \in \mathbb{N}$ such that $t^m + 1$ lies in the ideal generated by $p(t)$. But then if $p(\bar{T})\xi = 0$ we have $\bar{T}^m \xi = \xi$, that is, $\xi_{i+m} = \xi_i$. Hence for solutions of $p(\bar{T})\xi = 0$ instead of the condition $\xi_{-d+1} = \ldots = \xi_0 = \xi_{n+1} = \ldots = \xi_{n+d} = 0$ we can equally well consider the condition $\xi_{-d+1} = \ldots = \xi_0 = \xi_{n+m+1} = \ldots = \xi_{n+m+d} = 0$. But the latter condition determines H_{n+m}. Thus we get an isomorphism between H_{n+m} and H_n. Explicitly, the isomorphism maps $\sum_{j=1}^{n+m} \xi_j g_j \in H_{n+m}$ to $\sum_{j=1}^{n} \xi_j g_j \in H_n$. □

In particular, if X is finite and nonempty, the sequence $\{c_n\}_n$ is bounded. If d is the largest number in X, then using that $B(g, g_{n+d}) = 1$ for any $g \in G_n \backslash G_{n-1}$, we see that if $c_{n-1} = 0$ then H_n, which is nonzero and contained in $G_n \backslash G_{n-1}$, is not contained in H_{n+d}. It follows that the sequence $1 = c_n, c_{n+1}, \ldots, c_{n+d}$ cannot be increasing. Hence $c_k \le d$ for any $k \in \mathbb{N}$.

12.2 Entropy of Binary Shifts

In the notation of the previous section let X be a subset of $\mathbb{N}$ associated with a bitstream, and let $A(X)$ be the C*-algebra generated by symmetries s_n, $n \in \mathbb{Z}$, satisfying the commutation relations $s_i s_j = (-1)^{\chi(|i-j|)} s_j s_i$, where χ is the characteristic function of X. Let also τ_X be the canonical tracial state on $A(X)$. The shift $s_n \mapsto s_{n+1}$ defines an automorphism σ_X of $A(X)$ which is called the **binary shift** corresponding to X. In the present section we shall study entropic properties of such automorphisms.

We start with a simple estimate.

Proposition 12.2.1. *We have $\frac{1}{2}\log 2 \le ht(\sigma_X) \le \log 2$. If X is nonperiodic and φ is a σ_X-invariant state on $A(X)$ then $h_\varphi(\sigma_X) \le \frac{1}{2}\log 2$.*

Proof. The estimates for the topological entropy follow from Theorem 11.2.1. They will also follow from the argument below and Proposition 12.2.4.

As in the previous section denote by A_n the C*-algebra generated by $s_1, \ldots, s_n$. Then by the by now standard argument, see e.g. the proof of Proposition 10.4.5, we have

$$h_\varphi(\sigma_X) \leq \liminf_n \frac{1}{n} \log \operatorname{rank} A_n \quad \text{and} \quad ht(\sigma_X) \leq \limsup_n \frac{1}{n} \log \operatorname{rank} A_n.$$

By Theorem 12.1.4 the rank of A_n is not greater than 2^n, and if X is nonperiodic it is equal to $2^{n/2}$ for an infinite number of n's. This gives the required upper estimates for $ht(\sigma_X)$ and $h_\varphi(\sigma_X)$. $\square$

Our next goal is to show that generically a binary shift has zero entropy and different bitstreams give rise to nonconjugate systems.

As in the previous section let $G = \bigoplus_{\mathbb{Z}} \mathbb{Z}/2\mathbb{Z}$ and let $\omega_X = \omega$ denote the bicharacter on G defined by X. Let T denote the shift to the right on G, so that σ_X is the automorphism of $A(X) = C^*(G, \omega_X)$ defined by T. We shall identify a subset X of $\mathbb{N}$ with the point $(1_X(i))_{i \geq 1}$ in $Y = \prod_{i=1}^\infty \{0, 1\}$, where 1_X is the characteristic function of X. Let μ denote the product measure on Y with weights $(\frac{1}{2}, \frac{1}{2})$.

Lemma 12.2.2. *Let $\rho_X(g, h) = \omega_X(g, h)\omega_X(h, g)$ for $g, h \in G$. Then the sequence $\{\rho_X(T^n g, h)\}_{n=1}^\infty$ is uniformly distributed for $g, h \neq 0$ for μ-almost every X.*

Recall that a sequence $\{a_n\}_n$ such that $a_n = \pm 1$ is said to be uniformly distributed if $n^{-1} \sum_{m=1}^n a_m \to 0$ as $n \to \infty$.

Proof of Lemma 12.2.2. Let f denote the function on Y defined by

$$f(y) = (-1)^{y_1}, \quad y = (y_1, y_2, \ldots) \in Y.$$

Let S be the shift to the left on Y, $S(y_1, y_2, \ldots) = (y_2, y_3, \ldots)$. Then if $i > j$ we have

$$\rho_X(g_i, g_j) = (f \circ S^{i-j-1})(X),$$

where as before $g_i = 1$ in the i-th summand $(\mathbb{Z}/2\mathbb{Z})_i$. Let $g = g_{i_1} + \ldots + g_{i_k}$, $i_1 < \ldots < i_k$, and $h = g_{j_1} + \ldots + g_{j_l}$, $j_1 < \ldots < j_l$, be given elements in G. Replacing g by $T^{n_0} g$ for some large n_0 we may assume $i_s > j_t$ for all s, t. Let F denote the function on Y defined by $F(X) = \rho_X(g, h)$. Then

$$F(X) = \prod_{1 \leq s \leq k, 1 \leq t \leq l} (f \circ S^{i_s - j_t - 1})(X).$$

It follows that $\rho_X(T^n g, h) = (F \circ S^n)(X)$. By Birkhoff's ergodic theorem, see e.g. [72, Theorem 3.41], for almost all X we have

$$\frac{1}{n} \sum_{m=1}^n (F \circ S^m)(X) \to \int_Y F d\mu.$$

The function f takes the value ± 1 with probability $\frac{1}{2}$. Hence $\int_Y f d\mu = 0$, and similarly $\int_Y f \circ S^{i_k - j_1 - 1} d\mu = 0$. Now F is a product of $f \circ S^{i_k - j_1 - 1}$ and a function depending only on the first $i_k - j_1 - 1$ coordinates. Since μ is a product measure, we have $\int_Y F d\mu = 0$, whence $\{\rho_X(T^n g, h)\}_{n=1}^\infty$ is uniformly distributed. □

If we combine the above lemma with Theorems 11.4.1 and 11.4.2 we obtain the following result.

Theorem 12.2.3. *For almost all subsets X of $\mathbb{N}$ the binary shift σ_X has entropy $h_\tau(\sigma_X) = 0$ with respect to the trace $\tau = \tau_X$. For any such X the trace τ_X is the unique σ_X-invariant state on $A(X)$.* □

Though almost every binary shift has zero entropy, we shall see soon that in many cases the entropy is $\frac{1}{2} \log 2$.

Similarly to the discussion at the end of Sect. 11.5, we conclude that binary shifts provide many examples of systems for which the tensor product formula for the entropy fails. More precisely, we have the following result.

Proposition 12.2.4. *If X is nonperiodic then $h_{\tau \otimes \tau}(\sigma_X \otimes \sigma_X) = \log 2$.*

Proof. Similarly to the proof of Proposition 12.2.1 we have

$$h_{\tau \otimes \tau}(\sigma_X \otimes \sigma_X) \leq \liminf_n \frac{1}{n} \log \operatorname{rank} A_n \otimes A_n = \log 2.$$

The argument from the proof of Theorem 11.2.1 for the converse inequality now goes as follows. Let D denote the C*-subalgebra of $A(X) \otimes A(X)$ generated by the symmetries $s_n \otimes s_n$, $n \in \mathbb{Z}$. Then D is abelian, and the system $(D, (\tau \otimes \tau)|_D, (\sigma_X \otimes \sigma_X)|_D)$ is isomorphic to the Bernoulli shift with weights $(\frac{1}{2}, \frac{1}{2})$, hence it has entropy $\log 2$. Thus $h_{\tau \otimes \tau}(\sigma_X \otimes \sigma_X) \geq h_{(\tau \otimes \tau)|D}(\sigma_X \otimes \sigma_X|_D) = \log 2$. □

Denote by $M(X)$ the weak operator closure of $A(X)$ in the GNS-representation corresponding to $\tau = \tau_X$. Denote the extensions of τ_X and σ_X to $M(X)$ by the same symbols. Note that if X is nonperiodic then by Corollary 12.1.5 the algebra $M(X)$ is the hyperfinite II_1-factor.

Remark 12.2.5. In analogy with the classical situation if (M, τ, σ) is a W*-dynamical system we can say that a σ-invariant von Neumann subalgebra M_π is a Pinsker algebra if it satisfies the following conditions:

(i) $h_{\tau|M_\pi}(\sigma|_{M_\pi}) = 0$,

(ii) if B is a finite dimensional C*-subalgebra of M_π such that $h_\tau(B; \sigma) = 0$ then $B \subset M_\pi$.

In the abelian case such an algebra always exists. In the noncommutative case, however, this is false. Indeed, by Theorem 12.2.3 there exists a set X such that $h_\tau(\sigma_X) = 0$. Then $H_\tau(B; \sigma_X) = 0$ for all finite dimensional C*-subalgebras B of $M(X)$. Consider now the W*-dynamical system

$(M(X)\bar{\otimes}M(X), \tau \otimes \tau, \sigma_X \otimes \sigma_X)$. Then $h_{\tau\otimes\tau}(\sigma_X \otimes \sigma_X) = \log 2$ by Proposition 12.2.4. For each finite dimensional C*-subalgebra B of $M(X)$ we have $h_{\tau\otimes\tau}(B \otimes \mathbb{C}; \sigma_X \otimes \sigma_X) = 0$. Hence, if the Pinsker algebra for the system $(M(X)\bar{\otimes}M(X), \tau \otimes \tau, \sigma_X \otimes \sigma_X)$ existed, it would contain $M(X) \otimes \mathbb{C}$, and by symmetry $\mathbb{C} \otimes M(X)$. Thus it would coincide with $M(X)\bar{\otimes}M(X)$, a contradiction. ♦

Lemma 12.2.6. *Let $X_1, X_2 \subset \mathbb{N}$. Suppose there exist sequences $\{n_k\}_{k=1}^\infty \subset X_1$ and $\{m_k\}_{k=1}^\infty \subset \mathbb{N}$ such that $n_k, m_k \to \infty$ and $\mathbb{Z} \cap [n_k - m_k, n_k + m_k] \subset \mathbb{N}\backslash X_2$. Then the systems $(M(X_1), \tau_{X_1}, \sigma_{X_1})$ and $(M(X_2), \tau_{X_2}, \sigma_{X_2})$ are nonconjugate.*

Proof. Let s_n^1 and s_n^2 be the symmetries corresponding to X_1 and X_2, respectively. Then we have
$$\sigma_{X_1}^{n_k}(s_n^1)s_n^1 = -s_n^1\sigma_{X_1}^{n_k}(s_n^1)$$
for all n and k, so that $\|[\sigma_{X_1}^{n_k}(s_n^1), s_n^1]\|_2 = 2$. On the other hand, we have
$$\sigma_{X_2}^{n_k}(s_i^2)s_j^2 = s_j^2\sigma_{X_2}^{n_k}(s_i^2)$$
whenever $|i - j| \leq m_k$, from which it follows that
$$\lim_{k\to\infty} \|[\sigma_{X_2}^{n_k}(x), y]\|_2 = 0 \text{ for any } x, y \in M(X_2).$$
Thus σ_{X_1} and σ_{X_2} are nonconjugate. □

Theorem 12.2.7. *For almost all pairs (X_1, X_2) (with respect to the measure $\mu \times \mu$ on $Y \times Y$) the W^*-dynamical systems $(M(X_1), \tau_{X_1}, \sigma_{X_1})$ and $(M(X_2), \tau_{X_2}, \sigma_{X_2})$ are nonconjugate.*

Proof. Choose a strictly increasing sequence $\{m_k\}_k$ of natural numbers. For each $k \in \mathbb{N}$ let f_k be the function on $Y \times Y$ defined by
$$f_k(y', y'') = \begin{cases} 1, & \text{if } y'_{m_k+1} = 1, y''_l = 0 \text{ for } 1 \leq l \leq 2m_k + 1, \\ 0, & \text{otherwise.} \end{cases}$$

As before let S be the shift to the left on Y. Then for $n \geq m_k + 1$ the conditions $n \in X_1$ and $\mathbb{Z} \cap [n - m_k, n + m_k] \subset \mathbb{N}\backslash X_2$ mean exactly that $f_k(S^{n-m_k-1}X_1, S^{n-m_k-1}X_2) = 1$. For almost every pair (X_1, X_2) Birkhoff's ergodic theorem implies
$$\frac{1}{n}\sum_{m=1}^n f_k(S^n X_1, S^n X_2) \to \int_{Y\times Y} f_k d(\mu \times \mu) = 2^{-2m_k - 2} > 0.$$

Thus for almost all pairs (X_1, X_2) we have $f_k(S^{n-m_k-1}X_1, S^{n-m_k-1}X_2) = 1$ for infinitely many n's. In particular, for almost every pair (X_1, X_2) we can find an increasing sequence $\{n_k\}_k$ such that the assumptions of Lemma 12.2.6 are satisfied. This completes the proof of the theorem. □

Having proved generic results, we now consider special cases. We start with the periodic case.

Proposition 12.2.8. *Suppose $X \subset \mathbb{N}$ is periodic. Then $h_\tau(\sigma_X) = \log 2$.*

Proof. Let p be the period of $-X \cup X$. Put $w = s_0 s_p$. Then w lies in the center of $A(X)$. Since either w or iw is a symmetry, and $\tau(w) = 0$, the restriction of σ_X to the C*-algebra generated by $\sigma_X^n(w)$ is the Bernoulli shift with weights $(\frac{1}{2}, \frac{1}{2})$. Hence $h_\tau(\sigma_X) \geq \log 2$. On the other hand, by Proposition 12.2.1 we have $h_\tau(\sigma_X) \leq ht(\sigma_X) \leq \log 2$. Thus the proof is complete. □

We next consider binary shifts corresponding to finite sets. It turns out that these systems admit several dynamical characterizations.

Recall that an automorphism α of a C*-algebra A is asymptotically abelian if $\lim_n \|[\alpha^n(a), b]\| = 0$ for all $a, b \in A$. It is asymptotically abelian with locality if there is a dense *-subalgebra $\mathcal{A}$ such that for each pair $a, b \in \mathcal{A}$ we have $[\alpha^n(a), b] = 0$ for sufficiently large n, and $C^*(a, b)$ is finite dimensional. Recall also that a W*-dynamical system (M, τ, α) is called an entropic K-system if

$$\lim_{n \to \infty} h_\tau(\gamma; \alpha^n) = H_\tau(\gamma)$$

for any channel $\gamma \colon B \to M$.

Theorem 12.2.9. *Suppose $X \subset \mathbb{N}$ is nonperiodic. Then the following four conditions are equivalent:*

(i) *X is finite;*

(ii) *$(A(X), \sigma_X)$ is asymptotically abelian;*

(iii) *$(A(X), \sigma_X)$ is asymptotically abelian with locality;*

(iv) *$(M(X), \tau_X, \sigma_X)$ is an entropic K-system.*

Proof. Suppose X is finite, say $X \subset \{1, 2, \ldots, r\}$. Let $\mathcal{A} = \bigcup_{n=1}^\infty A_{[-n,n]}$, where $A_{[-n,n]}$ is generated by $s_{-n}, \ldots, s_n$. Then $\mathcal{A}$ is dense in $A(X)$. If $a, b \in A_{[-n,n]}$, let $m > 2n + r$. Then $[\sigma_X^m(a), b] = 0$, proving $(A(X), \sigma_X)$ is asymptotically abelian with locality. Thus (i) implies (iii).

Clearly (iii) implies (ii). Assume (ii). Since $\|[s_0, \sigma_X^n(s_0)]\| = \|[s_0, s_n]\| = 2$ if $n \in X$, it is immediate from asymptotic abelianness that there exists $r \in \mathbb{N}$ such that $n \notin X$ for $n > r$. Thus $X \subset \{1, \ldots, r\}$, and (i) follows.

We conclude by showing (i)⇔(iv). Assume $X \subset \{1, \ldots, r\}$. Let M_0 be the weak operator closure of $\bigcup_n A_{[-n,0]}$ in $M(X)$. Since $M(X)$ is a factor and $A_{[-n+r+1, n]} \subset \sigma_X^{-n}(M_0)' \cap \sigma_X^n(M_0)$, we conclude that $(M(X), \tau_X, \sigma_X)$ is an entropic K-system by Theorem 4.3.2. Note however that we have the following factorization property, which makes the proof of Theorem 4.3.2 in this case very simple. If $a \in A_{[-n,n]}$ and $b \in \sigma_X^m(A_{[-n,n]})$ with $m \geq 2n + r + 1$, then $ab = ba$ and $\tau(ab) = \tau(a)\tau(b)$.

Conversely, assume $(M(X), \tau_X, \sigma_X)$ is an entropic K-system. Let C be a finite dimensional abelian subalgebra of $A(X)$. Since

$$H_\tau(C, \sigma_X^n(C), \ldots, \sigma_X^{(2m-1)n}(C)) \leq m H_\tau(C, \sigma_X^n(C)),$$

so that $2h_\tau(C;\sigma_X^n) \leq H_\tau(C,\sigma_X^n(C))$, we conclude $H_\tau(C,\sigma_X^n(C)) \to 2H_\tau(C)$.

Let $C = C^*(s_0) = \mathbb{C}1 + \mathbb{C}s_0$. Then $H_\tau(C) = \log 2$, so $H_\tau(C,\sigma_X^n(C)) \geq \log 2$. If s_n and s_0 commute then C and $\sigma_X^n(C)$ are independent, so that $H_\tau(C,\sigma_X^n(C)) = 2\log 2$. On the other hand, if they anticommute, then C and $\sigma_X^n(C)$ generate the algebra of 2×2 matrices, so that $H_\tau(C,\sigma_X^n(C)) \leq \log 2$. To summarize,

$$H_\tau(C,\sigma_X^n(C)) = \begin{cases} \log 4, & \text{if } n \notin X, \\ \log 2, & \text{if } n \in X. \end{cases}$$

Since $H_\tau(C,\sigma_X^n(C)) \to 2H_\tau(C) = \log 4$, we conclude that there exists r such that $H_\tau(C,\sigma_X^n(C)) = \log 4$ whenever $n > r$. Thus $X \subset \{1,\ldots,r\}$. □

Note that we could alternatively use Theorem 4.3.5 to show that (iv) implies that the system $(M(X),\tau_X,\sigma_X)$ is strongly asymptotically abelian, and then deduce finiteness of X by the same argument as for the implication (ii)⇒(i) above.

Our next result shows that for X finite, dynamical entropy coincides with both topological entropy and mean entropy.

Theorem 12.2.10. *Let X be a finite nonempty subset of $\mathbb{N}$. Then*

$$h_\tau(\sigma_X) = ht(\sigma_X) = \lim_{n\to\infty} \frac{1}{n} H_\tau(A_n) = \frac{1}{2}\log 2.$$

Proof. If $X \subset \{1,\ldots,r\}$, then the algebras $A_n, \sigma_X^m(A_n), \ldots, \sigma_X^{km}(A_n)$ are τ-independent for $m \geq n+r+1$. By Proposition 10.4.5 we conclude that

$$h_\tau(\sigma_X) = \lim_{n\to\infty} \frac{1}{n} H_\tau(A_n).$$

On the other hand, by Theorem 12.1.4(i) we have $H_\tau(A_n) = \log \operatorname{rank} A_n$. Since $\operatorname{rank} A_n = 2^{n/2}$ for infinitely many n's by Theorem 12.1.4(iii), we get

$$h_\tau(\sigma_X) = \lim_{n\to\infty} \frac{1}{n} H_\tau(A_n) = \lim_{n\to\infty} \frac{1}{n}\log \operatorname{rank} A_n = \frac{1}{2}\log 2.$$

Since we always have

$$ht(\sigma_X) \leq \limsup_{n\to\infty} \frac{1}{n} \log \operatorname{rank} A_n,$$

we also get $ht(\sigma_X) = \frac{1}{2}\log 2$. □

Remark that by Theorem 12.2.10 the algebra $A(X)$ is UHF, hence τ_X is the unique tracial state on $A(X)$. By Theorem 11.4.2, if X is finite, there exist invariant states on $A(X)$ different from τ_X. Then by Theorem 9.3.3 we have $h_\varphi(\sigma_X) < \frac{1}{2}\log 2$ for any such state.

Corollary 12.2.11. *Suppose $X_i \subset \mathbb{N}$, $i = 1,2$, are finite. Then*

$$h_{\tau_{X_1}\otimes\tau_{X_2}}(\sigma_{X_1}\otimes\sigma_{X_2}) = \log 2 = h_{\tau_{X_1}}(\sigma_{X_1}) + h_{\tau_{X_2}}(\sigma_{X_2}).$$

Proof. Let $A_{in} \subset A(X_i)$ be the algebra generated by $s_1, \ldots, s_n \in A(X_i)$, $i = 1, 2$. Then

$$h_{\tau_{X_1} \otimes \tau_{X_2}}(\sigma_{X_1} \otimes \sigma_{X_2}) \leq \liminf_{n \to \infty} \frac{1}{n} H_{\tau_{X_1} \otimes \tau_{X_2}}(A_{1n} \otimes A_{2n})$$

$$= \lim_{n \to \infty} \frac{1}{n}(H_{\tau_{X_1}}(A_{1n}) + H_{\tau_{X_2}}(A_{2n})).$$

On the other hand, by Theorem 3.2.2(iv),

$$h_{\tau_{X_1} \otimes \tau_{X_2}}(\sigma_{X_1} \otimes \sigma_{X_2}) \geq h_{\tau_{X_1}}(\sigma_{X_1}) + h_{\tau_{X_2}}(\sigma_{X_2}).$$

Thus the result follows from the previous theorem. □

We next show that even when X is infinite the entropy can often be $\frac{1}{2} \log 2$.

Lemma 12.2.12. *Suppose X is nonperiodic and either contained in the even numbers or the odd numbers, then $h_\tau(\sigma_X) = \frac{1}{2} \log 2$.*

Proof. Assume first $X \subset \{1, 3, 5, \ldots\}$. Let $C_0 = C^*(s_0)$. Since $s_{2i}s_{2j} = s_{2j}s_{2i}$ for all $i, j \in \mathbb{Z}$, the algebras $\sigma_X^{2n}(C_0)$, $n \in \mathbb{Z}$, are independent. Thus σ_X^2 acts as the Bernoulli shift with weights $(\frac{1}{2}, \frac{1}{2})$ on the abelian C*-algebra C they generate, so that $h_{\tau|C}(\sigma_X^2|_C) = \log 2$. Hence

$$h_\tau(\sigma_X) = \frac{1}{2}h_\tau(\sigma_X^2) \geq \frac{1}{2}h_{\tau|C}(\sigma_X^2|_C) = \frac{1}{2}\log 2,$$

which together with Proposition 12.2.1 proves the lemma for X contained in the odd numbers.

Suppose next $X \subset \{2, 4, 6, \ldots\}$. Then $s_{2i+1}s_{2j} = s_{2j}s_{2i+1}$ for all $i, j \in \mathbb{Z}$. Letting $t_i = s_{2i}s_{2i+1}$ it follows that $t_i t_j = t_j t_i$ for all $i, j \in \mathbb{Z}$. Thus the C*-algebra $D = C^*(t_i \mid i \in \mathbb{Z})$ is abelian, and $\sigma_X^2|_D$ is the Bernoulli shift with weights $(\frac{1}{2}, \frac{1}{2})$. Now the argument is completed as in the odd case. □

Let us say that a W*-dynamical system (M, τ, σ) has **completely positive entropy** if $h_\tau(\gamma; \sigma) > 0$ for any nontrivial channel $\gamma \colon B \to M$ (that is, $\gamma(B) \neq \mathbb{C}1$). Clearly, an entropic K-system has completely positive entropy. Although by Theorem 12.2.9 the system $(M(X), \tau_X, \sigma_X)$ is an entropic K-system only if X is finite, we shall next show that it can have completely positive entropy even if X is infinite.

Proposition 12.2.13. *Suppose $q \geq 3$ is an odd integer and $X \subset \{q^n \mid n \geq 0\}$. Then $(M(X), \tau_X, \sigma_X)$ has completely positive entropy.*

Proof. Since $h_\tau(\gamma; \sigma_X^m) \leq m h_\tau(\gamma; \sigma_X)$ and $H_\tau(\gamma) > 0$ for nontrivial γ, it suffices to prove that

$$\lim_{n \to \infty} h_\tau(\gamma; \sigma_X^{2q^n}) = H_\tau(\gamma).$$

Note that if $x, y \in A_{[-m,m]}$ and $m \leq (q^n - 1)/2$ then x and $\sigma_X^{2kq^n}(y)$ commute for any $k \in \mathbb{N}$. Indeed, otherwise there exist i, j such that $|i|, |j| \leq m$ and $q^l = 2kq^n + i - j \in X$ for some l. It follows that $(2k-1)q^n + 1 \leq q^l \leq (2k+1)q^n - 1$, so that $2k - 1 < q^{l-n} < 2k+1$, which is impossible as q is an odd integer.

For any $\varepsilon > 0$ we can find a finite partition of unity $1 = \sum_i x_i$ in $A_{[-m,m]}$ for some m such that

$$H_\tau(\gamma) < H_\tau(\gamma; \{x_i\}_i) + \varepsilon.$$

Then for $q^n \geq 2m + 1$ the elements $x_{i_0}, \sigma_X^{2q^n}(x_{i_1}), \ldots, \sigma_X^{2kq^n}(x_{i_k})$ are independent, so that

$$H_\tau(\gamma, \sigma_X^{2q^n} \circ \gamma, \ldots, \sigma_X^{2kq^n} \circ \gamma; \{x_{i_0} \sigma_X^{2q^n}(x_{i_1}) \ldots \sigma_X^{2kq^n}(x_{i_k})\})$$

$$= \sum_{i_0,\ldots,i_k} \eta(\tau(x_{i_0} \sigma_X^{2q^n}(x_{i_1}) \ldots \sigma_X^{2kq^n}(x_{i_k}))) + (k+1) \sum_i S(\tau(\gamma(\cdot)x_i), \tau \circ \gamma)$$

$$= (k+1) \sum_i (\eta(\tau(x_i)) + S(\tau(\gamma(\cdot)x_i), \tau \circ \gamma)) > (k+1)(H_\tau(\gamma) - \varepsilon),$$

whence $h_\tau(\gamma; \sigma_X^{2q^n}) > H_\tau(\gamma) - \varepsilon$. Since ε was arbitrary and we always have $h_\tau(\gamma; \sigma_X^{2q^n}) \leq H_\tau(\gamma)$, the proof is complete. □

We thus get examples of systems which have completely positive entropy, but are not entropic K-systems. It turns out that among them there are uncountably many nonisomorphic ones.

Proposition 12.2.14. *Let $q \geq 2$ be an integer and $X_i \subset \{q^n \mid n \geq 0\}$, $i = 1, 2$. Suppose the set $X_1 \cap (\mathbb{N} \backslash X_2)$ is infinite. Then $(M(X_1), \tau_{X_1}, \sigma_{X_1})$ and $(M(X_2), \tau_{X_2}, \sigma_{X_2})$ are nonconjugate.*

Proof. Let $n_1 < n_2 < \ldots$ be such that $q^{n_k} \in X_1 \cap (\mathbb{N} \backslash X_2)$. Since $q^{n_k} \notin X_2$, we have

$$[q^{n_k} - q^{n_k - 1} + 1, q^{n_k} + q^{n_k - 1} - 1] \subset \mathbb{N} \backslash X_2.$$

Thus the result follows from Lemma 12.2.6. □

If we combine the above proposition with Lemma 12.2.12 and Proposition 12.2.13 we obtain the following.

Theorem 12.2.15. *There is an uncountable family of pairwise nonconjugate systems $(M(X), \tau_X, \sigma_X)$ with completely positive entropy such that $h_{\tau_X}(\sigma_X) = \frac{1}{2} \log 2$.* □

12.3 Notes

The study of binary shifts was initiated by Powers [171]. One should be aware that in the literature one usually understands by a binary shift the one-sided shift. The criterion for simplicity of the C*-algebra of a bitstream was obtained by Price [173]. The fact that the dimensions c_n of the centers of the algebras A_n have the form described in Theorem 12.1.4 was proved by Powers and Price [172]. The observation that any sequence $\{c_n\}_n$ with properties as in Theorem 12.1.4 arises from a bitstream was made by Price [175]. In our exposition of these results we also benefited from the work of Vik [220]. Proposition 12.1.6 is due to Enomoto, Nagisa, Watatani and Yoshida [64].

Some of the results for binary shifts can be generalized to automorphisms of the shift type on twisted group C*-algebras, see e.g. [39], [63], [38].

The entropy of the binary shifts corresponding to finite sets, Theorem 12.2.10, was computed by Choda [40]. A larger class of binary shifts with entropy $\frac{1}{2}\log 2$ was exhibited by Price [174]. The first example of a binary shift with zero entropy was given by Narnhofer, Størmer and Thirring [128]; see [69] for many such examples. It was the first example of a system for which the tensor product formula for the entropy fails. The fact that generically a binary shift has zero entropy, Theorem 12.2.3, was stated by Narnhofer and Thirring [134].

As was proved already by Powers [171], one-sided binary shifts corresponding to different nonperiodic sets are nonisomorphic. On the other hand, Lemma 12.2.6, proved by Golodets and Størmer [80], seems to be the only known sufficient condition for nonisomorphism of two-sided binary shifts. The observation that it implies that generically binary shifts are nonisomorphic, Theorem 12.2.7, is new. The rest of Sect. 12.2 is based on the work of Golodets and Størmer [80].

We have computed entropies of a large class of binary shifts, and the only values we have obtained are $0, \frac{1}{2}\log 2, \log 2$. The following problem thus naturally presents itself. Does there exist $X \subset \mathbb{N}$ such that $h_\tau(\sigma_X) \neq 0, \frac{1}{2}\log 2, \log 2$?

Let $(M(X), \tau_X, \sigma_X)$ be a binary shift. Consider the von Neumann subalgebra M_0 of $M(X)$ generated by s_n, $n \leq 0$. Then $M_0 \subset \sigma_X(M_0)$, $\cup_n \sigma_X^n(M_0)$ is dense in $M(X)$, $\cap_n \sigma_X^n(M_0) = \mathbb{C}1$. Thus $(M(X), \tau_X, \sigma_X)$ is what we called in Chap. 4 an algebraic K-system. Therefore there exist algebraic K-systems with zero entropy.

Theorem 12.2.15 shows that on the hyperfinite II_1-factor there exist uncountably many nonisomorphic systems with the same finite completely positive entropy. On the other hand, there are no examples of nonisomorphic entropic K-systems. In this respect a challenging open problem is whether binary shifts corresponding to different finite subsets are isomorphic. In Chap. 13 we shall construct uncountably many nonisomorphic entropic K-systems on the hyperfinite III_1-factor.

13
Bogoliubov Automorphisms

In this chapter we shall consider one of the basic models of quantum statistical mechanics, a system of noninteracting fermions. Our goal is to compute the topological entropy of such a system, as well as the dynamical entropy with respect to a natural class of states.

13.1 Canonical Anticommutation Relations

In the present section we collect basic facts about the CAR-algebra, the algebra of the canonical anticommutation relations. We refer the reader to [30, Chapter 5.2] for more details.

Let H be a Hilbert space. The **CAR-algebra** $\mathcal{A}(H)$ over H is a unital C*-algebra generated by elements $a(f)$ and $a^*(f)$, $f \in H$, such that the mapping $H \ni f \mapsto a^*(f)$ is linear[1], $a(f)^* = a^*(f)$ and

$$a^*(f)a(g) + a(g)a^*(f) = (f,g)1, \quad a(f)a(g) + a(g)a(f) = 0$$

for any $f, g \in H$.

If H is one-dimensional, $H = \mathbb{C}f$, $\|f\| = 1$, then the algebra $\mathcal{A}(\mathbb{C}f)$ is isomorphic to $\mathrm{Mat}_2(\mathbb{C})$, with matrix units defined by

$$e_{11}(f) = a(f)a^*(f), \ e_{22}(f) = a^*(f)a(f), \ e_{12}(f) = a(f), \ e_{21}(f) = a^*(f). \tag{13.1}$$

More generally, let $H = \mathbb{C}f \oplus K$, $\|f\| = 1$. Then the self-adjoint unitary $V(f) = a(f)a^*(f) - a^*(f)a(f)$ anticommutes with $a^*(f)$ and commutes with $a^*(g)$ for $g \in K$. Hence the map $a^*(g) \mapsto a^*(g)V(f)$ extends to an embedding of $\mathcal{A}(K)$ into $\mathcal{A}(H)$ such that the image commutes with $\mathcal{A}(\mathbb{C}f) \cong \mathrm{Mat}_2(\mathbb{C})$, and we thus get an isomorphism

---
[1] In the literature the opposite notation for the operators $a(f)$ and $a^*(f)$ is sometimes used, but we follow the conventions of [30].

$$\mathcal{A}(\mathbb{C}f \oplus K) \cong \mathrm{Mat}_2(\mathbb{C}) \otimes \mathcal{A}(K). \tag{13.2}$$

It follows that if H has dimension n then

$$\mathcal{A}(H) \cong \mathrm{Mat}_2(\mathbb{C})^{\otimes n}, \tag{13.3}$$

so $\mathcal{A}(H)$ is a full matrix algebra of dimension 2^{2n}. Explicitly, if $f_1, \ldots, f_n$ is an orthonormal basis in H then the matrix units $e_{ij}^{(k)}$ in the k-th factor $\mathrm{Mat}_2(\mathbb{C})$ are given by

$$e_{21}^{(k)} = a^*(f_k)V(f_1)\ldots V(f_{k-1}), \quad e_{12}^{(k)} = a(f_k)V(f_1)\ldots V(f_{k-1}),$$
$$e_{11}^{(k)} = a(f_k)a^*(f_k), \quad e_{22}^{(k)} = a^*(f_k)a(f_k). \tag{13.4}$$

Note that as $a^*(f)$ is a partial isometry when $\|f\| = 1$, we have $\|a^*(f)\| = 1$, and hence

$$\|a^*(g)\| = \|g\| \tag{13.5}$$

for any vector g.

Let U be a unitary operator on H. By the universal property of the CAR-algebra, it defines an automorphism α_U of $\mathcal{A}(H)$, $\alpha_U(a(f)) = a(Uf)$, called a **Bogoliubov automorphism**.

Let now A be an operator on H, $0 \leq A \leq 1$. It defines a state ω_A on $\mathcal{A}(H)$ by the formula

$$\omega_A(a^*(f_1)\ldots a^*(f_n)a(g_m)\ldots a(g_1)) = \delta_{nm}\det((Af_i, g_j))_{i,j},$$

called a **quasi-free state**.

If $H = \mathbb{C}f$, $\|f\| = 1$, and $A = \lambda 1$, then the quasi-free state ω_A on $\mathcal{A}(\mathbb{C}f) \cong \mathrm{Mat}_2(\mathbb{C})$ has density matrix

$$\begin{pmatrix} 1-\lambda & 0 \\ 0 & \lambda \end{pmatrix}.$$

More generally, if $H = \mathbb{C}f \oplus K$, $\|f\| = 1$, and $Af = \lambda f$, then under the isomorphism (13.2) we have

$$\omega_A = \mathrm{Tr}\left(\cdot \begin{pmatrix} 1-\lambda & 0 \\ 0 & \lambda \end{pmatrix}\right) \otimes \omega_{A|K}.$$

It follows that if H has dimension n and $f_1, \ldots, f_n$ is an orthonormal basis consisting of eigenvectors of A, $Af_k = \lambda_k f_k$, then under the isomorphism (13.3) we get

$$\omega_A = \bigotimes_{k=1}^n \mathrm{Tr}\left(\cdot \begin{pmatrix} 1-\lambda_k & 0 \\ 0 & \lambda_k \end{pmatrix}\right). \tag{13.6}$$

This gives a way to prove existence of quasi-free states: if A has pure point spectrum prove that the above product state is indeed the required quasi-free state, then approximate an arbitrary quasi-free state by quasi-free states corresponding to operators with pure point spectrum.

Finally note that if two operators A and U commute then the state ω_A is α_U-invariant.

13.2 Topological Entropy

Let U be a unitary operator on a separable Hilbert space H. In this section we compute the topological entropy of the Bogoliubov automorphism defined by U. To formulate the result recall, see C.7, that there exists a unique decomposition $H = H_s \oplus H_a$ such that $U_s = U|_{H_s}$ has singular spectral measure and $U_a = U|_{H_a}$ has absolutely continuous spectral measure with respect to the Lebesgue measure on $\mathbb{T}$. Furthermore, we have a direct integral decomposition

$$H_a = \int_\mathbb{T}^\oplus H_z d\mu(z), \quad U_a = \int_\mathbb{T}^\oplus z\, d\mu(z),$$

where μ is the normalized Lebesgue measure on $\mathbb{T}$. Denote by m_U the multiplicity function of U_a, $m_U(z) = \dim H_z$.

Theorem 13.2.1. *With the above notation, for the Bogoliubov automorphism α_U of the CAR-algebra $\mathcal{A}(H)$ we have*

$$ht(\alpha_U) = (\log 2) \int_\mathbb{T} m_U(z) d\mu(z).$$

In particular, $ht(\alpha_U) = 0$ if U has singular spectral measure.

Consider the simplest case when the unitary U is the bilateral shift, so it has homogeneous Lebesgue spectrum of multiplicity one. Then the theorem asserts that $ht(\alpha_U) = \log 2$.

To prove this let $\{f_n\}_{n \in \mathbb{Z}}$ be an orthonormal basis such that $Uf_n = f_{n+1}$. Then according to (13.3) and (13.4) the unital C*-subalgebra generated by $a^*(f_n)a(f_n)$, $n \in \mathbb{Z}$, is isomorphic to $C(\{0,1\}^\mathbb{Z})$, and the restriction of α_U to this subalgebra is just the shift automorphism with entropy $\log 2$. Hence $ht(\alpha_U) \geq \log 2$.

On the other hand, if Ω is a finite subset of $\mathcal{A}(H_{k,l})$, where $H_{k,l}$ is the space spanned by f_n, $n = k, \ldots, l$, then $\alpha_U^m(\Omega) \subset \mathcal{A}(H_{k,l+m})$, so that for any $\delta > 0$ and $n \in \mathbb{N}$

$$\operatorname{rcp}(\Omega \cup \alpha_U(\Omega) \cup \ldots \cup \alpha_U^{n-1}(\Omega), \delta) \leq \operatorname{rank} \mathcal{A}(H_{k,l+n-1}) = 2^{l-k+n}.$$

Hence $ht(\Omega, \delta; \alpha_U) \leq \log 2$. So indeed $ht(\alpha_U) = \log 2$.

Consider now the case of singular spectral measure. Then $ht(\alpha_U) = 0$. To show this we need a couple of technical lemmas.

Lemma 13.2.2. *Let U be a unitary operator with singular spectral measure, P a finite rank projection. Then for any $\varepsilon > 0$ there exists $n_0 \in \mathbb{N}$ such that for $n \geq n_0$ there is a projection P_n such that*

$$\operatorname{rank} P_n \leq \varepsilon n \quad \text{and} \quad \|(1-P_n)U^k P\| < \varepsilon \quad \text{for} \quad k = 0, \ldots, n.$$

Proof. Note that if $P = Q_1 + Q_2$ and Q_{in} is such that $\operatorname{rank} Q_{in} \le \varepsilon n/2$ and $\|(1-Q_{in})U^k Q_i\| < \varepsilon/2$, $i = 1, 2$, then the projection $P_n = Q_{1n} \vee Q_{2n}$ satisfies the conditions of the lemma. Thus it suffices to prove the lemma for rank one projections. Replacing H by a subspace we may further assume that P is the projection onto the space spanned by a cyclic vector for U. Then we can identify the Hilbert space with $L^2(\mathbb{T}, \nu)$, where ν is a singular probability measure, such that P becomes the projection onto the constant functions and U becomes the operator of multiplication by the function z.

Choose $N \in \mathbb{N}$ such that $\sqrt{2}/N < \varepsilon$. Since ν is singular, there exist $m \in \mathbb{N}$ and disjoint arcs $I_1, \ldots, I_m$ on $\mathbb{T}$ of length not greater than $1/N^2 m$ each such that
$$\nu(\mathbb{T} \setminus (I_1 \cup \ldots \cup I_m)) < \frac{\varepsilon^2}{2}.$$

Consider now any n of the form $n = Nml$, $l \in \mathbb{N}$. Subdivide each arc I_s into l arcs I_{st}, $1 \le t \le l$, of length not greater than $1/N^2 ml$ each. Let P_n be the projection onto the subspace spanned by the characteristic functions of these arcs. Then $\operatorname{rank} P_n \le ml = n/N$. On each arc I_{st} choose a point z_{st}. Then for $w \in I_{st}$ and $k = 0, \ldots, n$ we have
$$|w^k - z_{st}^k| \le \frac{n}{N^2 ml} = \frac{1}{N},$$
so that
$$\left\| z^k - \sum_{s,t} z_{st}^k \mathbb{1}_{I_{st}} \right\|_2^2 = \nu(\mathbb{T} \setminus (\cup_{s,t} I_{st})) + \sum_{s,t} \|(z^k - z_{st}^k) \mathbb{1}_{I_{st}}\|_2^2 < \frac{\varepsilon^2}{2} + \frac{1}{N^2} < \varepsilon^2.$$

Hence $\|z^k - P_n z^k\|_2 < \varepsilon$, that is, $\|(1 - P_n)U^k P\| < \varepsilon$.

Now for any $n \in \mathbb{N}$ set
$$P_n = P_{([n/Nm]+1)Nm}.$$
Then $\|(1 - P_n)U^k P\| < \varepsilon$ for $k = 0, \ldots, n$, and
$$\operatorname{rank} P_n \le \left(\left[\frac{n}{Nm} \right] + 1 \right) m \le n \left(\frac{1}{N} + \frac{m}{n} \right),$$
so that $\operatorname{rank} P_n < \varepsilon n$ if n is sufficiently large. □

Lemma 13.2.3. *For any $m \in \mathbb{N}$ and $\delta > 0$ there exists $\varepsilon > 0$ such that if P and Q are projections in $B(H)$ such that $\operatorname{rank} P = m$ and $\|(1-Q)P\| < \varepsilon$, then for any x in the unit ball of $\mathcal{A}(PH) \subset \mathcal{A}(H)$ there exists y in the unit ball of $\mathcal{A}(QH) \subset \mathcal{A}(H)$ such that $\|x - y\| < \delta$.*

Proof. This follows from norm-continuity of the map $g \mapsto a^*(g)$. In fact, by (13.5) we know that

$$\|a^*(g) - a^*(Qg)\| = \|(1-Q)g\|$$

for any $g \in H$. □

Let U be a unitary operator with singular spectral measure. Let Ω be a finite subset of the unit ball of $\mathcal{A}(PH)$ for a projection P of rank m. Then $\alpha_U^k(\Omega)$ is contained in the unit ball of $\mathcal{A}(U^k P U^{-k} H)$. For $\delta > 0$ choose $\varepsilon < \delta$ in accordance with Lemma 13.2.3. Choose a projection P_n according to Lemma 13.2.2. Then $\alpha_U^k(\Omega) \subset_\delta \mathcal{A}(P_n H)$ for $k = 0, \ldots, n$. Consequently

$$\mathrm{rcp}(\Omega \cup \alpha_U(\Omega) \cup \ldots \cup \alpha_U^{n-1}(\Omega), \delta) \leq \mathrm{rank}\,\mathcal{A}(P_n H) = 2^{\mathrm{rank}\,P_n} \leq 2^{\varepsilon n} \leq 2^{\delta n},$$

whence $ht(\Omega, \delta; \alpha_U) \leq \delta \log 2$. Thus $ht(\alpha_U) = 0$.

It turns out that the general result follows essentially from the above cases, thanks to the following characterization of the Lebesgue integral.

Proposition 13.2.4. *Let h be a function defined on the set of all unitary operators on a separable Hilbert space such that:*

(i) $h(U)$ *depends only on the conjugacy class of U;*

(ii) *if $U_a \cong W^N$, where W is the bilateral shift and $N \in \mathbb{N}$, then $h(U) = N$;*

(iii) *if P is a projection commuting with U then $h(U|_{PH}) \leq h(U)$;*

(iv) *if $\{P_n\}_n$ is a sequence of projections commuting with U such that $P_n \to 1$ strongly, then $h(U|_{P_n H}) \to h(U)$;*

(v) $h(U^n) = nh(U)$ *for any $n \in \mathbb{N}$.*

Then

$$h(U) = \int_\mathbb{T} m_U(z) d\mu(z).$$

Proof. Denote $\int_\mathbb{T} m_U(z) d\mu(z)$ by $\mu(U)$. Let us also write $U \preceq V$ if U is conjugate to the restriction of V to an invariant subspace.

In view of (iv) it suffices to consider unitary operators with bounded multiplicity function m_U. Moreover, using (iv) and regularity of the measure we may assume that the sets $m_U^{-1}(\{k\})$, $k \geq 0$, are closed. Then we can find a sequence $\{U(n)\}_n$ of unitary operators such that $U_s \cong U(n)_s$, $m_U \leq m_{U(n)}$, $\mu(U(n)) \to \mu(U)$, and $m_{U(n)}$ is the sum of the characteristic functions of p_n arcs of length $2\pi/q_n$ each, where $q_n \in \mathbb{N}$.

Observe next that if I is a measurable subset of $\mathbb{T}$ such that the map $I \ni z \mapsto z^k$ is injective with image J, then the operator of multiplication by the function $z \mapsto z^k$ on $L^2(I, d\mu)$ is conjugate to the operator of multiplication by the function $z \mapsto z$ on $L^2(J, d\mu)$. It follows that $U(n)_a^{q_n} \cong W^{p_n}$, where W is the bilateral shift, that is, the operator of multiplication by the function z on $L^2(\mathbb{T}, d\mu)$. More generally, the above observation implies that for any unitary operator V we have

$$m_{V^k}(z) = \sum_{w: w^k = z} m_V(w). \qquad (13.7)$$

232   13 Bogoliubov Automorphisms

We have $U_a^{q_n} \preceq U(n)_a^{q_n} \cong W^{p_n}$. Let $V(n)$ be a unitary operator such that $V(n)^{p_n} \cong U^{q_n}$. Since $V(n)_a^{p_n} \preceq W^{p_n}$, by (13.7) we have $m_{V(n)} \leq 1$. Therefore

$$V(n)^{p_n} \cong U^{q_n} \preceq U(n)^{q_n}, \quad V(n)_a \preceq W \quad \text{and} \quad U(n)_a^{q_n} \cong W^{p_n}.$$

Applying μ we get

$$\frac{p_n}{q_n}\mu(V(n)) = \mu(U) \leq \mu(U(n)) = \frac{p_n}{q_n}.$$

Since $\mu(U(n)) \to \mu(U)$, by the above inequality we get $p_n/q_n \to \mu(U)$ and $\mu(V(n)) \to 1$ (unless $\mu(U) = 0$, in which case we have $h(U) = 0$, as $h(U) \leq h(U(n)) = p_n/q_n$). Since $m_{V(n)} \leq 1$, it follows that $m_{V(n)} \to 1$ in measure. Then by (iv) and (ii) we get $h(V(n)_a) \to 1$. Since $V(n)_a \preceq V(n) \preceq W \oplus V(n)_s$, it follows that $h(V(n)) \to 1$. Since

$$h(U) = \frac{p_n}{q_n} h(V(n)),$$

we conclude that $p_n/q_n \to h(U)$. Thus $h(U) = \mu(U)$.   □

Proof of Theorem 13.2.1. We have to check that the function which takes the value $(\log 2)^{-1} ht(\alpha_U)$ at U satisfies the conditions of Proposition 13.2.4.

Properties (i), (iii) and (v) are immediate consequences of general properties of topological entropy. Property (iv) follows from the fact that any finite subset in $\mathcal{A}(H)$ can be approximated by a finite subset in $\mathcal{A}(P_n H)$ for all sufficiently large n's.

It remains to prove (ii). So assume that U is such that there exists an orthonormal basis $\{f_{i,n} \mid i = 1, \ldots, N, n \in \mathbb{Z}\}$ in H_a such that $Uf_{i,n} = f_{i,n+1}$. Then, similarly to the argument after the formulation of Theorem 13.2.1, the C*-algebra generated by $a^*(f_{i,n})a(f_{i,n})$, $i = 1, \ldots, N$, $n \in \mathbb{Z}$, is isomorphic to $C(\{0,1\}^{\mathbb{Z}})^{\otimes N}$, and the restriction of α_U to this subalgebra is the tensor product of the shift automorphisms, so that $ht(\alpha_U) \geq N \log 2$.

Denote by $P_{k,l}$ the projection onto the space spanned by $f_{i,n}$, $i = 1, \ldots, N$, $n = k, \ldots, l$. Let P be a finite rank projection in $B(H_s)$, Ω a finite subset of the unit ball of $\mathcal{A}(PH_s \oplus P_{k,l} H_a)$, and $\delta > 0$. Choose $\varepsilon < \delta$ in accordance with Lemma 13.2.3 applied to $m = \text{rank}\, P + \text{rank}\, P_{k,l}$. For $n \in \mathbb{N}$ find a projection $P_n \in B(H_s)$ according to Lemma 13.2.2. Then $\alpha_U^j(\Omega)$ is contained in the unit ball of $\mathcal{A}(U_s^j P U_s^{-j} H_s \oplus P_{k+j,l+j} H_a)$, so that $\alpha_U^j(\Omega) \subset_\delta \mathcal{A}(P_n H_s \oplus P_{k,l+n-1} H_a)$ for $j = 0, \ldots, n-1$. Consequently

$$\text{rcp}(\Omega \cup \alpha_U(\Omega) \cup \ldots \cup \alpha_U^{n-1}(\Omega), \delta) \leq 2^{\text{rank}\, P_n + \text{rank}\, P_{k,l+n-1}} \leq 2^{\delta n + N(l-k+n)},$$

whence $ht(\Omega, \delta; \alpha_U) \leq (\delta + N) \log 2$. Thus $ht(\alpha_U) \leq N \log 2$, and property (ii) is proved. This completes the proof of the theorem.   □

The following example of an asymptotically abelian system with non-tracial states of maximal entropy was promised in Chap. 9. Let us first introduce one more notion. The **even CAR-algebra** is the fixed point subalgebra $\mathcal{A}(H)_e$ of the CAR-algebra $\mathcal{A}(H)$ for the Bogoliubov automorphism α_{-1}

corresponding to the operator -1. It is generated by operators of the form $a^{\#}(f)a^{\#}(g)$, where $a^{\#}(f)$ is any of the operators $a^*(f)$ and $a(f)$.

Example 13.2.5. It easy to see that the restriction α of the Bogoliubov automorphism α_U to the even CAR-algebra is asymptotically abelian if and only if $(U^n f, g) \to 0$ as $n \to \infty$ for any $f, g \in H$. If in addition U has singular spectrum then by Theorem 13.2.1 we have $ht(\alpha) = 0$, while there are many nontracial α-invariant states (for example, quasi-free states corresponding to scalars $\lambda \in (0, 1/2)$). Unitaries with such properties can be obtained using Riesz products. We shall briefly recall the construction.

Let $q > 3$ be a real number, $\{n_k\}_{k=1}^\infty$ a sequence of positive integers such that $n_{k+1} \geq q n_k$, $\{a_k\}_{k=1}^\infty$ a sequence of real numbers such that $a_k \in (-1, 1)$, $a_k \to 0$ as $k \to \infty$, $\sum_k a_k^2 = \infty$. Then the sequence of measures

$$\frac{1}{2\pi}\left[\prod_{k=1}^n (1 + a_k \cos n_k t)\right] dt$$

on $[0, 2\pi]$ converges weakly* to a probability measure μ with Fourier coefficients

$$\hat{\mu}(n) = \mu(e^{int}) = \begin{cases} \prod_{k=1}^\infty \left(\frac{a_k}{2}\right)^{|\varepsilon_k|}, & \text{if } n = \sum_k \varepsilon_k n_k \text{ with } \varepsilon_k \in \{-1, 0, 1\}, \\ 0, & \text{otherwise.} \end{cases}$$

The measure μ is singular by [238, Theorem V.7.6]. We see also that $\hat{\mu}(n) \to 0$ as $|n| \to \infty$. Thus the operator U of multiplication by e^{it} on $L^2([0, 2\pi], d\mu)$ has the desired properties. ◆

13.3 Classical Bernoullian Subsystems

In this section we begin the computation of the dynamical entropy of Bogoliubov automorphisms with respect to quasi-free states. Consider first the case when U is the bilateral shift, so there exists an orthonormal basis $\{f_n\}_{n\in\mathbb{Z}}$ such that $Uf_n = f_{n+1}$, and $A = \lambda 1$, $0 \leq \lambda \leq 1$, is a scalar operator. We shall write ω_λ instead of $\omega_{\lambda 1}$ for the quasi-free state corresponding to $\lambda 1$. Then by (13.3)-(13.6) the C*-algebra generated by $a^*(f_n)a(f_n)$, $n \in \mathbb{Z}$, is isomorphic to $C(\{0, 1\}^\mathbb{Z})$, and by restricting ω_λ and α_U to this subalgebra we obtain the classical Bernoulli system with weights $(1-\lambda, \lambda)$ and entropy $\eta(\lambda) + \eta(1-\lambda)$. Since this subalgebra lies in the centralizer of the state, we get

$$h_{\omega_\lambda}(\alpha_U) \geq \eta(\lambda) + \eta(1-\lambda).$$

In fact, it is not difficult to see that equality holds, which allows us to conclude along the lines of the proof of Theorem 13.2.1 that

$$h_{\omega_\lambda}(\alpha_U) = (\eta(\lambda) + \eta(1-\lambda)) \int_{\mathbb{T}} m_U(z) d\mu(z)$$

for any unitary U. Since we shall in the next section prove a more general formula without relying on this particular case, we are not going into details. Our goal in this section is rather to prove an analogue of the above inequality when A is close to a scalar, which is the main technical result we shall need later.

Proposition 13.3.1. *For given $\varepsilon > 0$ and C, $0 < C < 1$, there exists $\delta > 0$ such that if $\operatorname{Sp} A \subset (\lambda_0 - \delta, \lambda_0 + \delta)$ for some $\lambda_0 \in (0, C)$, and the spectrum of U has a homogeneous Lebesgue component, so that there exists a unit vector f such that $\{U^n f\}_{n \in \mathbb{Z}}$ is an orthonormal system, then*

$$h_{\omega_A}(\alpha_U) \geq \eta(\lambda_0) + \eta(1-\lambda_0) - \varepsilon.$$

The same inequality is true for the restrictions of ω_A and α_U to the even CAR-algebra.

Consider the matrix units $e_{ij}(f)$ defined by (13.1). Let $\mathcal{P}$ be the algebra spanned by $p_1 = e_{11}(f) = a(f)a^*(f)$ and $p_2 = 1 - p_1$. For $n \in \mathbb{N}$ we want to estimate

$$H_{\omega_A}(\mathcal{P}, \alpha_U(\mathcal{P}), \ldots, \alpha_U^{n-1}(\mathcal{P}); \{p_{i_0} \alpha_U(p_{i_1}) \ldots \alpha_U^{n-1}(p_{i_{n-1}})\}_{i_0, \ldots, i_{n-1} = 1, 2}).$$

For this we need first of all to estimate the action of the modular group of ω_A on p_1 and p_2.

Assuming that $\operatorname{Ker} A = \operatorname{Ker}(1 - A) = 0$ set

$$B = \frac{A}{1 - A}.$$

Then ω_A is a KMS$_{-1}$-state for the one-parameter group $\{\sigma_t = \alpha_{B^{it}}\}_t$ of Bogoliubov automorphisms. If A has pure point spectrum this follows from (13.3)-(13.6), since the product-state in (13.6) is a KMS$_{-1}$-state for the one-parameter group

$$\bigotimes_{k=1}^n \operatorname{Ad} \begin{pmatrix} (1-\lambda_k)^{it} & 0 \\ 0 & \lambda_k^{it} \end{pmatrix},$$

which is exactly the above group of Bogoliubov automorphisms. In the general case we can approximate A by operators with pure point spectrum and argue by continuity. Another possibility is to use the explicit description of the GNS-representation, see [30, Example 5.2.20]. It follows that the extension of σ_t to $\pi_{\omega_A}(\mathcal{A}(H))''$ is the modular group of the state.

Lemma 13.3.2. *If $\operatorname{Sp} A \subset (0, 1)$ and*

$$\left\| \left(\frac{A}{1-A}\right)^{1/2} f - \left(\frac{\lambda}{1-\lambda}\right)^{1/2} f \right\| < \delta$$

for some $\delta > 0$ and f, $\|f\| = 1$, where $\lambda = (Af, f)$, then

$$\|\lambda_j^{1/2}\sigma_{-i/2}(e_{kj}(f)) - \lambda_k^{1/2} e_{kj}(f)\|_{\omega_A} \leq \sqrt{2}(\lambda_j\lambda_k)^{1/4}\delta^{1/2} \quad \text{for} \ k, j = 1, 2,$$

where $\lambda_1 = 1 - \lambda$, $\lambda_2 = \lambda$.

Proof. We shall write e_{kj} for $e_{kj}(f)$. By definition we have

$$\omega_A(e_{11}) = \lambda_1, \quad \omega_A(e_{22}) = \lambda_2, \quad \omega_A(e_{12}) = \omega_A(e_{21}) = 0.$$

By the KMS-condition, for any operators x and y,

$$\omega_A(\sigma_{-i/2}(x)^*\sigma_{-i/2}(y)) = \omega_A(\sigma_{i/2}(x^*)\sigma_{-i/2}(y))$$
$$= \omega_A(\sigma_{-i/2}(y)\sigma_{-i/2}(x^*)) = \omega_A(yx^*).$$

Using also that $\omega_A(\sigma_{-i/2}(x)^*y) = \omega_A(\sigma_{i/2}(x^*)y) = \omega_A(x^*\sigma_{-i/2}(y))$, we get

$$\|\lambda_j^{1/2}\sigma_{-i/2}(e_{kj}) - \lambda_k^{1/2} e_{kj}\|_{\omega_A}^2 = 2(\lambda_j\lambda_k)^{1/2}((\lambda_j\lambda_k)^{1/2} - \omega_A(e_{jk}\sigma_{-i/2}(e_{kj}))).$$

So we must prove that $\omega_A(e_{jk}\sigma_{-i/2}(e_{kj}))$ is close to $(\lambda_j\lambda_k)^{1/2}$ to within δ. Since

$$\lambda_1 - \omega_A(e_{11}\sigma_{-i/2}(e_{11})) = \omega_A(e_{11}\sigma_{-i/2}(e_{22})) = \lambda_2 - \omega_A(e_{22}\sigma_{-i/2}(e_{22}))$$

and

$$\omega_A(e_{12}\sigma_{-i/2}(e_{21})) = \omega_A(e_{21}\sigma_{-i/2}(e_{12}))$$

by the KMS-condition, we must show that $\omega_A(e_{11}\sigma_{-i/2}(e_{22}))$ is close to zero and $\omega_A(e_{21}\sigma_{-i/2}(e_{12}))$ is close to $\lambda^{1/2}(1-\lambda)^{1/2}$ to within δ. Let $B = \dfrac{A}{1-A}$ and $\beta = \dfrac{\lambda}{1-\lambda}$. Then

$$\sigma_{-i/2}(e_{21}) = \sigma_{-i/2}(a^*(f)) = a^*(B^{1/2}f),$$

so

$$\|\sigma_{-i/2}(e_{21}) - \beta^{1/2}e_{21}\| = \|B^{1/2}f - \beta^{1/2}f\| < \delta.$$

Hence, since $e_{11}e_{21} = 0$,

$$|\omega_A(e_{11}\sigma_{-i/2}(e_{22}))| = |\omega_A(e_{11}(\sigma_{-i/2}(e_{21}) - \beta^{1/2}e_{21})\sigma_{-i/2}(e_{12}))| < \delta$$

and since $\omega_A(e_{11}) = 1 - \lambda$,

$$|\omega_A(e_{12}\sigma_{-i/2}(e_{21})) - \lambda^{1/2}(1-\lambda)^{1/2}| = |\omega_A(e_{12}(\sigma_{-i/2}(e_{21}) - \beta^{1/2}e_{21}))| < \delta,$$

which proves the lemma. $\square$

The previous lemma will allow us to estimate the correction term of the entropy. It will also allow us to estimate the classical term thanks to the following approximate factorization result.

13 Bogoliubov Automorphisms

Lemma 13.3.3. *Let $\{e_{ij}\}_{i,j=1}^n$ be a system of matrix units in a von Neumann algebra M with $\sum_k e_{kk} = 1$, and let ω be a faithful normal state on M. Then for any $x \in M$ commuting with the matrix units we have*

$$|\omega(e_{kk}x) - \lambda_k \omega(x)| \le \sqrt{2} \sum_j \|\lambda_j^{1/2} \sigma_{-i/2}^\omega(e_{kj}) - \lambda_k^{1/2} e_{kj}\|_\omega \|x\|_\omega^\#,$$

where $\|x\|_\omega^\# = (\omega(x^*x) + \omega(xx^*))^{1/2}$ and $\lambda_k = \omega(e_{kk})$.

Proof. We may assume that ω is the vector state defined by a cyclic and separating vector ξ. Let J be the corresponding modular conjugation. Then

$$\omega(e_{kk}x) = (e_{kk}x\xi, \xi) = (e_{jk}x\xi, e_{jk}\xi) = (Je_{jk}\xi, Je_{jk}x\xi).$$

Using that x commutes with e_{jk}, and JxJ commutes with e_{kj}, we also have

$$(e_{kj}\xi, Je_{jk}x\xi) = (e_{kj}, JxJJe_{jk}\xi) = (e_{kj}Jx^*\xi, Je_{jk}\xi)$$

and

$$(e_{kj}Jx^*\xi, e_{kj}\xi) = (e_{kj}\xi, e_{kj}Jx\xi) = (e_{jj}\xi, Jx\xi) = (x\xi, Je_{jj}\xi).$$

It follows that

$$\lambda_j \omega(e_{kk}x) = \lambda_j^{1/2}((\lambda_j^{1/2}Je_{jk} - \lambda_k^{1/2}e_{kj})\xi, Je_{jk}x\xi)$$
$$+ \lambda_k^{1/2}(e_{kj}Jx^*\xi, (\lambda_j^{1/2}Je_{jk} - \lambda_k^{1/2}e_{kj})\xi) + \lambda_k(x\xi, Je_{jj}\xi).$$

Since $\sigma_{-i/2}^\omega(y)\xi = Jy^*\xi$ for $y \in M$, we then get

$$|\lambda_j \omega(e_{kk}x) - \lambda_k(x\xi, Je_{jj}\xi)|$$
$$\le (\lambda_j^{1/2}\|x\|_\omega + \lambda_k^{1/2}\|x^*\|_\omega)\|\lambda_j^{1/2}\sigma_{-i/2}^\omega(e_{kj}) - \lambda_k^{1/2}e_{kj}\|_\omega$$
$$\le \sqrt{2}\|x\|_\omega^\# \|\lambda_j^{1/2}\sigma_{-i/2}^\omega(e_{kj}) - \lambda_k^{1/2}e_{kj}\|_\omega.$$

Summing up the above inequalities over all j's we obtain the desired estimate. $\square$

Proof of Proposition 13.3.1. First choose $\delta_1 > 0$ such that

$$|\eta(\lambda) + \eta(1-\lambda) - \eta(\lambda_0) - \eta(1-\lambda_0)| < \frac{\varepsilon}{3} \tag{13.8}$$

if $\lambda \in (\lambda_0 - \delta_1, \lambda_0 + \delta_1)$, $\lambda_0 \in (0, C)$.

For $n \in \mathbb{N}$ we want to estimate

$$H_{\omega_A}(\mathcal{P}, \alpha_U(\mathcal{P}), \ldots, \alpha_U^{n-1}(\mathcal{P}); \{p_{i_0}\alpha_U(p_{i_1})\ldots\alpha_U^{n-1}(p_{i_{n-1}})\}_{i_0,\ldots,i_{n-1}=1,2}),$$

where $\mathcal{P}$ is spanned by $p_1 = a(f)a^*(f)$ and $p_2 = a^*(f)a(f)$.

Start with the correction term, which is by definition

13.3 Classical Bernoullian Subsystems

$$\sum_{k=0}^{n-1}\sum_{j=1}^{2} S(\omega_A(\cdot\,\sigma_{-i/2}(\alpha_U^k(p_j)))|_{\alpha_U^k(\mathcal{P})}, \omega_A|_{\alpha_U^k(\mathcal{P})})$$

$$= n\sum_{j=1}^{2} S(\omega_A(\cdot\,\sigma_{-i/2}(p_j))|_{\mathcal{P}}, \omega_A|_{\mathcal{P}})$$

$$= n\sum_{j=1}^{2}\sum_{k=1}^{2} \omega_A(p_k\sigma_{-i/2}(p_j))(\log\omega_A(p_k\sigma_{-i/2}(p_j)) - \log\omega_A(p_k))$$

$$= n\sum_{k=1}^{2}\left(\eta(\omega_A(p_k)) - \sum_{j=1}^{2}\eta(\omega_A(p_k\sigma_{-i/2}(p_j)))\right).$$

Since $\omega_A(p_k\sigma_{-i/2}(p_j))$ can be made arbitrarily close to $\delta_{kj}\omega_A(p_k)$ by Lemma 13.3.2, we can find $\delta_2 \in (0, \delta_1)$ such that if $\operatorname{Sp} A \subset (\lambda_0 - \delta_2, \lambda_0 + \delta_2)$, $\lambda_0 \in (0, C)$, then

$$n\sum_{j=1}^{2} S(\omega_A(\cdot\,\sigma_{-i/2}(p_j))|_{\mathcal{P}}, \omega_A|_{\mathcal{P}}) > -n\frac{\varepsilon}{3}. \tag{13.9}$$

Turning to the classical term, by Lemma 4.3.4 there exists $\varepsilon_1 > 0$ such that if

$$|\omega_A(p_i x) - \omega_A(p_i)\omega_A(x)| \leq \varepsilon_1 \|x\|, \quad i = 1, 2, \tag{13.10}$$

for any x commuting with p_1 and p_2, then

$$\sum_{i=1}^{2}\sum_{j\in J}\eta(\omega_A(p_i x_j)) \geq \sum_{i=1}^{2}\eta(\omega_A(p_i)) + \sum_{j\in J}\eta(\omega_A(x_j)) - \frac{\varepsilon}{3}$$

for any partition of unity $1 = \sum_{j\in J} x_j$ commuting with p_1 and p_2. By Lemmas 13.3.2 and 13.3.3 applied to $\{e_{kj}(f)\}_{k,j=1}^{2}$, there exists $\delta_3 \in (0, \delta_2)$ such that if $\operatorname{Sp} A \subset (\lambda_0 - \delta_3, \lambda_0 + \delta_3)$, $\lambda_0 \in (0, C)$, then (13.10) is satisfied. Applying the above inequality recursively we then get

$$\sum_{i_0,\ldots,i_{n-1}=1}^{2}\eta(\omega_A(p_{i_0}\alpha_U(p_{i_1})\ldots\alpha_U^{n-1}(p_{i_{n-1}})))$$

$$\geq n\sum_{i=1}^{2}\eta(\omega_A(p_i)) - (n-1)\frac{\varepsilon}{3} = n(\eta(\lambda) + \eta(1-\lambda)) - (n-1)\frac{\varepsilon}{3}, \tag{13.11}$$

where $\lambda = (Af, f) \in (\lambda_0 - \delta_3, \lambda_0 + \delta_3)$.

It follows from (13.8), (13.9) and (13.11) that if we take $\delta = \delta_3$ then for any $n \in \mathbb{N}$

$$H_{\omega_A}(\mathcal{P}, \alpha_U(\mathcal{P}), \ldots, \alpha_U^{n-1}(\mathcal{P}); \{p_{i_0}\alpha_U(p_{i_1})\ldots\alpha_U^{n-1}(p_{i_{n-1}})\})$$

$$> n(\eta(\lambda_0) + \eta(1-\lambda_0)) - (3n-1)\frac{\varepsilon}{3}$$

as long as $\operatorname{Sp} A \subset (\lambda_0 - \delta, \lambda_0 + \delta)$ for some $\lambda_0 \in (0, C)$. Hence

$$h_{\omega_A}(\alpha_U) \geq h_{\omega_A}(\mathcal{P}; \alpha_U) \geq \eta(\lambda_0) + \eta(1-\lambda_0) - \varepsilon.$$

Since $\mathcal{P}$ is a subalgebra of the even CAR-algebra, the same inequalities hold for the restrictions of α_U and ω_A to $\mathcal{A}(H)_e$. □

13.4 Dynamical Entropy

In this section we shall prove a formula for the entropy of a Bogoliubov automorphism of the CAR-algebra with respect to a quasi-free state. To formulate the result consider a unitary U on a separable Hilbert space H and an operator A, $0 \leq A \leq 1$, commuting with U. Let $U_a = U|_{H_a}$ be the part of U with absolutely continuous spectral measure. Then, see App. C, there exists a direct integral decomposition

$$H_a = \int_{\mathbb{T}}^{\oplus} H_z d\mu(z), \quad U_a = \int_{\mathbb{T}}^{\oplus} z\, d\mu(z), \quad A|_{H_a} = \int_{\mathbb{T}}^{\oplus} A_z d\mu(z),$$

where μ is the normalized Lebesgue measure on the torus $\mathbb{T}$.

Theorem 13.4.1. *With the above notation, for the Bogoliubov automorphism α_U of the CAR-algebra $\mathcal{A}(H)$ and the quasi-free state ω_A defined by A we have*

$$h_{\omega_A}(\alpha_U) = \int_{\mathbb{T}} \operatorname{Tr}(\eta(A_z) + \eta(1 - A_z)) d\mu(z).$$

A particular case, where one can get an explicit direct integral decomposition, is given in the following corollary.

Corollary 13.4.2. *Let I be an open subset of $\mathbb{R}$, ω a real locally absolutely continuous function on I, ρ a measurable function on I, $0 \leq \rho \leq 1$. Let U and A be the operators on $L^2(I, dx)$ of multiplication by the functions $e^{i\omega}$ and ρ, respectively. Then*

$$h_{\omega_A}(\alpha_U) = \frac{1}{2\pi} \int_I (\eta(\rho(x)) + \eta(1 - \rho(x))) |\omega'(x)| dx.$$

Proof. Consider the sets

$$I_1 = \{x \in I \,|\, \omega'(x) = 0\} \quad \text{and} \quad I_2 = \{x \in I \,|\, \omega'(x) \neq 0\}.$$

For any compact subset $X \subset I_1$ the set $\omega(X)$ has measure zero. Hence the spectrum of the restriction of U to $L^2(X, dx)$ has measure zero. It follows that the restriction of U to $L^2(I_1, dx)$ has singular spectral measure and thus does not contribute to the entropy.

On the other hand, if $x \in I_2$ then $\omega(y) \neq \omega(x)$ for any y close to x, $y \neq x$, so that for every $z \in \mathbb{T}$ the set

$$Y(z) = \{x \in I_2 \,|\, e^{i\omega(x)} = z\}$$

is at most countable. For any $f \in L^1(I_2, |\omega'(x)|dx)$ we have

$$\int_{I_2} f(x)|\omega'(x)|dx = 2\pi \int_{\mathbb{T}} \sum_{x \in Y(z)} f(x) d\mu(z),$$

see e.g. [67, Theorem 3.2.3] for the case when ω is Lipschitzian, and use [67, Theorem 3.1.8] and [67, Theorem 2.10.43] to extend the result to the general case. It follows that if for a function f and $z \in \mathbb{T}$ we put

$$f_z = (2\pi)^{1/2}|\omega'|^{-1/2}f|_{Y(z)},$$

then the map $f \mapsto (f_z)_z$ defines an isometry between the spaces $L^2(I_2, dx)$ and $\int_{\mathbb{T}}^{\oplus} \ell_2(Y(z)) d\mu(z)$. Since $I \backslash (I_1 \cup I_2)$ has measure zero, we thus obtain a direct integral decomposition of U_a and $A|_{H_a}$ such that A_z is the operator of multiplication by $\rho|_{Y(z)}$. Hence

$$\int_{\mathbb{T}} \text{Tr}(\eta(A_z) + \eta(1 - A_z)) d\mu(z) = \int_{\mathbb{T}} \sum_{x \in Y(z)} (\eta(\rho(x)) + \eta(1 - \rho(x))) d\mu(z)$$

$$= \frac{1}{2\pi} \int_{I_2} (\eta(\rho(x)) + \eta(1 - \rho(x))) |\omega'(x)| dx,$$

which proves the corollary. □

Turning to the proof of Theorem 13.4.1 we shall first establish the lower bound for the entropy. The main technical step in this direction has been done in the previous section, and our goal now is to extend that result to all unitaries with absolutely continuous spectrum. This extension is based on the following two results, which provide tools for changing the spectrum of a unitary operator without changing the entropy.

Let us first make a few remarks about the even CAR-algebra.

If K and L are mutually orthogonal subspaces of H then $\mathcal{A}(K)$ and $\mathcal{A}(L)_e$ commute, and the C*-algebra they generate decomposes into the tensor product $\mathcal{A}(K) \otimes \mathcal{A}(L)_e$. Indeed, it suffices to consider the case when K is finite dimensional, and then this is true as $\mathcal{A}(K)$ is a full matrix algebra. If in addition K and L are invariant subspaces for an operator A, $0 \le A \le 1$, then

$$\omega_A|_{\mathcal{A}(K) \otimes \mathcal{A}(L)_e} = \omega_A|_{\mathcal{A}(K)} \otimes \omega_A|_{\mathcal{A}(L)_e}.$$

If moreover $H = K \oplus L$, the subalgebra $\mathcal{A}(K) \otimes \mathcal{A}(L)_e$ of $\mathcal{A}(H)$ is the fixed point algebra for the involutive Bogoliubov automorphism $\alpha_{1 \oplus -1}$ preserving ω_A. In particular, there exists an ω_A-preserving conditional expectation onto $\mathcal{A}(K) \otimes \mathcal{A}(L)_e$, and by monotonicity and superadditivity of entropy, Theorem 3.2.2(v),(iv), we get

$$h_{\omega_A}(\alpha_U) \geq h_{\omega_A|\mathcal{A}(K) \otimes \omega_A|\mathcal{A}(L)_e}(\alpha_U|_{\mathcal{A}(K)} \otimes \alpha_U|_{\mathcal{A}(L)_e})$$
$$\geq h_{\omega_A|\mathcal{A}(K)}(\alpha_U|_{\mathcal{A}(K)}) + h_{\omega_A|\mathcal{A}(L)_e}(\alpha_U|_{\mathcal{A}(L)_e}). \quad (13.12)$$

Lemma 13.4.3. *Let U_n be a unitary operator on a Hilbert space H_n, $n \in \mathbb{N}$, and $\{z_n\}_{n=1}^\infty \subset \mathbb{T}$. Consider two unitary operators U' and U'' on the space $H = \oplus_{n=1}^\infty H_n$,*

$$U' = \bigoplus_{n=1}^\infty U_n \quad \text{and} \quad U'' = \bigoplus_{n=1}^\infty z_n U_n.$$

Then $h_\omega(\alpha_{U'}) = h_\omega(\alpha_{U''})$ for any $\alpha_{U'}$- and $\alpha_{U''}$-invariant state ω on $\mathcal{A}(H)$. The same holds for the restrictions of the automorphisms to the even CAR-algebra $\mathcal{A}(H)_e$.

Proof. Consider the unitary operator $V = \oplus_{n=1}^\infty z_n 1$. For each $n \in \mathbb{N}$ choose an increasing sequence $\{H_{nk}\}_{k=1}^\infty$ of finite-dimensional subspaces of H_n such that $\cup_k H_{nk}$ is dense in H_n. Set

$$K_n = H_{1n} \oplus \ldots \oplus H_{nn}.$$

Then K_n is finite-dimensional, $K_n \subset K_{n+1}$, and $\cup_n K_n$ is dense in H. Since $VK_n = K_n$ and $\alpha_{U''} = \alpha_V \circ \alpha_{U'} = \alpha_{U'} \circ \alpha_V$, we have $\alpha_{U''}^k(\mathcal{A}(K_n)) = \alpha_{U'}^k(\mathcal{A}(K_n))$ for any $k \in \mathbb{Z}$, whence $h_\omega(\mathcal{A}(K_n); \alpha_{U''}) = h_\omega(\mathcal{A}(K_n); \alpha_{U'})$. As $\{\mathcal{A}(K_n)\}_n$ is an increasing sequence with dense union in $\mathcal{A}(H)$, it follows that $h_\omega(\alpha_{U''}) = h_\omega(\alpha_{U'})$.

Similarly, using the algebras $\mathcal{A}(K_n)_e$ instead of $\mathcal{A}(K_n)$, we get the result for the even CAR-algebra. $\square$

Lemma 13.4.4. *Let X_1, X_2 be measurable subsets of $\mathbb{T}$ such that $\mu(X_1) = \mu(X_2) > 0$. Then there exist a countable measurable partition $X_1 = \sqcup_{n=1}^\infty Y_n$ of X_1 and a sequence of numbers $\{z_n\}_{n=1}^\infty \subset \mathbb{T}$ such that $X_2 = \sqcup_{n=1}^\infty z_n Y_n$ modulo a set of measure zero.*

Proof. It suffices to prove that there exists a subset $Y \subset X_1$ of positive measure and $z \in \mathbb{T}$ such that $zY \subset X_2$. Then an application of Zorn's lemma finishes the proof.

Denoting by $\mathbb{1}_{X_1}$ and $\mathbb{1}_{X_2}$ the characteristic functions of the sets X_1 and X_2 we have

$$\iint_{\mathbb{T} \times \mathbb{T}} \mathbb{1}_{X_1}(w) \mathbb{1}_{X_2}(zw) d\mu(w) d\mu(z) = \mu(X_1)\mu(X_2) > 0.$$

Hence there exists z such that

$$\int_\mathbb{T} \mathbb{1}_{X_1}(w) \mathbb{1}_{X_2}(zw) d\mu(w) > 0,$$

that is, $zX_1 \cap X_2$ has positive measure. $\square$

Now we can extend Proposition 13.3.1 to arbitrary unitaries.

13.4 Dynamical Entropy

Proposition 13.4.5. *For $\varepsilon > 0$ and C, $0 < C < 1$, choose $\delta > 0$ as in Proposition 13.3.1. Let U be a unitary with absolutely continuous spectral measure and a direct integral decomposition*

$$H = \int_{\mathbb{T}}^{\oplus} H_z d\mu(z), \quad U = \int_{\mathbb{T}}^{\oplus} z \, d\mu(z).$$

Set $X = \{z \in \mathbb{T} \mid H_z \neq 0\}$. Let A, $0 \leq A \leq 1$ be an operator commuting with U such that $\operatorname{Sp} A \subset (\lambda_0 - \delta, \lambda_0 + \delta)$ for some $\lambda_0 \in (0, C)$. Then

$$h_{\omega_A}(\alpha_U) \geq \mu(X)(\eta(\lambda_0) + \eta(1 - \lambda_0) - \varepsilon).$$

The same inequality is true for the entropy of the restriction of α_U to $\mathcal{A}(H)_e$.

Proof. Let $\{n_k\}_{k=1}^{\infty} \subset \mathbb{N}$ be a sequence such that $\mu(X) = \sum_k n_k^{-1}$. By Lemma 13.4.4 there exist a measurable partition $X = \sqcup_{k,m=1}^{\infty} X_{km}$ and a set $\{z_{km}\}_{k,m=1}^{\infty} \subset \mathbb{T}$ such that for all $k \in \mathbb{N}$

$$\exp\left(2\pi i \left[0, \frac{1}{n_k}\right]\right) = \bigsqcup_m z_{km} X_{km}$$

modulo sets of measure zero. Let $H_{km} = \int_{X_{km}}^{\oplus} H_z d\mu(z)$ be the spectral subspace of U corresponding to the set X_{km}. Set $H_k = \oplus_m H_{km}$, and define a unitary operator U_k on H_k by

$$U_k = \bigoplus_m z_{km} U|_{H_{km}}.$$

By construction the multiplicity function of U_k is positive on $\exp(2\pi i [0, n_k^{-1}])$, so that $U_k^{n_k}$ has Lebesgue spectrum. By Lemma 13.4.3 and Proposition 13.3.1 we get

$$h_{\omega_A|\mathcal{A}(H_k)_e}(\alpha_U|_{\mathcal{A}(H_k)_e}) = \frac{1}{n_k} h_{\omega_A|\mathcal{A}(H_k)_e}(\alpha_{U_k^{n_k}}|_{\mathcal{A}(H_k)_e})$$

$$\geq \frac{1}{n_k}(\eta(\lambda_0) + \eta(1 - \lambda_0) - \varepsilon).$$

By (13.12) for any $n \in \mathbb{N}$ we get

$$h_{\omega_A}(\alpha_U) \geq \sum_{k=1}^{n} h_{\omega_A|\mathcal{A}(H_k)_e}(\alpha_U|_{\mathcal{A}(H_k)_e})$$

$$\geq \left(\sum_{k=1}^{n} \frac{1}{n_k}\right)(\eta(\lambda_0) + \eta(1 - \lambda_0) - \varepsilon).$$

Letting $n \to \infty$ we obtain the required estimate.

The same proof works for the restriction of the automorphism to the even part of the algebra. □

13 Bogoliubov Automorphisms

We are now ready to proof a half of Theorem 13.4.1.

Proof of the lower bound for the entropy in Theorem 13.4.1. Since $h_{\omega_A}(\alpha_U) \geq h_{\omega_A|A(H_a)}(\alpha_U|_{A(H_a)})$ by (13.12), to prove the result it suffices to consider unitaries with absolutely continuous spectral measure.

Assuming that U is such a unitary, let us first show that if $h_{\omega_A}(\alpha_U) < \infty$ then A_z has pure point spectrum for almost all $z \in \mathbb{T}$.

Fix $\delta_0 \in (0, 1/2)$ and take $\varepsilon \in (0, \eta(\delta_0))$. Let δ be as in the formulation of Proposition 13.4.5 with $C = 1 - \delta_0$. For any Borel subset X of $\mathbb{R}$, let $\mathbb{1}_X(A)$ be the spectral projection of A corresponding to X. Then

$$\mathbb{1}_X(A) = \int_\mathbb{T}^\oplus \mathbb{1}_X(A_z) d\mu(z).$$

Define a measurable function φ_X on $\mathbb{T}$,

$$\varphi_X(z) = \begin{cases} 1, & \text{if } \mathbb{1}_X(A_z) \neq 0, \\ 0, & \text{otherwise.} \end{cases}$$

By Proposition 13.4.5 if X is a Borel subset of $(\lambda_0 - \delta, \lambda_0 + \delta)$ for some $\lambda_0 \in (\delta_0, 1 - \delta_0)$ then

$$h_{\omega_A}(\alpha_U|_{A(\mathbb{1}_X(A)H)_e}) \geq (\eta(\lambda_0) + \eta(1 - \lambda_0) - \varepsilon) \int_\mathbb{T} \varphi_X(z) d\mu(z)$$

$$\geq \eta(1 - \delta_0) \int_\mathbb{T} \varphi_X(z) d\mu(z), \tag{13.13}$$

where we have used that $\eta(\lambda_0) + \eta(1 - \lambda_0) \geq \eta(\delta_0) + \eta(1 - \delta_0)$ in the second inequality. Let $t_0 = \delta_0 < t_1 < \ldots < t_m = 1 - \delta_0$, $t_k - t_{k-1} < \delta$. Then, similarly to the proof of Proposition 13.4.5, using (13.12) we obtain from (13.13) the inequality

$$h_{\omega_A}(\alpha_U) \geq \eta(1 - \delta_0) \int_\mathbb{T} \sum_{k=1}^m \varphi_{(t_{k-1}, t_k]}(z) d\mu(z).$$

If we let $\max(t_k - t_{k-1}) \to 0$ then the integrand in the expression above converges to infinity at every point $z \in \mathbb{T}$ such that $(\delta_0, 1 - \delta_0) \cap \operatorname{Sp} A_z$ is infinite. Therefore if $h_{\omega_A}(\alpha_U) < \infty$, then the intersection $(\delta_0, 1 - \delta_0) \cap \operatorname{Sp} A_z$ is finite for almost all $z \in \mathbb{T}$. Since δ_0 is arbitrary, it follows that A_z has pure point spectrum for almost all $z \in \mathbb{T}$.

Thus to get the lower bound for the entropy we may assume that A_z has pure point spectrum for almost all $z \in \mathbb{T}$, since otherwise both the entropy and the integral in the formulation of the theorem are infinite. But then, see C.6, we have

$$H = \bigoplus_{n=1}^N L^2(X_n, d\mu),$$

where X_n is a measurable subset of $\mathbb{T}$, $N \in \mathbb{N} \cup \{\infty\}$, and U and A act on $L^2(X_n, d\mu)$ as multiplications by functions z and $\lambda_n(z)$, respectively. We must prove that

$$h_{\omega_A}(\alpha_U) \geq \sum_{n=1}^{N} \int_{X_n} (\eta(\lambda_n(z)) + \eta(1 - \lambda_n(z)))d\mu(z).$$

Using (13.12) once again, we see that it suffices to estimate $h_{\omega_A}(\alpha_U|_{\mathcal{A}(H)_e})$ assuming $N = 1$. As in the proof above, fixing $\delta_0 > 0$, $\varepsilon > 0$ and choosing $t_0 = \delta_0 < t_1 < \ldots < t_m = 1 - \delta_0$, we get that

$$h_{\omega_A}(\alpha_U|_{\mathcal{A}(H)_e}) \geq \sum_{k=1}^{m} \int_{\{z \in X_1 \mid t_{k-1} < \lambda_1(z) \leq t_k\}} (\eta(t_k) + \eta(1-t_k) - \varepsilon)d\mu(z)$$

if $\max(t_k - t_{k-1})$ is small enough. Letting $\max(t_k - t_{k-1}) \to 0$, we obtain

$$h_{\omega_A}(\alpha_U|_{\mathcal{A}(H)_e}) \geq \int_{\{z \in X_1 \mid \delta_0 < \lambda_1(z) \leq 1-\delta_0\}} (\eta(\lambda_1(z)) + \eta(1-\lambda_1(z)))d\mu(z) - \varepsilon.$$

Since δ_0 and ε were arbitrarily small, this completes the proof of the lower bound for the entropy. □

To obtain the upper bound we first need to establish a few properties of the von Neumann entropy of quasi-free states.

Lemma 13.4.6. *Let H be finite-dimensional, ω_A a quasi-free state on $\mathcal{A}(H)$. Then*
(i) $S(\omega_A) = \mathrm{Tr}(\eta(A) + \eta(1-A))$;
(ii) *if $H = H_1 \oplus H_2$ then $S(\omega_A) \leq S(\omega_A|_{\mathcal{A}(H_1)}) + S(\omega_A|_{\mathcal{A}(H_2)})$.*

Proof. Part (i) is an immediate consequence of (13.6). To prove (ii) note that $\omega_A|_{\mathcal{A}(H_i)}$ is the quasi-free state defined by the operator $P_i A P_i \in B(H_i)$, where P_i is the projection onto H_i. Since η is an operator concave function by B.4, we have

$$P_1 \eta(A) P_1 + P_2 \eta(A) P_2 \leq \eta(P_1 A P_1 + P_2 A P_2) = \eta(P_1 A P_1) + \eta(P_2 A P_2)$$

by B.2, whence

$$\mathrm{Tr}(\eta(A)) \leq \mathrm{Tr}(\eta(P_1 A P_1)) + \mathrm{Tr}(\eta(P_2 A P_2)).$$

Similarly we get

$$\mathrm{Tr}(\eta(1-A)) \leq \mathrm{Tr}(\eta(P_1 - P_1 A P_1)) + \mathrm{Tr}(\eta(P_2 - P_2 A P_2)),$$

and (ii) is proved. Moreover, using B.5 we see that the equality in (ii) holds if and only if H_1 and H_2 are invariant subspaces for A. □

Proof of the upper bound for the entropy in Theorem 13.4.1. Let $V \in B(K)$ be the infinite direct sum of bilateral shifts, so that V is the operator with countably multiple homogeneous Lebesgue spectrum. Consider the unitary $U \oplus V$ and the operator $A \oplus 0$ on $H \oplus K$. Then by (13.12)

$$h_{\omega_{A\oplus 0}}(\alpha_{U\oplus V}) \geq h_{\omega_A}(\alpha_U).$$

On the other hand, the integral in the formulation of the theorem does not change if we replace U by $U \oplus V$ and A by $A \oplus 0$. So in establishing the upper bound for the entropy we may replace U by $U \oplus V$ and thus assume that U_a has countably multiple homogeneous Lebesgue spectrum. Furthermore, we may assume that A_z has pure point spectrum for almost all $z \in \mathbb{T}$, since otherwise the integral is infinite. Then, see again C.6, we can represent H_a as the sum of a countably many copies of $L^2(\mathbb{T}, d\mu)$ in such a way that U and A act on the n-th copy as multiplications by functions z and $\lambda_n(z)$, respectively. In view of Proposition 3.2.4 to obtain the upper bound it suffices to consider a finite number of copies of $L^2(\mathbb{T})$. Thus we assume that

$$H_a = \bigoplus_{m=1}^{m_0} L^2(\mathbb{T}, d\mu), \quad U_a = \bigoplus_{m=1}^{m_0} z, \quad A|_{H_a} = \bigoplus_{m=1}^{m_0} \lambda_m(z),$$

and we have to prove that

$$h_{\omega_A}(\alpha_U) \leq \sum_{m=1}^{m_0} \int_{\mathbb{T}} (\eta(\lambda_m(z)) + \eta(1 - \lambda_m(z))) d\mu(z).$$

Let H_0 be the m_0-dimensional subspace of H spanned by the constant functions in each copy of $L^2(\mathbb{T})$. Then $H_a = \oplus_{n \in \mathbb{Z}} U^n H_0$. For $n \in \mathbb{N}$ set $H_n = \oplus_{k=0}^n U^k H_0$. We assert that

$$h_{\omega_A}(\alpha_U) \leq \liminf_{n \to \infty} \frac{1}{n} S(\omega_A|_{\mathcal{A}(H_{n-1})}). \tag{13.14}$$

The proof is similar to the argument in the proof of Theorem 13.2.1 showing that the topological entropy has property (ii) from Proposition 13.2.4. Denote by $P_{k,l}$ the projection onto the space $\oplus_{i=k}^l U^i H_0$. Let P be a finite rank projection in $B(H_s)$. Let $\gamma \colon B \to \mathcal{A}(PH_s \oplus P_{k,l}H_a)$ be a channel. We want to estimate $h_{\omega_A}(\gamma; \alpha_U)$.

Fix $\delta > 0$. Choose $\varepsilon < \delta$ in accordance with Lemma 13.2.3 applied to $m = \operatorname{rank} P + \operatorname{rank} P_{k,l}$. For $n \in \mathbb{N}$ find a projection $P_n \in B(H_s)$ according to Lemma 13.2.2. Denote by E_n a conditional expectation onto the finite dimensional algebra $\mathcal{A}(P_n H_s \oplus P_{k,l+n-1} H_a)$. Then for any x in the unit ball of B the element $(\alpha_U^j \circ \gamma)(x)$ is contained in the unit ball of the algebra $\mathcal{A}(U_s^j P U_s^{-j} H_s \oplus P_{k+j,l+j} H_a)$, so that

$$(\alpha_U^j \circ \gamma)(x) \in_\delta \mathcal{A}(P_n H_s \oplus P_{k,l+n-1} H_a)$$

for $j = 0, \ldots, n-1$, whence $\|\alpha_U^j \circ \gamma - E_n \circ \alpha_U^j \circ \gamma\| < \delta$. By Proposition 3.1.11

$$H_{\omega_A}(\gamma, \alpha_U \circ \gamma, \ldots, \alpha_U^{n-1} \circ \gamma) \leq H_{\omega_A}(E_n \circ \gamma, \ldots, E_n \circ \alpha_U^{n-1} \circ \gamma) + n\varepsilon_1$$
$$\leq H_{\omega_A}(\mathcal{A}(P_n H_s \oplus P_{k,l+n-1} H_a)) + n\varepsilon_1,$$

where ε_1 depends only on δ and the dimension of B, and $\varepsilon_1(\delta, \dim B) \to 0$ as $\delta \to 0$. Using Lemma 13.4.6(ii) we then estimate

$$\begin{aligned}
H_{\omega_A}(\mathcal{A}(P_nH_s \oplus P_{k,l+n-1}H_a)) &\leq S(\omega_A|_{\mathcal{A}(P_nH_s\oplus P_{k,l+n-1}H_a)}) \\
&\leq S(\omega_A|_{\mathcal{A}(P_nH_s)}) + S(\omega_A|_{\mathcal{A}(P_{k,l+n-1}H_a)}) \\
&\leq \log \operatorname{rank} \mathcal{A}(P_nH_s) + S(\omega_A|_{\mathcal{A}(P_{k,l+n-1}H_a)}) \\
&= \operatorname{rank} P_n \log 2 + S(\omega_A|_{\mathcal{A}(H_{l-k+n})}) \\
&\leq n\delta \log 2 + S(\omega_A|_{\mathcal{A}(H_{l-k+n})}).
\end{aligned}$$

We therefore get

$$h_{\omega_A}(\gamma; \alpha_U) \leq \varepsilon_1 + \delta \log 2 + \liminf_{n\to\infty} \frac{1}{n} S(\omega_A|_{\mathcal{A}(H_{n-1})}).$$

Since δ was arbitrarily small, and $\varepsilon_1 = \varepsilon_1(\delta, \dim B) \to 0$ as $\delta \to 0$, we conclude that $h_{\omega_A}(\gamma; \alpha_U)$ is not larger then the right hand side in (13.14). Since any channel into $\mathcal{A}(H)$ can be approximated by a channel of the above form, we thus get (13.14).

It is possible to show that the limit in (13.14) exists and coincides with the required upper bound. But it is easier to argue as follows. Applying Lemma 13.4.6(ii) again we obtain

$$S(\omega_A|_{\mathcal{A}(H_{n-1})}) \leq \sum_{k=0}^{n-1} S(\omega_A|_{\mathcal{A}(U^kH_0)}) = nS(\omega_A|_{\mathcal{A}(H_0)}).$$

The state $\omega_A|_{\mathcal{A}(H_0)}$ is the quasi-free state corresponding to the operator $P_{0,0}AP_{0,0}$, where $P_{0,0}$ is the projection onto the space H_0. Notice that the compression of the operator of multiplication by a function $\lambda(z)$ to the space of constant functions is the scalar $\int_\mathbb{T} \lambda(z)d\mu(z)$. Hence $\omega_A|_{\mathcal{A}(H_0)}$ is the quasi-free state corresponding to the operator with eigenvalues $\lambda_m = \int_\mathbb{T} \lambda_m(z)d\mu(z)$, $m = 1, \ldots, m_0$. Therefore by Lemma 13.4.6(i) and (13.14) we get

$$h_{\omega_A}(\alpha_U) \leq S(\omega_A|_{\mathcal{A}(H_0)}) = \sum_{m=1}^{m_0} (\eta(\lambda_m) + \eta(1-\lambda_m)).$$

Apply the above inequality to the operator U^n. Notice that if V is the operator of multiplication by the function z on $L^2(\exp(2\pi i[(k-1)/n, k/n])$ and B is the operator of multiplication by a function $\lambda(z)$, then V^n can be identified with the operator of multiplication by z on $L^2(\mathbb{T})$ in a way such that B becomes the operator of multiplication by the function

$$\exp(2\pi i t) \mapsto \lambda\left(\exp\left(2\pi i \left(\frac{t}{n} + \frac{k-1}{n}\right)\right)\right), \quad t \in [0,1].$$

Using that $h_{\omega_A}(\alpha_U) = n^{-1} h_{\omega_A}(\alpha_{U^n})$, we therefore conclude that

$$h_{\omega_A}(\alpha_U) \leq \frac{1}{n} \sum_{m=1}^{m_0} \sum_{k=1}^{n} (\eta(\lambda_{mnk}) + \eta(1 - \lambda_{mnk})), \qquad (13.15)$$

where

$$\lambda_{mnk} = \int_0^1 \lambda_m \left(\exp\left(2\pi i \left(\frac{t}{n} + \frac{k-1}{n} \right) \right) \right) dt = n \int_{(k-1)/n}^{k/n} \lambda_m \left(e^{2\pi i t} \right) dt.$$

Now observe that if g is a bounded measurable function and f a continuous function, then

$$\lim_{n \to \infty} \frac{1}{n} \sum_{k=1}^{n} f\left(n \int_{(k-1)/n}^{k/n} g(t) dt \right) = \int_0^1 f(g(t)) dt.$$

To show this define a linear operator F_n on $L^1(0,1)$ by

$$(F_n h)(\tau) = n \int_{(k-1)/n}^{k/n} h(t) dt \quad \text{for} \quad \tau \in \left[\frac{k-1}{n}, \frac{k}{n} \right].$$

Then $F_n \to \mathrm{id}$ in the pointwise norm topology. Indeed, since $\|F_n\| = 1$, it suffices to prove the assertion for continuous functions, for which it is obvious. Thus $F_n g \to g$ in mean, hence in measure. By virtue of uniform continuity of f we conclude that $f \circ F_n g \to f \circ g$ in measure, whence $\int_0^1 f \circ F_n g \, dt \to \int_0^1 f \circ g \, dt$.

Applying this observation to (13.15) we get the required upper bound for the entropy. □

Remark 13.4.7. The formula in Theorem 13.4.1 is also valid for the restriction of the automorphism to the even CAR-algebra. Moreover, it is valid for the gauge-invariant part of the algebra, which is by definition the fixed point algebra for the Bogoliubov automorphisms corresponding to the scalar operators $e^{i\theta} 1$, $\theta \in \mathbb{R}$. Indeed, the key estimate for the lower bound of the entropy in Proposition 13.3.1 uses a subalgebra $\mathcal{P}$ of the gauge-invariant algebra, and the rest of the proof works with minimal changes. On the other hand, the upper bound for the entropy is obtained directly from Theorem 13.4.1, since there exist ω_A-preserving conditional expectations onto the even and gauge-invariant subalgebras. ♦

Let us show next that Bogoliubov automorphisms allow us to construct a large class of entropic K-systems.

Example 13.4.8. Let U be the operator of multiplication by e^{it} on $H = L^2(0, 2\pi)$, A the operator of multiplication by $e^t(1+e^t)^{-1}$. Let ω and τ_θ, $\theta \in \mathbb{R}$, be the quasi-free state and the Bogoliubov automorphism of the even CAR-algebra $\mathcal{A}(H)_e$ corresponding to A and $e^{i\theta} U$, respectively. Set $M = \pi_\omega(\mathcal{A}(H)_e)''$. Extend ω and τ_θ to M by continuity. Then
(i) M is the hyperfinite III$_1$-factor;

(ii) the systems (M, ω, τ_θ), $\theta \in \mathbb{R}$, are entropic K-systems with the same finite entropy;

(iii) if π, θ_1 and θ_2 are linearly independent over $\mathbb{Q}$ then $(M, \omega, \tau_{\theta_1})$ and $(M, \omega, \tau_{\theta_2})$ are nonconjugate.

Part (i) follows from the fact that A has continuous spectrum. We shall only sketch the argument. We claim that the centralizer of the state ω_A on $N = \pi_{\omega_A}(\mathcal{A}(H))''$ is trivial. Moreover, the modular operator has continuous spectrum on the orthogonal complement of the cyclic vector. To show this one can use the explicit description of the GNS-representation in [30, Example 5.2.20], which gives an identification of the space of the representation with $\mathfrak{F}(H) \otimes \mathfrak{F}(H)$ and the cyclic vector with $\Omega \otimes \Omega$, where $\mathfrak{F}(H) = \oplus_{n \geq 0} \wedge^n H$ is the Fermi-Fock space and $\Omega \in \wedge^0 H \cong \mathbb{C}$ is the vacuum vector. One can check that under this identification $\Delta_{\omega_A}^{it}$ is $\Gamma(B^{it}) \otimes \Gamma(B^{it})$, where $B = A(1-A)^{-1}$ and $\Gamma(U) = \oplus_{n \geq 0} U^{\otimes n}$. This implies that the cyclic vector is the only eigenvector of $\Delta_{\omega_A}^{it}$ for $t \neq 0$, and hence the centralizer of ω_A on N is trivial. It follows that the centralizer of ω on M is also trivial. Hence M is a factor of type III$_1$ by [205, Theorem 29.9]. It is hyperfinite since $\mathcal{A}(H)_e$ is an AF-algebra.

To show (ii) consider the closed subspace H_0 of H spanned by the functions e^{-int}, $n \geq 0$. Let M_0 be the W*-subalgebra of M generated by $\mathcal{A}(H_0)_e$. Then $M_0 \subset \tau_\theta(M_0)$, $\cup_{n \in \mathbb{N}}(\tau_\theta^{-n}(M_0)' \cap \tau_\theta^n(M_0)) \supset \cup_{n \in \mathbb{N}} \mathcal{A}(U^n H_0 \ominus U^{-n} H_0)_e$ is weakly dense in M, and $\cap_n \tau_\theta^n(M_0) = \mathbb{C}1$ since M is a factor. Hence (M, ω, τ_θ) is an entropic K-system by Theorem 4.3.2. Theorem 13.4.1 shows that $h_\omega(\tau_\theta)$ is finite and does not depend on θ. Note this follows also from the elementary considerations at the beginning of Sect. 13.2 and from Lemma 13.4.3, since U is the bilateral shift.

To show nonconjugacy notice that

$$U = \left(\frac{A}{1-A}\right)^i,$$

so that $\tau_0 = \sigma_1^\omega$. Denote by γ_θ the Bogoliubov automorphism of M corresponding to the operator $e^{i\theta}1$. Then $\tau_\theta = \gamma_\theta \circ \tau_0$. Assume now that α is an automorphism of M such that $\alpha \circ \tau_{\theta_1} \circ \alpha^{-1} = \tau_{\theta_2}$ and $\omega \circ \alpha = \omega$ (remark that the second equality follows in fact from the first one since τ_θ is ergodic and hence ω is the unique normal τ_θ-invariant state). Since α preserves the state, it commutes with the modular group. In particular, it commutes with τ_0. Hence $\alpha \circ \gamma_{\theta_1} \circ \alpha^{-1} = \gamma_{\theta_2}$. If π, θ_1 and θ_2 are linearly independent, for any θ we can find a sequence $\{m_n\}_n$ of natural numbers such that

$$e^{im_n \theta_1} \to 1 \quad \text{and} \quad e^{im_n \theta_2} \to e^{i\theta}.$$

It follows that $\gamma_{\theta_1}^{m_n} \to \text{id}$ and $\gamma_{\theta_2}^{m_n} \to \gamma_\theta$ in the pointwise strong operator topology. Thus the equality $\alpha \circ \gamma_{\theta_1} \circ \alpha^{-1} = \gamma_{\theta_2}$ implies that γ_θ is the identity automorphism for any θ. This is however true only for $\theta = \pi n$, $n \in \mathbb{Z}$. Hence the systems $(M, \omega, \tau_{\theta_1})$ and $(M, \omega, \tau_{\theta_2})$ are nonconjugate. ♦

13.5 Notes

Quasi-free states were introduced by Shale and Stinespring [194] and have together with Bogoliubov actions been studied extensively since then, see [30].

Consider the Hilbert space $H = L^2(\mathbb{T}, d\mu)$. Let A be the operator of multiplication by a function ρ, $0 \le \rho \le 1$. Denote by H_n be the subspace of H spanned by $1, z, \ldots, z^n$. Then it is natural to call

$$s(\omega_A) = \lim_{n\to\infty} \frac{1}{n} S(\omega_A|_{\mathcal{A}(H_{n-1})})$$

the mean entropy of the quasi-free state ω_A. It is related to mean entropy of states on spin lattice systems as follows. Denote $z^n \in L^2(\mathbb{T})$ by e_n. Even though there is no canonical isomorphism of $\mathcal{A}(H)$ onto $B = \mathrm{Mat}_2(\mathbb{C})^{\otimes \mathbb{Z}}$, there is a canonical isomorphism of the even CAR-algebra $\mathcal{A}(H)_e$ onto B^β, where

$$\beta = \bigotimes_{\mathbb{Z}} \mathrm{Ad}\begin{pmatrix} 1 & 0 \\ 0 & -1 \end{pmatrix}.$$

Namely, denoting by $\{e_{ij}^{(n)}\}$ the matrix units in the n-th factor of B we formally put

$$a(e_n) \mapsto \ldots \otimes \begin{pmatrix} 1 & 0 \\ 0 & -1 \end{pmatrix} \otimes \begin{pmatrix} 1 & 0 \\ 0 & -1 \end{pmatrix} \otimes e_{12}^{(n)} \otimes 1 \otimes 1 \otimes \ldots,$$

which induces a well-defined map $\mathcal{A}(H)_e \to B^\beta$. Using this isomorphism, given an α_{-1}-invariant state φ on $\mathcal{A}(H)$ we get a state on B^β. Composing it with the conditional expectation $B \to B^\beta$, $x \mapsto (x + \beta(x))/2$, we get a state $\tilde\varphi$ on B. The map $\varphi \mapsto \tilde\varphi$ is a one-to-one correspondence between α_{-1}-invariant states on $\mathcal{A}(H)$ and β-invariant states on B. One can easily check that $s(\varphi) = s(\tilde\varphi)$, where $s(\varphi)$ is defined as above.

We have

$$s(\omega_A) = \int_{\mathbb{T}} (\eta(\rho)(z) + \eta(1 - \rho(z))) d\mu(z).$$

This was proved by Fannes [65]. As was observed later, see [5], the result also follows from Szegö's theorem. By virtue of Corollary 13.4.2 and Remark 13.4.7 we can then conclude that $s(\tilde\omega_A) = h_{\tilde\omega_A}(\alpha)$, where α is the shift to the right on B. As we discussed in Chap. 9, one expects that $s(\psi) = h_\psi(\alpha)$ for any α-invariant state ψ on $B = \mathrm{Mat}_2(\mathbb{C})^{\otimes \mathbb{Z}}$. Corollary 13.4.2 therefore says that this is the case for states derived from quasi-free states, and provides a natural extension of this fact. From this point of view the formula in Corollary 13.4.2 appeared and was proved in a particular case already in the paper of Connes, Narnhofer and Thirring [50]. Explicitly it appeared in the work of Narnhofer and Thirring [130]. In the more general setting of Theorem 13.4.1 the formula for the dynamical entropy was conjectured by Connes for the tracial quasi-free state $\omega_{1/2}$, and by Størmer and Voiculescu [212] for an arbitrary quasi-free state ω_A. Størmer and Voiculescu proved the formula under the additional assumption that $A|_{H_a}$ has pure point spectrum. We use their argument in Sect. 13.2 to compute the topological entropy. In certain cases of continuous spectrum, covered by Corollary 13.4.2, the formula was then proved by Park and Shin [155] and Narnhofer and Thirring [133]. See also [14] for related results. In full generality Theorem 13.4.1 was proved by Neshveyev [139].

Example 13.2.5 is due to the authors [140]. Example 13.4.8 is from [139]. A careful analysis of the modular group of a quasi-free state allows one to prove a more general nonisomorphism result [77].

As we remarked in Notes to Chap. 3, the notion of entropy makes sense for an action of an arbitrary discrete amenable group. If G is a countable abelian group and $U: G \to B(H)$ a unitary representation of G on H, then we can define an action α_U of G on $\mathcal{A}(H)$ by Bogoliubov automorphisms. The formula for the entropy from Theorem 13.4.1 makes sense for such an action with obvious changes: we have to desintegrate the representation over the dual group $\hat{G}$ and use the Haar measure on $\hat{G}$ instead of the Lebesgue measure on $\mathbb{T}$. Particular cases of this formula were proved in [155], [21], [149], [78]. As was remarked in [139], a combination of the methods of [78] and [139] shows that the formula is valid at least for all torsion-free abelian groups.

Similar results hold for the canonical commutation relations (CCR). For a Hilbert space H, the CCR-algebra $\mathcal{U}(H)$ is the universal C*-algebra generated by unitaries $W(f)$, $f \in H$, such that

$$W(f)W(g) = e^{i\,\mathrm{Im}(f,g)/2} W(f+g).$$

In other words, $\mathcal{U}(H)$ is the twisted group C*-algebra of the discrete abelian group H corresponding to the cocycle $\omega(f,g) = \exp(i\,\mathrm{Im}(f,g)/2)$. A unitary operator U on H defines an automorphism α_U of $\mathcal{A}(H)$ by $\alpha_U(W(f)) = W(Uf)$. A positive operator A defines a quasi-free state ω_A on $\mathcal{U}(H)$ by

$$\omega_A(W(f)) = e^{-(\|f\|^2 + 2(Af,f))/4}.$$

Then, similarly to Theorem 13.4.1, we have, see [139],

$$h_{\omega_A}(\alpha_U) = \int_{\mathbb{T}} \mathrm{Tr}(\eta(A_z) - \eta(1 + A_z)) d\mu(z).$$

By Theorem 11.2.1 we may also conclude that $ht(\alpha_U) = 0$ if the spectrum of U consists of a finite number of roots of unity, and $ht(\alpha_U) = +\infty$ otherwise.

Finally note that automorphisms of Bogoliubov type arise from representations of the group $U(\infty)$, and a similar formula for the entropy holds in that case [28].

14
Free Products

In this chapter we shall study dynamical systems on free products of C^*-algebras. These algebras are in many respects the most noncommutative C^*-algebras. From our experience so far with entropy we should expect that shifts on infinite free products have zero entropy. We shall see that this is indeed the case. More generally, we shall show that a free product automorphism is not more chaotic than its free factors, so that the entropy of a free product automorphism is the supremum of the entropies of the factors.

14.1 Free Products of Algebras and Maps

Let A_1 and A_2 be unital C^*-algebras. Consider the **full**, or maximal, **free product** C^*-algebra $A_1 \hat{*} A_2$ which is generated by copies of A_1 and A_2 and has the following universal property: any unital representations of A_1 and A_2 on a Hilbert space H extend to a representation of $A_1 \hat{*} A_2$.

Lemma 14.1.1. *For states φ_i on A_i, $i = 1, 2$, there exists a state $\varphi = \varphi_1 * \varphi_2$ on $A_1 \hat{*} A_2$ uniquely determined by the following condition:*

$$\varphi(a_1 \ldots a_m) = 0 \quad \text{if} \quad a_j \in \operatorname{Ker} \varphi_{i_j}, \quad i_j \neq i_{j+1} \quad (1 \leq j \leq m-1). \tag{14.1}$$

The C^*-algebra $A = \pi_\varphi(A_1 \hat{*} A_2)$ is called the **reduced free product** of A_1 and A_2 with respect to φ_1 and φ_2. We write

$$(A, \varphi) = (A_1, \varphi_1) * (A_2, \varphi_2).$$

Proof of Lemma 14.1.1. We shall explicitly describe the GNS-representation of φ. Let (H_i, ξ_i, π_i) be the GNS-triple associated with φ_i. Set $H_i^\circ = H_i \ominus \mathbb{C}\xi_i$. Consider the Hilbert space H, with a distinguished unit vector ξ, defined by

$$H = \mathbb{C}\xi \oplus \bigoplus_{\substack{m=1 \\ i_j \neq i_{j+1}}}^{\infty} \bigoplus_{i_1, \ldots, i_m} H_{i_1}^\circ \otimes \ldots \otimes H_{i_m}^\circ.$$

We write $(H, \xi) = (H_1, \xi_1) * (H_2, \xi_2)$.

Consider the closed subspace $H(i)$ of H spanned by ξ and the Hilbert spaces $H_{i_1}^\circ \otimes \ldots \otimes H_{i_m}^\circ$ with $i_1 \neq i$. Define a unitary $U_i \colon H_i \otimes H(i) \to H$ by

$$U_i(\xi_i \otimes \xi) = \xi,$$
$$U_i(\xi_i \otimes \eta) = \eta \text{ for } \eta \in H_{i_1}^\circ \otimes \ldots \otimes H_{i_m}^\circ,$$
$$U_i(\zeta \otimes \xi) = \zeta \text{ for } \zeta \in H_i^\circ,$$
$$U_i(\zeta \otimes \eta) = \zeta \otimes \eta \text{ for } \zeta \in H_i^\circ,\ \eta \in H_{i_1}^\circ \otimes \ldots \otimes H_{i_m}^\circ.$$

Define a representation π of $A_1 \hat{*} A_2$ on H by setting $\pi(a) = U_i(\pi_i(a) \otimes 1)U_i^*$ for $a \in A_i$. Note that if $a \in \operatorname{Ker} \varphi_i$, $\zeta_1 \otimes \ldots \otimes \zeta_m \in H_{i_1}^\circ \otimes \ldots \otimes H_{i_m}^\circ$ and $i_1 \neq i$, then

$$\pi(a)(\zeta_1 \otimes \ldots \otimes \zeta_m) = \pi_i(a)\xi_i \otimes \zeta_1 \otimes \ldots \otimes \zeta_m.$$

If $i_1 = i$, then $\pi_i(a)\zeta_1 = \tilde{\zeta}_1 + \lambda \xi_i$ for some $\tilde{\zeta}_1 \in H_i^\circ$ and $\lambda \in \mathbb{C}$, and

$$\pi(a)(\zeta_1 \otimes \ldots \otimes \zeta_m) = \tilde{\zeta}_1 \otimes \zeta_2 \otimes \ldots \otimes \zeta_m + \lambda \zeta_2 \otimes \ldots \otimes \zeta_m.$$

It is now easy to see that the state $\omega_\xi \circ \pi$ satisfies (14.1), where $\omega_\xi = (\cdot\, \xi, \xi)$. It is also easy to check that ξ is cyclic, so (H, ξ, π) is the GNS-triple corresponding to $\varphi_1 * \varphi_2$. □

Remark 14.1.2. If φ_1 and φ_2 are pure, then $\varphi = \varphi_1 * \varphi_2$ is also pure. Indeed, the projection $p_i \in B(H_i)$ onto $\mathbb{C}\xi_i$ belongs to the weak operator closure of A_i in $B(H_i)$. Under the representation of $B(H_i)$ on H this projection becomes the projection onto $H(i)$. Hence $p_1 p_2$ is the projection onto $\mathbb{C}\xi$. Since it belongs to $\pi_\varphi(A)''$, and ξ is cyclic, we conclude that $\pi_\varphi(A)'' = B(H)$. ◆

If α_i is an automorphism of A_i, $i = 1, 2$, then there exists a unique automorphism $\alpha_1 * \alpha_2$ of $A_1 \hat{*} A_2$ extending α_1 and α_2. It is clear that if in addition α_i preserves a state φ_i, then $\varphi_1 * \varphi_2$ is $\alpha_1 * \alpha_2$-invariant, so we have a well-defined automorphism of the corresponding reduced free product. We denote it again by $\alpha_1 * \alpha_2$. Using the above description of the GNS-representation it is easy to construct a unitary implementation of $\alpha_1 * \alpha_2$ from unitary implementations of α_1 and α_2.

It will be convenient for us to have not only free products of automorphisms, but also free products of unital completely positive maps.

Theorem 14.1.3. *Let $\theta_i \colon A_i \to B_i$ be a unital completely positive map, ψ_i a state on B_i, $\psi_i \circ \theta_i = \varphi_i$, $i = 1, 2$. Assume the GNS-representations corresponding to the states ψ_i and φ_i, $i = 1, 2$, are faithful. Set*

$$(A, \varphi) = (A_1, \varphi_1) * (A_2, \varphi_2) \quad \text{and} \quad (B, \psi) = (B_1, \psi_1) * (B_2, \psi_2).$$

Then there exists a unital completely positive map $\theta = \theta_1 * \theta_2 \colon A \to B$ such that

14.1 Free Products of Algebras and Maps

$$\theta(a_1 \ldots a_m) = \theta_{i_1}(a_1) \ldots \theta_{i_m}(a_m) \quad \text{if} \quad a_j \in \operatorname{Ker} \varphi_{i_j}, \ i_j \neq i_{j+1}. \tag{14.2}$$

If, moreover, θ_i extends to a normal unital completely positive map $\pi_{\varphi_i}(A_i)'' \to \pi_{\psi_i}(B_i)''$ for $i = 1, 2$, then θ extends to a normal unital completely positive map $\pi_\varphi(A)'' \to \pi_\psi(B)''$.

Proof. By Stinespring's dilation A.2 there exist a representation of A_i on a Hilbert space H_i and an isometry $V_i \colon H_{\psi_i} \to H_i$ such that $\theta_i(a) = V_i^* a V_i$ for $a \in A_i$. Set $\xi_i = V_i \xi_{\psi_i}$. Then ξ_i defines φ_i, so that we can identify H_{φ_i} with $\overline{A_i \xi_i}$.

Assume first that $H_i = H_{\varphi_i}$, $i = 1, 2$. Then using the description of the GNS-representations corresponding to φ and ψ from Lemma 14.1.1 we get an obvious isometry $V = V_1 * V_2 \colon H_\psi \to H_\varphi$, and thus a normal unital completely positive map $\theta \colon B(H_\varphi) \to B(H_\psi)$, $\theta(x) = V^* x V$.

To check (14.2) consider a vector $\zeta \in H^\circ_{\psi_{k_1}} \otimes \ldots \otimes H^\circ_{\psi_{k_n}} \subset H_\psi$. If $i_m \neq k_1$, then both sides of (14.2) applied to the vector ζ give

$$V_{i_1}^* a_1 \xi_{i_1} \otimes \ldots \otimes V_{i_m}^* a_m \xi_{i_m} \otimes \zeta.$$

If $i_m = k_1$, we can assume that ζ is of the form $\zeta_1 \otimes \zeta_2$ for some $\zeta_1 \in H^\circ_{\psi_{i_m}}$ and $\zeta_2 \in H^\circ_{\psi_{k_2}} \otimes \ldots \otimes H^\circ_{\psi_{k_n}}$. Then $a_m V_{i_m} \zeta_1 = \tilde{\zeta}_1 + \lambda \xi_{i_m}$, where $\tilde{\zeta}_1 \in H^\circ_{\varphi_{i_m}}$. Hence $\theta_{i_m}(a_m) \zeta_1 = V_{i_m}^* \tilde{\zeta}_1 + \lambda \xi_{\psi_{i_m}}$ and

$$\theta(a_1 \ldots a_m) \zeta$$
$$= V^* a_1 \ldots a_{m-1} (\tilde{\zeta}_1 \otimes V \zeta_2 + \lambda V \zeta_2)$$
$$= V_{i_1}^* a_1 \xi_{i_1} \otimes \ldots \otimes V_{i_{m-1}}^* a_{m-1} \xi_{i_{m-1}} \otimes V_{i_m}^* \tilde{\zeta}_1 \otimes \zeta_2 + \lambda \theta(a_1 \ldots a_{m-1}) \zeta_2$$
$$= \theta(a_1 \ldots a_{m-1})(V_{i_m}^* \tilde{\zeta}_1 \otimes \zeta_2 + \lambda \zeta_2)$$
$$= \theta(a_1 \ldots a_{m-1}) \theta_{i_m}(a_m) \zeta.$$

So in this case the equality $\theta(a_1 \ldots a_m) \zeta = \theta_{i_1}(a_1) \ldots \theta_{i_m}(a_m) \zeta$ follows by induction on m.

Without assuming $H_i = H_{\varphi_i}$, consider now $(H, \xi) = (H_1, \xi_1) * (H_2, \xi_2)$. Then we still have representations of A_1 and A_2 on H, and thus a representation of $A_1 \hat{*} A_2$. We also have an isometry $V = V_1 * V_2 \colon H_\psi \to H$, and thus a unital completely positive map $B(H) \to B(H_\psi)$, $x \mapsto V^* x V$, which is a normal extension of the free product of the maps $B(H_i) \to B(H_{\psi_i})$, $x \mapsto V_i^* x V_i$, $i = 1, 2$. So what we have to prove is that the representation of $A_1 \hat{*} A_2$ on H factorizes through A.

Set $K_i = H_i \ominus H_{\varphi_i}$. Then K_i is an A_i-invariant subspace of H_i°. Let L_i be the subspace of H spanned by the spaces $K_i \otimes H^\circ_{k_1} \otimes \ldots \otimes H^\circ_{k_n}$. Then we have

$$H = H_\varphi \oplus \overline{(A_1 \hat{*} A_2) L_1} \oplus \overline{(A_1 \hat{*} A_2) L_2}.$$

Let $a = a_{i_1} \ldots a_{i_m} \in A_1 \hat{*} A_2$, $a_j \in \operatorname{Ker} \varphi_{i_j}$, $i_j \neq i_{j+1}$. Then $a L_1 \subset L_1^\perp$ unless $m = 1$ and $i_m = 1$. Let $E_1 = \operatorname{id}_{A_1} * \varphi_2 \colon A \to A_1$, which exists by the case considered above. Then by definition $E_1(\pi_\varphi(a)) = 0$ unless $m = 1$ and $i_m = 1$. It follows that for any $\zeta, \eta \in L_1$ we have $(a\zeta, \eta) = (E_1(\pi_\varphi(a))\zeta, \eta)$. Hence for any $a, b, c \in A_1 \hat{*} A_2$ and $\zeta, \eta \in L_1$ we have $(ab\zeta, c\eta) = (E_1(\pi_\varphi(c^*ab))\zeta, \eta)$, so that the functional $(\cdot b\zeta, c\eta)$ on $A_1 \hat{*} A_2$ factorizes through A. Hence the representation of $A_1 \hat{*} A_2$ on $\overline{(A_1 \hat{*} A_2) L_1}$ factorizes through A (moreover, what we have proved is that the representation of A on $\overline{(A_1 \hat{*} A_2) L_1}$ is induced from the representation of A_1 on L_1 via the conditional expectation $E_1 \colon A \to A_1$). The same is true for $\overline{(A_1 \hat{*} A_2) L_2}$. So we get a representation of A on $B(H)$, and the proof of the first part of the theorem is complete.

The second part follows by observing that if θ_1 and θ_2 extend to normal maps, then the representation of A_i on H_i extends to a normal representation of $\pi_{\varphi_i}(A_i)''$, $i = 1, 2$, which together with normality of the conditional expectations E_1 and E_2 implies that the representation of A on H extends to a normal representation of $\pi_\varphi(A)''$. □

14.2 Free Shifts

Let φ be a state on a C*-algebra A. We call the shift α to the right on the infinite free product of (A, φ) with itself the **free shift** defined by (A, φ). To be more precise, for each $n \in \mathbb{N}$ fix a copy (A_n, φ_n) of (A, φ) and an isomorphism $\pi_n \colon (A, \varphi) \to (A_n, \varphi_n)$. Set

$$(B, \psi) = *_{n \in \mathbb{Z}} (A_n, \varphi_n).$$

We write $(B, \psi) = (A, \varphi)^{*\mathbb{Z}}$. Then α is the unique automorphism of B such that $\alpha(\pi_n(a)) = \pi_{n+1}(a)$ for $a \in A$.

Our aim in this section is to prove that free shifts have zero entropy. The proof is based on the following result, which gives a good feeling of how far free product algebras are from abelian algebras.

Theorem 14.2.1. *Let $(A, \psi) = *_{i \in I}(A_i, \varphi_i)$ be an infinite free product, D a unital abelian C*-algebra, $P \colon A \to D$ a unital positive map, and μ a state on D. Then for any $\varepsilon > 0$ there exists a subset $J \subset I$ such that $|J| < (6/\varepsilon)^2$ and*

$$\|P(a) - \psi(a)1\|_\mu \le \varepsilon \|a\| \quad \text{for any} \quad a \in A_i, \ i \notin J.$$

Proof. Consider first the case when D is finite dimensional and P can be extended to a normal map $\pi_\psi(A)'' \to D$. In other words, $D \cong \mathbb{C}^N$, and P is given by normal states $\psi_1, \ldots, \psi_N$ on $\pi_\psi(A)''$. Let also μ be defined by a probability distribution $(\lambda_1, \ldots, \lambda_N)$.

Let $(H_i, \xi_i) = (H_{\varphi_i}, \xi_{\varphi_i})$ and $(H, \xi) = *_i(H_i, \xi_i)$, so that ξ is the cyclic vector defining ψ. Since ψ_k is normal, there exist vectors $\zeta_{kn} \in H$, $n \in \mathbb{N}$, such that

$$\psi_k = \sum_n \omega_{\zeta_{kn}},$$

where as usual $\omega_\zeta(x) = (x\zeta, \zeta)$. Denoting by $H(i)$ the closed subspace of H spanned by ξ and the subspaces $H^\circ_{i_1} \otimes \ldots \otimes H^\circ_{i_m}$ with $i_1 \neq i$, we have an orthogonal decomposition

$$\zeta_{kn} = \lambda_{kn}\xi + \sum_i \zeta_{kni}, \quad \zeta_{kni} \in H(i)^\perp.$$

Let J consist of all $i \in I$ such that

$$\sum_{k,n} \lambda_k \|\zeta_{kni}\|^2 > \left(\frac{\varepsilon}{6}\right)^2.$$

Since $\sum_{k,n,i} \lambda_k\|\zeta_{kni}\|^2 = \sum_{k,n}\lambda_k(\|\zeta_{kn}\|^2 - |\lambda_{kn}|^2) \le \sum_k \lambda_k \psi_k(1) = 1$, we have $|J| < (6/\varepsilon)^2$. Fix $i \notin J$, and let $a \in \operatorname{Ker}\varphi_i$. Then $aH(i) \subset H(i)^\perp$. Since $\zeta_{kn} - \zeta_{kni} \in H(i)$, we thus have

$$(a(\zeta_{kn} - \zeta_{kni}), \zeta_{kn} - \zeta_{kni}) = 0.$$

It follows that

$$|\psi_k(a)| = \left|\sum_n (a\zeta_{kn}, \zeta_{kn})\right|$$

$$= \left|\sum_n \Big((a\zeta_{kni}, \zeta_{kn}) + (a\zeta_{kn}, \zeta_{kni}) - (a\zeta_{kni}, \zeta_{kni})\Big)\right|$$

$$\le \left(2\sum_n \|\zeta_{kni}\|\,\|\zeta_{kn}\| + \sum_n \|\zeta_{kni}\|^2\right)\|a\|$$

$$\le 3\left(\sum_n \|\zeta_{kni}\|^2\right)^{1/2} \|a\|,$$

where in the last step we used the Cauchy-Schwarz inequality and that $\sum_n \|\zeta_{kni}\|^2 \le \sum_n \|\zeta_{kn}\|^2 = 1$. Since $i \notin J$, it follows that

$$\|P(a)\|_\mu^2 = \sum_k \lambda_k |\psi_k(a)|^2 \le 9\sum_{k,n} \lambda_k \|\zeta_{kni}\|^2 \|a\|^2 \le \left(\frac{\varepsilon}{2}\right)^2 \|a\|^2.$$

Hence for any $a \in A_i$ we get

$$\|P(a) - \psi(a)1\|_\mu = \|P(a - \psi(a)1)\|_\mu \le \frac{\varepsilon}{2}\|a - \psi(a)1\| \le \varepsilon\|a\|.$$

Assume now that D is finite dimensional, but P is not necessarily normal. If the statement of the theorem is wrong, there exist $J \subset I$ and elements $a_j \in A_j$, $j \in J$, such that $|J| \ge (6/\varepsilon)^2$ and

$$\|P(a_j) - \psi(a_j)1\|_\mu > \varepsilon\|a_j\|, \quad j \in J.$$

But then, since P can be approximated by normal maps in the pointwise norm topology, we get a contradiction with the previous case.

Consider the general case. Replacing D by $\pi_\mu(D)''$ we may assume that D is a von Neumann algebra and μ is a faithful normal state. Choose an increasing net $\{D_k\}_k$ of finite dimensional C*-subalgebras of D such that $\cup_k D_k$ is dense in D. Let $E_k\colon D \to D_k$ be the μ-preserving conditional expectation. Set

$$J_k = \{i \in I \mid \|(E_k \circ P)(a) - \psi(a)1\|_\mu > \varepsilon\|a\| \text{ for some } a \in A_i\}.$$

Then $|J_k| < (6/\varepsilon)^2$ for any k. So there exists k_0 such that $|J_{k_0}|$ is maximal. Since $E_{k_0} = E_{k_0} \circ E_k$ for $k \succ k_0$, and E_{k_0} is a contraction in the L^2-norm, we have $J_{k_0} \subset J_k$ if $k \succ k_0$. Hence $J_k = J_{k_0}$. In other words, for any $a \in A_i$, $i \notin J_{k_0}$, and $k \succ k_0$ we have $\|(E_k \circ P)(a) - \psi(a)1\|_\mu \le \varepsilon\|a\|$. Since $E_k \to \mathrm{id}$ in the pointwise strong operator topology, we get $\|P(a) - \psi(a)1\|_\mu \le \varepsilon\|a\|$. □

Using Lemma 3.1.2 and the above theorem it is now easy to show that the dynamical entropy of a free shift is zero. But we can prove a stronger statement. Let us first observe the following.

Corollary 14.2.2. *Under the assumptions of the theorem for any $\varepsilon > 0$ there exists a finite subset $J \subset I$ such that*

$$\|P(a) - \psi(a)1\|_\mu \le \varepsilon\|a\| \quad \text{for every } a \text{ in } *_{i \in I \setminus J}(A_i, \varphi_i).$$

Proof. Assume this does not hold. Then there exists a sequence $\{J_n\}_n$ of disjoint finite subsets of I, and elements a_n in $(B_n, \psi_n) = *_{i \in J_n}(A_i, \varphi_i)$ such that

$$\|P(a_n) - \psi(a_n)1\|_\mu > \varepsilon\|a_n\|.$$

Since $*_{n \in \mathbb{N}}(B_n, \psi_n)$ can be considered as a subalgebra of $*_{i \in I}(A_i, \varphi_i)$, this contradicts the theorem. □

Corollary 14.2.3. *Let α be the free shift on $(B, \psi) = (A, \varphi)^{*\mathbb{Z}}$, β an automorphism of an abelian C*-algebra D. Then any $(\alpha \otimes \beta)$-invariant state on $B \otimes D$ is of the form $\psi \otimes \mu$, where μ is a β-invariant state on D.*

In particular, any stationary coupling of (B, ψ, α) with an abelian dynamical system is trivial, so that $h_\psi(\alpha) = h_\psi^{ST}(\alpha) = 0$.

Proof. Given an $(\alpha \otimes \beta)$-invariant state λ on $B \otimes D$, set $\mu = \lambda|_D$ and consider the unital completely positive map $P\colon B \to \pi_\mu(D)''$ such that

$$\mu(P(x)y) = \lambda(x \otimes y) \quad \text{for } x \in B,\ y \in D.$$

Then $\beta \circ P = P \circ \alpha$. In particular, $\|(P \circ \alpha^n)(x) - \psi(x)1\|_\mu$ does not depend on $n \in \mathbb{Z}$. By the previous corollary we conclude that $P(x) = \psi(x)1$ for every x lying in a finite free product, hence for all $x \in B$. Therefore $\lambda = \psi \otimes \mu$. □

For exact algebras we have an even stronger result.

Theorem 14.2.4. *Let A be an exact C^*-algebra, φ a state on A with faithful GNS-representation. Let α be the free shift on $(B, \psi) = (A, \varphi)^{*\mathbb{Z}}$. Then $ht(\alpha) = 0$.*

This will be proved in the next section. Notice that (B, ψ) depends only on the image of A in the GNS-representation. So the assumption of faithfulness of the representation is in fact redundant if we use the fact that quotients of exact algebras are exact.

14.3 Free Product Automorphisms

Our goal is to prove the following result.

Theorem 14.3.1. *Let $(A_i, \varphi_i, \alpha_i)$ be a C^*-dynamical system, $i = 1, 2$. Assume the GNS-representations corresponding to φ_1 and φ_2 are faithful. Set $(A, \varphi) = (A_1, \varphi_1) * (A_2, \varphi_2)$. Then for the automorphism $\alpha_1 * \alpha_2$ of A we have*
$$ht(\alpha_1 * \alpha_2) = \max\{ht(\alpha_1), ht(\alpha_2)\}.$$

This theorem implies in particular that reduced free products of exact algebras are exact. We shall first prove this for matrix algebras by explicitly constructing approximations.

Lemma 14.3.2. *Let H_i be a finite dimensional Hilbert space, $\xi_i \in H_i$ a unit vector, $\Omega_i \subset B(H_i)$ a finite self-adjoint subset containing the unit, $i = 1, 2$. Set $(H, \xi) = (H_1, \xi_1) * (H_2, \xi_2)$. Then for any $N \in \mathbb{N}$ there exist a Hilbert space K and unital completely positive maps $\theta \colon B(H) \to B(K)$ and $\gamma \colon B(K) \to B(H)$ such that*

(i) $\dim K < (\max_i |\Omega_i|)^{(N-1)^2}(\dim H_1 + \dim H_2)$;

(ii) $\|(\gamma \circ \theta)(a) - a\| \leq 3N^{-1/2}\|a\|$ *for* $a \in \Omega_1 \cup \Omega_2$.

Proof. Consider the filtration
$$\mathbb{C}\xi_i = E_i^0 \subset E_i^1 \subset \ldots \subset E_i^{N-1} \subset E_i^N = H_i,$$
where E_i^k is spanned by $\Omega_i E_i^{k-1}$ for $1 \leq k \leq N-1$, and set $F_i^k = E_i^k \ominus E_i^{k-1}$, $1 \leq k \leq N$. Let K be the subspace of H spanned by ξ and all $F_{i_1}^{k_1} \otimes \ldots \otimes F_{i_{d+1}}^{k_{d+1}}$, $d \geq 0$, with $k_1 + \ldots + k_d < N$. As $F_i^k \subset E_i^{N-1}$ for $k \leq N-1$ and $\dim E_i^{N-1} \leq |\Omega_i|^{N-1}$, we have (i) satisfied.

We define $\theta \colon B(H) \to B(K)$ to be the compression map. The map γ will be defined by an isometry $V \colon H \to K \otimes H$. Set $V\xi = \xi \otimes \xi$. For a vector $\zeta = \zeta_1 \otimes \ldots \otimes \zeta_m \in F_{i_1}^{k_1} \otimes \ldots \otimes F_{i_m}^{k_m}$, let $d \geq 0$ be the largest number such that $d < m$ and $k_1 + \ldots + k_d < N$. Then we set

$$V\zeta = \sum_{l=1}^{d} \left(\frac{k_l}{N}\right)^{1/2} (\zeta_1 \otimes \ldots \otimes \zeta_l) \otimes (\zeta_{l+1} \otimes \ldots \otimes \zeta_m)$$
$$+ \left(\frac{N - k_1 - \ldots - k_d}{N}\right)^{1/2} (\zeta_1 \otimes \ldots \otimes \zeta_{d+1}) \otimes (\zeta_{d+2} \otimes \ldots \otimes \zeta_m),$$

where we consider $\zeta_1 \otimes \ldots \otimes \zeta_l$ as an element of K, and $\zeta_{l+1} \otimes \ldots \otimes \zeta_m$ as an element of H, and similarly when l is replaced by $d+1$. Note that if $m = d+1$, then $\zeta_{d+2} \otimes \ldots \otimes \zeta_m$ should be understood as ξ. Now define $\gamma \colon B(K) \to B(H)$ by $\gamma(x) = V^*(x \otimes 1)V$.

To estimate $\|(\gamma \circ \theta)(a) - a\|$ we consider V as an operator $H \to H \otimes H$. If $a \in \Omega_i$, then $aF_i^k \subset E_i^{k+1}$, and $aF_i^k \perp E_i^{k-2}$ as $a^*E_i^{k-2} \subset E_i^{k-1}$. So $aF_i^k \subset F_i^{k-1} \oplus F_i^k \oplus F_i^{k+1}$ (with $F_i^0 = \mathbb{C}\xi_i$ and $F_i^{-1} = F_i^{N+1} = 0$). Thus we can decompose a into the sum of operators a^-, a^0, a^+ such that $a^0 F_i^k \subset F_i^k$ and $a^{\pm} F_i^k \subset F_i^{k\pm 1}$. These operators are of norm not greater than $\|a\|$. Since each subspace $F_{i_1}^{k_1} \otimes \ldots \otimes F_{i_m}^{k_m}$ is a^0-invariant by definition of the representations of $B(H_1)$ and $B(H_2)$ on H, see the proof of Lemma 14.1.1, we have $Va^0 = (a^0 \otimes 1)V$. To estimate $Va^+ - (a^+ \otimes 1)V$ it will be convenient to introduce two more isometries $V_i \colon H \to H \otimes H$, $i = 1, 2$.

Set $V_i \xi = \xi \otimes \xi$. For $\zeta = \zeta_1 \otimes \ldots \otimes \zeta_m \in F_{i_1}^{k_1} \otimes \ldots \otimes F_{i_m}^{k_m}$, set $V_i \zeta = V\zeta$ if $i_1 \neq i$. If $i_1 = i$, let $d' \geq 0$ be the largest number such that $d' < m$ and $k_1 + \ldots + k_{d'} < N + 1$, and set

$$V_i \zeta = \left(\frac{k_1 - 1}{N}\right)^{1/2} \zeta_1 \otimes (\zeta_2 \otimes \ldots \otimes \zeta_m)$$
$$+ \sum_{l=2}^{d'} \left(\frac{k_l}{N}\right)^{1/2} (\zeta_1 \otimes \ldots \otimes \zeta_l) \otimes (\zeta_{l+1} \otimes \ldots \otimes \zeta_m)$$
$$+ \left(\frac{N - k_1 - \ldots - k_{d'} + 1}{N}\right)^{1/2} (\zeta_1 \otimes \ldots \otimes \zeta_{d'+1}) \otimes (\zeta_{d'+2} \otimes \ldots \otimes \zeta_m).$$

Then $(a^+ \otimes 1)V = V_i a^+$ for $a \in \Omega_i$, so that

$$\|Va^+ - (a^+ \otimes 1)V\| \leq \|V - V_i\| \|a^+\| \leq \|V - V_i\| \|a\|.$$

We claim that $\|V - V_i\| \leq \sqrt{2/N}$. It is enough to check this on each subspace $F_{i_1}^{k_1} \otimes \ldots \otimes F_{i_m}^{k_m}$ with $i_1 = i$. Consider the case $d' \neq d$. Then $d' = d + 1$ and $k_1 + \ldots + k_{d+1} = N$, so that

$$V\zeta - V_i\zeta = \left(\left(\frac{k_1}{N}\right)^{1/2} - \left(\frac{k_1 - 1}{N}\right)^{1/2}\right) \zeta_1 \otimes (\zeta_2 \otimes \ldots \otimes \zeta_m)$$
$$- \left(\frac{1}{N}\right)^{1/2} (\zeta_1 \otimes \ldots \otimes \zeta_{d+2}) \otimes (\zeta_{d+3} \otimes \ldots \otimes \zeta_m).$$

Thus $V - V_i$ is the sum of two operators with orthogonal ranges and of norm not greater than $N^{-1/2}$ each. The case $d' = d$ is even easier.

14.3 Free Product Automorphisms   259

In a similar way we get $\|Va^- - (a^- \otimes 1)V\| \leq \sqrt{2/N}\|a\|$, whence

$$\|(\gamma \circ \theta)(a) - a\| = \|V^*((a \otimes 1)V - Va)\| \leq 2\sqrt{2/N}\|a\| \leq 3N^{-1/2}\|a\|,$$

and the proof of the lemma is complete. □

Now if we have two exact C*-algebras A_1 and A_2, we can approximate $A_1 * A_2$ by free products of matrix algebras, and then approximate the latter product by matrix algebras. For the first step we need however state preserving approximations. This is always possible to achieve by the following lemma.

Lemma 14.3.3. *Let $\Omega \subset B(H)$ be a finite self-adjoint subset containing the unit, $\xi \in H$ a unit vector, $\theta \colon B(H) \to \mathrm{Mat}_m(\mathbb{C})$ and $\gamma \colon \mathrm{Mat}_m(\mathbb{C}) \to B(H)$ unital completely positive maps such that*

$$\|(\gamma \circ \theta)(a) - a\| < \varepsilon\|a\| \quad \text{for} \ a \in \Omega.$$

Then for any $N \in \mathbb{N}$ there exist a Hilbert space H_0, a unit vector $\xi_0 \in H_0$ and normal unital completely positive maps $\theta_0 \colon B(H) \to B(H_0)$ and $\gamma_0 \colon B(H_0) \to B(H)$ such that

(i) $\dim H_0 \leq |\Omega|^{N-1} + m$;

(ii) $\omega_{\xi_0} \circ \theta_0 = \omega_\xi$, $\omega_\xi \circ \gamma_0 = \omega_{\xi_0}$;

(iii) $\|(\gamma_0 \circ \theta_0)(a) - a\| < (\varepsilon + 3N^{-1/2})\|a\|$ for $a \in \Omega$.

Proof. By A.7 we can assume that θ is normal. Then the representation of $B(H)$ given by the Stinespring dilation of θ is also normal, hence it is an amplification of the representation of $B(H)$ on H. In other words, there exist a Hilbert space H' and an isometry $W \colon \ell_2^m \to H \otimes H'$ such that $\theta(x) = W^*(x \otimes 1)W$ for $x \in B(H)$. Let ξ' be a unit vector in H'. Since we have the Fubini map $\mathrm{id} \otimes \omega_{\xi'} \colon B(H \otimes H') \to B(H)$, it suffices to prove the lemma for the Hilbert space $H \otimes H'$, the vector $\xi \otimes \xi'$ and the maps $\gamma' \colon \mathrm{Mat}_m(\mathbb{C}) \to B(H \otimes H')$, $\gamma'(x) = \gamma(x) \otimes 1$, and $\theta' \colon B(H \otimes H') \to \mathrm{Mat}_m(\mathbb{C})$, $\theta'(x) = W^*xW$. Thus by replacing H by $H \otimes H'$ and by setting $K = W\ell_2^m$ we can assume that K is a subspace of H of dimension m, and $\theta \colon B(H) \to B(K)$ is the compression map.

Then the map $\gamma \colon B(K) \to B(H)$ is given by an isometry $V \colon H \to K \otimes L$, so $\gamma(x) = V^*(x \otimes 1)V$.

As in the proof of Lemma 14.3.2 consider the filtration

$$\mathbb{C}\xi = E^0 \subset E^1 \subset \ldots \subset E^{N-1} \subset E^N = H,$$

where E^k is spanned by ΩE^{k-1} for $1 \leq k \leq N-1$, and set $F^k = E^k \ominus E^{k-1}$, $0 \leq k \leq N$. Let H_0 be the space spanned by E^{N-1} and K, so that

$$\dim H_0 \leq \dim E^{N-1} + \dim K \leq |\Omega|^{N-1} + m.$$

We set $\xi_0 = \xi$, and define $\theta_0 \colon B(H) \to B(H_0)$ to be the compression map.

To define γ_0 consider the Hilbert space $L_0 = L \oplus \mathbb{C}\xi_L$, where ξ_L is a unit vector. Define an isometry $V_0\colon H \to H_0 \otimes L_0$ by

$$V_0\zeta = \left(\frac{k}{N}\right)^{1/2} V\zeta + \left(\frac{N-k}{N}\right)^{1/2} \zeta \otimes \xi_L, \quad \zeta \in F^k, \quad 0 \le k \le N.$$

Then set $\gamma_0(x) = V_0^*(x \otimes 1)V_0$. Since $V_0\xi = \xi_0 \otimes \xi_L$, we obviously have $\omega_\xi \circ \gamma_0 = \omega_{\xi_0}$. So it remains to get an estimate for $\|(\gamma_0 \circ \theta_0)(a) - a\|$ for $a \in \Omega$.

Let $p_0\colon H \to H_0$ be the projection, so that $\theta_0(x) = p_0 x p_0$. Note that the image of V_0 is contained in the direct sum of the image of V and $H_0 \otimes \xi_L$. Hence $V_0^* = V_0^* V V^*$ on $H_0 \otimes L$. Thus for $\zeta \in F^k$ and $a \in \Omega$ we have

$V_0^*(\theta_0(a) \otimes 1)V_0\zeta$

$$= V_0^* \left(\left(\frac{k}{N}\right)^{1/2} VV^*(\theta_0(a) \otimes 1)V\zeta + \left(\frac{N-k}{N}\right)^{1/2} \theta_0(a)\zeta \otimes \xi_L\right)$$

$$= V_0^* \left(\left(\frac{k}{N}\right)^{1/2} V(\gamma \circ \theta)(a)\zeta + \left(\frac{N-k}{N}\right)^{1/2} p_0 a\zeta \otimes \xi_L\right).$$

Consider the operator $T_a\colon H \to H_0 \otimes L_0$ given by

$$T_a\zeta = \left(\frac{k}{N}\right)^{1/2} Va\zeta + \left(\frac{N-k}{N}\right)^{1/2} p_0 a\zeta \otimes \xi_L, \quad \zeta \in F^k.$$

Then for $\zeta = \sum_{k=0}^N \zeta_k$, $\zeta_k \in F^k$, we have

$$\|(\gamma_0 \circ \theta_0)(a)\zeta - V_0^* T_a\zeta\| = \left\|V_0^* V \sum_k \left(\frac{k}{N}\right)^{1/2} ((\gamma \circ \theta)(a) - a)\zeta_k\right\|$$

$$\le \|(\gamma \circ \theta)(a) - a\| \left\|\sum_k \left(\frac{k}{N}\right)^{1/2} \zeta_k\right\|.$$

Since

$$\left\|\sum_k \left(\frac{k}{N}\right)^{1/2} \zeta_k\right\|^2 = \sum_k \frac{k}{N}\|\zeta_k\|^2 \le \sum_k \|\zeta_k\|^2 = \|\zeta\|^2,$$

we see that $\|(\gamma_0 \circ \theta_0)(a) - V_0^* T_a\| < \varepsilon\|a\|$. On the other hand, the same proof as in Lemma 14.3.2 shows that $\|V_0 a - T_a\| \le 3N^{-1/2}\|a\|$, whence

$$\|(\gamma_0 \circ \theta_0)(a) - a\| < (\varepsilon + 3N^{-1/2})\|a\|.$$

In more detail, decompose $a = a^+ + a^0 + a^-$ so that $a^0 F^k \subset F^k$, $a^\pm F^k \subset F^{k\pm 1}$, and introduce contractions $V_0^\pm\colon H \to H_0 \otimes L_0$ by letting $V_0^- = V_0$ on F^N, $V_0^+ = V_0$ on F^0 and

$$V_0^\pm \zeta = \left(\frac{k \mp 1}{N}\right)^{1/2} V\zeta + \left(\frac{N-k\pm 1}{N}\right)^{1/2} p_0\zeta \otimes \xi_L, \quad \zeta \in F^k,$$

in all other cases. Then $V_0 a^0 = T_{a^0}$, $V_0^\pm a^\pm = T_{a^\pm}$ and $\|V_0 - V_0^\pm\| \le \sqrt{2/N}$. □

By combining the last two lemmas we get the following key result.

Lemma 14.3.4. *Let H_i be a Hilbert space, $\xi_i \in H_i$ a unit vector, $\Omega_i \subset B(H_i)$ a finite self-adjoint subset containing the unit, $\theta_i \colon B(H_i) \to \mathrm{Mat}_{m_i}(\mathbb{C})$ and $\gamma_i \colon \mathrm{Mat}_{m_i}(\mathbb{C}) \to B(H_i)$ unital completely positive maps such that*

$$\|(\gamma_i \circ \theta_i)(a) - a\| < \varepsilon \|a\| \quad \text{for} \ \ a \in \Omega_i, \ i = 1, 2.$$

Set $(H, \xi) = (H_1, \xi_1) * (H_2, \xi_2)$. Then for any $N \in \mathbb{N}$ there exist unital completely positive maps $\theta \colon B(H) \to \mathrm{Mat}_m(\mathbb{C})$ and $\gamma \colon \mathrm{Mat}_m(\mathbb{C}) \to B(H)$ such that

(i) $m < (\max_i |\Omega_i|)^{N(N-1)}(m_1 + m_2 + 2)$;
(ii) $\|(\gamma \circ \theta)(a) - a\| < (\varepsilon + 6N^{-1/2})\|a\|$ *for $a \in \Omega_1 \cup \Omega_2$.*

Proof. Assume that $\theta_i \colon B(H_i) \to \mathrm{Mat}_{m_i}(\mathbb{C})$ is normal and there exists a vector ζ_i in $\ell_2^{m_i}$ such that $\omega_{\zeta_i} \circ \theta_i = \omega_{\xi_i}$ and $\omega_{\xi_i} \circ \gamma_i = \omega_{\zeta_i}$, $i = 1, 2$. Let $(L, \zeta) = (\ell_2^{m_1}, \zeta_1) * (\ell_2^{m_2}, \zeta_2)$. Then by Theorem 14.1.3 and Remark 14.1.2 we have well-defined normal unital completely positive maps $\theta_1 * \theta_2 \colon B(H) \to B(L)$ and $\gamma_1 * \gamma_2 \colon B(L) \to B(H)$. On the other hand, by Lemma 14.3.2 applied to $\theta_i(\Omega_i) \subset \mathrm{Mat}_{m_i}(\mathbb{C})$ there exist unital completely positive maps $\tilde{\theta} \colon B(L) \to \mathrm{Mat}_m(\mathbb{C})$ and $\tilde{\gamma} \colon \mathrm{Mat}_m(\mathbb{C}) \to B(L)$ such that $m < (\max_i |\Omega_i|)^{(N-1)^2}(m_1 + m_2)$ and

$$\|(\tilde{\gamma} \circ \tilde{\theta} \circ \theta_i)(a) - \theta_i(a)\| \le 3N^{-1/2}\|a\| \quad \text{for} \ \ a \in \Omega_i, \ i = 1, 2.$$

Set $\theta = \tilde{\theta} \circ (\theta_1 * \theta_2)$ and $\gamma = (\gamma_1 * \gamma_2) \circ \tilde{\gamma}$. Then for $a \in \Omega_i$

$$\|(\gamma \circ \theta)(a) - a\| \le \|(\gamma_1 * \gamma_2)((\tilde{\gamma} \circ \tilde{\theta} \circ \theta_i)(a) - \theta_i(a))\| + \|(\gamma_i \circ \theta_i)(a) - a\|$$
$$< (\varepsilon + 3N^{-1/2})\|a\|.$$

In the general case we first have to modify the maps θ_i and γ_i to be normal and state preserving. By Lemma 14.3.3 this is possible by increasing ε to $\varepsilon + 3N^{-1/2}$ and m_i to $m_i + |\Omega_i|^{N-1}$, which gives the result. □

Proof of Theorem 14.3.1. Set $\alpha = \alpha_1 * \alpha_2$. Since $A_i \subset A$, we have $ht(\alpha) \ge ht(\alpha_i)$. By Theorem 6.2.6 to compute the entropy of α it is enough to consider finite subsets Ω of A of the form $\Omega = \Omega_1 \cup \Omega_2$, where Ω_i is a self-adjoint subset of A_i consisting of unitaries. Let $k = \max_i |\Omega_i|$. Fix $\varepsilon > 0$ and choose $N \in \mathbb{N}$ such that $6N^{-1/2} \le \varepsilon$. Then by Lemma 14.3.4 for any $n \in \mathbb{N}$ we have

$$\mathrm{rcp}(\Omega \cup \ldots \cup \alpha^{n-1}(\Omega), 2\varepsilon) \le (nk)^{N(N-1)}(\mathrm{rcp}(\Omega_1 \cup \ldots \cup \alpha_1^{n-1}(\Omega_1), \varepsilon)$$
$$+ \mathrm{rcp}(\Omega_2 \cup \ldots \cup \alpha_2^{n-1}(\Omega_2), \varepsilon) + 2),$$

whence $ht(\Omega, 2\varepsilon; \alpha) \le \max_i ht(\Omega_i, \varepsilon; \alpha_i)$, and the theorem is proved. □

The above proof cannot be used to compute the dynamical entropy of free product automorphisms. But we can at least conclude the following.

Corollary 14.3.5. *Let (X_i, μ_i, T_i) be an ergodic dynamical system, $i = 1, 2$. Consider the free product automorphism α of $(A, \varphi) = (L^\infty(X_1), \mu_1) * (L^\infty(X_2), \mu_2)$ defined by T_1 and T_2. Then*

$$h_\varphi(\alpha) = \max\{h_{\mu_1}(T_1), h_{\mu_2}(T_2)\}.$$

Proof. Since by Theorem 14.1.3 there exist φ-preserving conditional expectations onto the subalgebras $L^\infty(X_i, \mu_i)$ of A, we clearly have the inequality $\geq$. To show the opposite inequality, by the Jewett-Krieger theorem, Theorem 1.2.3, we may assume that X_i is a compact metric space, T_i is a homeomorphism of X_i, and μ_i is the unique T_i-invariant probability measure on X_i, so that $h_{top}(T_i) = h_{\mu_i}(T)$. Let $B \subset A$ be the C*-subalgebra generated by $C(X_1)$ and $C(X_2)$. Then by Theorem 14.3.1

$$h_{\varphi|B}(\alpha|B) \leq ht(\alpha|B) = \max\{h_{\mu_1}(T_1), h_{\mu_2}(T_2)\}.$$

Since $\pi_\varphi(B)$ is weakly operator dense in $\pi_\varphi(A)$, by Theorem 3.2.2(ii) we have $h_{\varphi|B}(\alpha|B) = h_\varphi(\alpha)$, and we get the result. $\square$

We shall next prove Theorem 14.2.4 by using the following trick, which allows us to relate free shifts to free product automorphisms. First recall, see Chap. 8, that if we have an action β of a discrete group G on a C*-algebra D then any state ψ on D extends to a state $\bar\psi$ on $D \rtimes_{r,\beta} G$ defined by $\bar\psi(x u_g) = \delta_{g,e} \psi(x)$. The restriction of $\bar\psi$ to $C_r^*(G) \subset D \rtimes_{r,\beta} G$ gives the canonical trace τ on $C_r^*(G)$.

Lemma 14.3.6. *Let G be a discrete group. For each $g \in G$ let $\pi_g \colon (A, \varphi) \to (A_g, \varphi_g)$ be an isomorphism. Set $(D, \psi) = *_{g \in G}(A_g, \varphi_g)$. Let $\beta \colon G \to \mathrm{Aut}(D)$ be the action defined by the left translations of G on itself, so $\beta_g(\pi_h(a)) = \pi_{gh}(a)$. Then there exists an isomorphism*

$$(D \rtimes_{r,\beta} G, \bar\psi) \cong (C_r^*(G), \tau) * (A, \varphi)$$

such that $\pi_g(a) \mapsto u_g a u_g^*$ and $u_g \mapsto u_g$.

Proof. It is easy to see that the isomorphism exists at the level of full free and crossed products. So we just have to check that it is state preserving. The kernel of $\bar\psi$ is spanned by u_g, $g \neq 0$, and elements of the form $\pi_{g_1}(a_1) \ldots \pi_{g_n}(a_n) u_g$, where $a_1, \ldots, a_n \in \mathrm{Ker}\,\varphi$ and $g_1, \ldots, g_n, g \in G$, $g_i \neq g_{i+1}$. Thus we have to check that

$$(\tau * \varphi)((u_{g_1} a_1 u_{g_1}^*) \ldots (u_{g_n} a_n u_{g_n}^*) u_g) = 0.$$

Since $\tau(u_{g_i}^* u_{g_{i+1}}) = 0$, this is immediate by definition of $\tau * \varphi$. $\square$

Proof of Theorem 14.2.4. Let $G = \bigoplus_\mathbb{Z} \mathbb{Z}/2\mathbb{Z}$, and T be the shift automorphism of G. Let α_T be the corresponding automorphism of $C^*(G)$, $\alpha_T(u_g) = u_{Tg}$. Denote by g_n the unit in the n-th copy of $\mathbb{Z}/2\mathbb{Z}$, so $Tg_n = g_{n+1}$.

Set $(D, \psi) = (A, \varphi)^{*G}$. That is, we consider copies of (A, φ) indexed by the elements of G, and define (D, ψ) as their free product. Identify the free product $(B, \psi) = (A, \varphi)^{*\mathbb{Z}}$ with

$$(A, \varphi)^{*\{g_n \mid n \in \mathbb{Z}\}} \subset (D, \psi) \subset (D \rtimes_\beta G, \bar\psi),$$

where β is the action defined by the left translations of G on itself as described in Lemma 14.3.6. By that lemma $(D \rtimes_\beta G, \bar\psi)$ is isomorphic to $(C^*(G), \tau) * (A, \varphi)$. Thus we get an embedding of (B, ψ) into $(C^*(G), \tau) * (A, \varphi)$ which intertwines α with $\alpha_T * \mathrm{id}_A$. By Theorem 14.3.1

$$ht(\alpha) \le ht(\alpha_T * \mathrm{id}_A) = ht(\alpha_T) = \log 2.$$

Since the automorphism α^n is the free shift defined by $(A, \varphi)^{*n}$, we get $ht(\alpha^n) \le \log 2$ for any $n \in \mathbb{N}$, and thus $ht(\alpha) = 0$. □

More generally, given a permutation σ of a set I, we can define in an obvious way an automorphism α_σ of $(A, \varphi)^{*I}$, which we call a free permutation automorphism.

Corollary 14.3.7. *Let φ be a state on an exact C^*-algebra A with faithful GNS-representation, σ a permutation of a set I. Then for the corresponding free permutation automorphism α_σ of $(B, \psi) = (A, \varphi)^{*I}$ we have $ht(\alpha_\sigma) = 0$.*

Proof. The free product $(A, \varphi)^{*I}$ decomposes into free products corresponding to orbits of σ. Hence by Theorem 14.3.1 (or rather by its extension to free products with an arbitrary number of factors) it suffices to prove the corollary assuming that σ acts transitively. If I is infinite, then α_σ is a free shift, so $ht(\alpha_\sigma) = 0$ by Theorem 14.2.4. If I is finite, then $\alpha_\sigma^n = \mathrm{id}$ for some $n \in \mathbb{N}$, and the equality $ht(\alpha_\sigma) = 0$ is immediate. □

14.4 Notes

Reduced free products of C*-algebras were introduced by Avitzour [12] and Voiculescu [221], see the books [228] and [90] for an exposition of the theory. Reduced free products of completely positive maps were first considered by Choda [42], but her proof of existence of such maps was incomplete. A complete proof of Theorem 14.1.3 was given by Blanchard and Dykema [25].

The dynamical entropy of free shifts was computed by Størmer [206], [209], who in particular proved Theorem 14.2.1 and Corollary 14.2.2. The definition of entropy in terms of stationary couplings did not exist at that moment. Otherwise the result would follow from triviality of stationary couplings of free shifts with abelian systems stated in Corollary 14.2.3, which was already obtained by Avitzour [12]. Partial cases of Theorem 14.2.4 and Corollary 14.3.7 were proved by Brown and Choda [35] and Dykema [56]. In full generality these results were proved by Brown, Dykema and Shlyakhtenko [36]. The connection between free and crossed products, Lemma 14.3.6, used in our proof of Theorem 14.2.4 was apparently first employed by Phillips [161].

Complementing the results of Sect. 14.2 showing how far free shifts are from abelian systems, it is worth to mention a result of Popa [168] that the free shift of the von Neumann algebra of the free group with infinitely many generators has no nontrivial invariant abelian von Neumann subalgebras. As we were informed by Ueda, a similar result is true for arbitrary free products.

Voiculescu [226] introduced another dynamical invariant, dynamical free entropy dimension, which is apparently more suitable for automorphisms of the free shift type. Unfortunately, the presently known properties of free entropy dimension are not sufficient to compute this invariant in interesting cases.

Entropic properties of free product automorphisms were first studied by Choda [42], [45]. Theorem 14.3.1 was obtained by Brown, Dykema and Shlyakhtenko [36], extending and partially building upon the work of Dykema and Shlyakhtenko [58]. In fact they proved a slightly stronger result allowing amalgamation over a finite dimensional subalgebra. The proof of Theorem 14.3.1 we presented is due to Ozawa [153]. Theorem 14.3.1 implies in particular that reduced free products of exact C*-algebras are exact, which is a result of Dykema [57]. The original proof of Brown, Dykema and Shlyakhtenko, although longer than Ozawa's (which is closer to the technique used by Dykema in [57]), is to a certain extent more transparent. The main idea is to embed the reduced free product of two C*-algebras A_1 and A_2 into a reduced free product of the Toeplitz algebra and $(A_1 \oplus A_2) \otimes \mathcal{O}_2$, where $\mathcal{O}_2$ is the Cuntz algebra. This reduces the proof to the case when one of the free factors is the Toeplitz algebra with the trivial automorphism. Then the reduced free product has a crossed product type structure, which makes it possible to compute the entropy.

Neither of the two available proofs of Theorem 14.3.1 seems to be applicable to dynamical entropy. In [46] a variant of completely positive approximation entropy is introduced, and it is stated that Theorem 14.3.1 holds for that entropy. However, by virtue of Lemma 14.3.3 that entropy is always zero.

One of the interesting consequences of Theorem 14.3.1 obtained in [36] is that every nuclear separable purely infinite simple C*-algebra has automorphisms with arbitrary topological entropy. The reason is briefly that any such algebra absorbs the Cuntz algebra $\mathcal{O}_\infty$. So it suffices to prove the result for $\mathcal{O}_\infty$. For this one uses a representation of $\mathcal{O}_\infty$ as the tensor product of $\mathcal{O}_\infty$ and a reduced free product of the Toeplitz algebra with an abelian algebra.

A

Completely Positive Maps

A.1. A linear map $\theta\colon A \to B$ between C*-algebras is called **completely positive** if
$$\theta \otimes \mathrm{id}\colon A \otimes \mathrm{Mat}_n(\mathbb{C}) \to B \otimes \mathrm{Mat}_n(\mathbb{C})$$
is positive for any $n \in \mathbb{N}$. Such a map is automatically bounded since any positive linear functional on a C*-algebra is bounded.

An operator $T = \sum_{i,j} b_{ij} \otimes e_{ij} \in B \otimes \mathrm{Mat}_n(\mathbb{C})$, where the e_{ij}'s are the usual matrix units in $\mathrm{Mat}_n(\mathbb{C})$, is positive if and only if $\sum_{i,j} b_i^* b_{ij} b_j \geq 0$ for any $b_1, \ldots, b_n \in B$. Since every positive element in $A \otimes \mathrm{Mat}_n(\mathbb{C})$ is the sum of n elements of the form $\sum_{i,j} a_i^* a_j \otimes e_{ij}$, we conclude that $\theta\colon A \to B$ is completely positive if and only if
$$\sum_{i,j=1}^n b_i^* \theta(a_i^* a_j) b_j \geq 0$$
for any $n \in \mathbb{N}$, $a_1, \ldots, a_n \in A$ and $b_1, \ldots, b_n \in B$.

A.2. If $\theta\colon A \to B$ is completely positive and $B \subset B(H)$, then there exist a Hilbert space K, a representation $\pi\colon A \to B(K)$ and a bounded operator $V\colon H \to K$ such that $\theta(a) = V^* \pi(a) V$ and $\overline{\pi(A)VH} = K$.

The construction is as follows. Consider the algebraic tensor product $A \odot H$, and define a sesquilinear form on it by
$$(a \odot \xi, c \odot \zeta) = (\theta(c^*a)\xi, \zeta).$$

Complete positivity of θ means exactly that this form is positive definite. Thus taking the quotient by the space $N = \{x \in A \odot H \mid (x,x) = 0\}$ we get a pre-Hilbert space. Denote its completion by K. Then define a representation $\pi\colon A \to B(K)$ by
$$\pi(a)[c \odot \zeta] = [ac \odot \zeta],$$
where $[x]$ denotes the image of x in $(A \odot H)/N$. Note that boundedness of π follows from complete positivity of θ and the inequality $\sum_{i,j} c_i^* a^* a c_j \otimes e_{ij} \leq$

$\|a\|^2 \sum_{i,j} c_i^* c_j \otimes e_{ij}$. If A is unital, set $V\xi = [1 \odot \xi]$ for $\xi \in H$. In the general case choose an approximate unit $\{e_i\}_i$ in A, set $V_i\xi = [e_i \odot \xi]$, and consider any weak operator limit point V of the V_i's. Note that since $\|V_i\| \leq \|\theta(e_i^2)\|^{1/2} \leq \|\theta\|^{1/2}$, such a point exists. Moreover, it is unique since it is determined by the condition $(V\xi, [a \odot \zeta]) = (\theta(a^*)\xi, \zeta)$. Then $V^*([a \odot \zeta]) = \theta(a)\zeta$ and $\pi(a)V\zeta = [a \odot \zeta]$, whence $V^*\pi(a)V = \theta(a)$.

It follows from the construction that $\|V\| \leq \|\theta\|^{1/2}$, and if A is unital then $\|V\| \leq \|\theta(1)\|^{1/2}$. The equality $\theta(a) = V^*\pi(a)V$ shows that the opposite inequalities hold. Hence $\|V\| = \|\theta\|^{1/2}$, and if A is unital then $\|\theta\| = \|\theta(1)\|$. In particular, if θ is contractive then $VV^* \leq 1$, so that

$$\theta(a)^*\theta(a) = V^*\pi(a)^*VV^*\pi(a)V \leq V^*\pi(a^*a)V = \theta(a^*a),$$

that is, θ is a **Schwarz map**.

The triple (K, π, V) is called the **Stinespring dilation** of θ. It is easy to see that it is unique up to an isomorphism. Note also that any map of the form $a \mapsto V^*\pi(a)V$, where π is a representation and V is a bounded operator, is completely positive.

A.3. If $\theta \colon A \to B$ is a positive linear map, and either A or B is abelian, then θ is completely positive.

Assume B is abelian. Then it suffices to prove that $\chi \circ \theta$ is completely positive for any character χ of B. In other words, we have to check that any positive linear functional on A is a completely positive mapping. But this is easy to see as the GNS-representation gives the Stinespring dilation of such a mapping.

Assume now that A is abelian. It is obvious that any positive linear map $\mathbb{C} \to B$ is completely positive. Hence any positive linear map $\mathbb{C}^n \to B$ is completely positive. It follows that if A is AF, then θ is completely positive. In the general case consider the second transpose $\theta^{**} \colon A^{**} \to B^{**}$ of θ. Since A^{**} is an inductive limit of finite dimensional abelian C*-algebras, we conclude that $\theta = \theta^{**}|_A$ is completely positive.

A.4. Let $\theta \colon A \to B$ be a contractive completely positive map. Let $A^\sim$ and $B^\sim$ be the C*-algebras obtained by adjoining units to A and B, respectively. Then θ extends to a unital completely positive map $\tilde\theta \colon A^\sim \to B^\sim$.

To see this, consider a faithful unital representation of $B^\sim$ on a Hilbert space H, and let (K, π, V) be the corresponding Stinespring dilation of θ. By A.2 we have $\|V\| \leq 1$. Consider the Hilbert space $\tilde K = K \oplus H$, the unital representation $\tilde\pi \colon A^\sim \to B(\tilde K)$ such that $\tilde\pi(a) = \pi(a) \oplus 0$ for $a \in A$, and define

$$\tilde V \colon H \to \tilde K, \quad \tilde V \xi = V\xi \oplus (1 - V^*V)^{1/2}\xi.$$

Then $\tilde V$ is an isometry, so that the completely positive map $A^\sim \ni a \mapsto \tilde V^* \tilde\pi(a) \tilde V \in B(H)$ is unital. It is easy to see that it coincides with θ on A, so that it is the required map $\tilde\theta$.

Note that if B is unital, then since $1_B B^\sim 1_B = B$, the map $\theta\colon A \to B$ extends to a unital completely positive map $A^\sim \to B$.

A.5. Let $\theta\colon A \to B$ be a contractive completely positive map. Then for any $x, y \in A$ we have

$$\|\theta(x^*y) - \theta(x)^*\theta(y)\| \le \|\theta(x^*x) - \theta(x)^*\theta(x)\|^{1/2}\|\theta(y^*y) - \theta(y)^*\theta(y)\|^{1/2}.$$

By A.4 we may assume that A, B and θ are unital. Then the operator V in the Stinespring dilation of θ is an isometry. Thus, without loss of generality we may assume that θ is the compression map $B(K) \to pB(K)p$, $a \mapsto pap$, where $p \in B(K)$ is a projection. Set $T(a) = (1-p)ap$. Then for any $a, c \in B(K)$

$$\theta(a^*c) - \theta(a)^*\theta(c) = pa^*cp - pa^*pcp = T(a)^*T(c),$$

so that

$$\|\theta(x^*y) - \theta(x)^*\theta(y)\|^2$$
$$= \|T(x)^*T(y)\|^2 = \|T(y)^*T(x)T(x)^*T(y)\|$$
$$\le \|T(x)T(x)^*\|\,\|T(y)^*T(y)\| = \|T(x)^*T(x)\|\,\|T(y)^*T(y)\|$$
$$= \|\theta(x^*x) - \theta(x)^*\theta(x)\|\,\|\theta(y^*y) - \theta(y)^*\theta(y)\|.$$

A.6. Let $\theta\colon A \to \mathrm{Mat}_n(\mathbb{C})$ be a linear map, $\theta(a) = \sum_{i,j} \theta_{ij}(a)e_{ij}$. Consider the linear functional φ_θ on $A \otimes \mathrm{Mat}_n(\mathbb{C})$ defined by $\varphi_\theta(\sum_{i,j} a_{ij} \otimes e_{ij}) = \sum_{i,j} \theta_{ij}(a_{ij})$. Then θ is completely positive if and only if φ_θ is positive.

Assume φ_θ is positive. Consider the corresponding GNS-representation of $A \otimes \mathrm{Mat}_n(\mathbb{C})$. It is necessarily of the form $a \otimes x \mapsto \pi(a) \otimes x \in B(H \otimes \ell_2^n)$ for some representation $\pi\colon A \to B(H)$. Let $\xi = \sum_i \xi_i \otimes e_i \in H \otimes \ell_2^n$ be the cyclic vector defining φ_θ, so that $\theta_{ij}(a) = (\pi(a)\xi_j, \xi_i)$. Consider the operator $V\colon \ell_2^n \to H$ defined by $Ve_i = \xi_i$. Then $V^*\zeta = \sum_i (\zeta, \xi_i)e_i$, whence

$$V^*\pi(a)Ve_j = \sum_i (\pi(a)\xi_j, \xi_i)e_i = \theta(a)e_j.$$

Thus θ is completely positive.

Conversely, assume θ is completely positive. Let (H, π, V) be its Stinespring dilation, where $V\colon \ell_2^n \to H$. Then φ_θ is defined by the representation $\pi \otimes \mathrm{id}$ of $A \otimes \mathrm{Mat}_n(\mathbb{C})$ on $H \otimes \ell_2^n$ and the vector $\xi = \sum_i Ve_i \otimes e_i$, that is, $\varphi_\theta(a \otimes x) = ((\pi(a) \otimes x)\xi, \xi)$. Hence φ_θ is positive.

Thus we get a one-to-one correspondence between completely positive maps $A \to \mathrm{Mat}_n(\mathbb{C})$ and positive linear functionals on $A \otimes \mathrm{Mat}_n(\mathbb{C})$.

A.7. If M is a von Neumann algebra, then any completely positive map $\theta\colon M \to \mathrm{Mat}_n(\mathbb{C})$ can be approximated in the pointwise norm topology by normal completely positive maps. Moreover, if θ is unital then the approximations can also be chosen unital.

The first assertion is an immediate consequence of A.6. The second one is true since if $\gamma\colon M \to \operatorname{Mat}_n(\mathbb{C})$ is a completely positive map such that $\gamma(1)$ is close to 1, then $\gamma(1)^{-1/2}\gamma(\cdot)\gamma(1)^{-1/2}$ is a unital completely positive map, which is close to γ.

A.8. If A is a C*-subalgebra of a C*-algebra B, then any completely positive map $\theta\colon A \to B(H)$ can be extended to a completely positive map $B \to B(H)$.

If H is finite dimensional, the result follows from A.6. In the general case first note that we may assume that B is unital and then by A.4 that A contains the unit of B. Choose a net $\{p_i\}_i$ of finite rank projections in $B(H)$ such that $p_i \nearrow 1$. For each i extend the map $A \ni a \mapsto p_i\theta(a)p_i \in p_iB(H)p_i$ to B. Since A contains the unit of B, the norms of such extensions are not greater than the norm of θ. Then take any pointwise weak operator limit point of these extensions.

This is a weak form of Arveson's extension theorem. A more general form assumes only that B is a unital C*-algebra and $A \subset B$ is an operator system, that is, a self-adjoint subspace containing the unit. For a proof see e.g. [214, Theorem XV.1.1].

A.9. A linear map $\gamma\colon \operatorname{Mat}_n(\mathbb{C}) \to A$ is completely positive if and only if the operator $T_\gamma = \sum_{i,j=1}^n \gamma(e_{ij}) \otimes e_{ij} \in A \otimes \operatorname{Mat}_n(\mathbb{C})$ is positive.

Since $\sum_{i,j} e_{ij} \otimes e_{ij} = (\sum_i e_{1i} \otimes e_{1i})^*(\sum_j e_{1j} \otimes e_{1j})$ is positive, positivity of T_γ is a necessary condition for complete positivity of γ. Conversely, assume T_γ is positive. Then T_γ is the sum of n elements of the form $\sum_{i,j} a_i^* a_j \otimes e_{ij}$. Thus it suffices to show that the map $\operatorname{Mat}_n(\mathbb{C}) \to A$, $e_{ij} \mapsto a_i^* a_j$, is completely positive. To show this, identify A with $A \otimes \mathbb{C}e_{11} \subset A \otimes \operatorname{Mat}_n(\mathbb{C})$, and write the map as $\operatorname{Mat}_n(\mathbb{C}) \ni x \mapsto V^*(1 \otimes x)V \in A \otimes \operatorname{Mat}_n(\mathbb{C})$, where $V = \sum_i a_i \otimes e_{i1}$.

Thus we get a one-to-one correspondence between completely positive maps $\operatorname{Mat}_n(\mathbb{C}) \to A$ and positive elements in $A \otimes \operatorname{Mat}_n(\mathbb{C})$.

A.10. If $\{A_i\}_i$ is an increasing net of unital C*-subalgebras of a unital C*-algebra A such that $\cup_i A_i$ is norm-dense in A, then any unital completely positive map $\gamma\colon \operatorname{Mat}_n(\mathbb{C}) \to A$ can be approximated in norm by unital completely positive maps $\operatorname{Mat}_n(\mathbb{C}) \to \cup_i A_i$. If in addition A is a strongly operator dense subalgebra of a von Neumann algebra M, then any unital completely positive map $\gamma\colon \operatorname{Mat}_n(\mathbb{C}) \to M$ can be approximated in the pointwise strong operator topology by unital completely positive maps $\operatorname{Mat}_n(\mathbb{C}) \to \cup_i A_i$.

By A.9 there exist completely positive maps $\gamma_i\colon \operatorname{Mat}_n(\mathbb{C}) \to A_i$ such that $\|T_{\gamma_i}\| \le \|T_\gamma\|$ and $T_{\gamma_i} \to T_\gamma$ in norm (resp., in the strong operator topology). In the first case we just have to modify the γ_i's to unital completely positive maps as we did in A.7. In the second case $\gamma_i(1) \to 1$ only in the strong operator topology. Fix a state φ on $\operatorname{Mat}_n(\mathbb{C})$. For $\varepsilon > 0$ define

$$\gamma_{i,\varepsilon}(x) = (\gamma_i(1) + \varepsilon 1)^{-1/2}(\gamma_i(x) + \varepsilon\varphi(x)1)(\gamma_i(1) + \varepsilon 1)^{-1/2}.$$

Then $\gamma_{i,\varepsilon}\colon \operatorname{Mat}_n(\mathbb{C}) \to A_i$ is a unital completely positive map, and for any fixed $\varepsilon > 0$ we have $\gamma_{i,\varepsilon}(x) \to (1+\varepsilon)^{-1}(\gamma(x) + \varepsilon\varphi(x)1)$ in the strong operator topology. Thus the required unital completely positive maps can be chosen from $\{\gamma_{i,\varepsilon}\}_{i,\varepsilon}$.

A.11. If A is a unital C*-algebra, I a two-sided ideal in A, and $\gamma\colon \mathrm{Mat}_n(\mathbb{C}) \to A/I$ a unital completely positive map, then there exists a lifting of γ to a unital completely positive map $\bar\gamma\colon \mathrm{Mat}_n(\mathbb{C}) \to A$.

By A.9 there exists a not necessarily unital lifting $\tilde\gamma$. Let $\tilde\gamma(1) = 1+a$, and let a_1 and a_2 be positive elements such that $a_1 a_2 = 0$ and $a = a_1 - a_2$. Let φ be a state on $\mathrm{Mat}_n(\mathbb{C})$. Then define
$$\bar\gamma(x) = (1+a_1)^{-1/2}(\tilde\gamma(x) + \varphi(x)a_2)(1+a_1)^{-1/2}.$$

A.12. If $\theta\colon A \to \mathrm{Mat}_n(\mathbb{C})$ and $\gamma\colon \mathrm{Mat}_n(\mathbb{C}) \to B$ are completely positive maps such that $\gamma \circ \theta$ is unital, then there exist $m \le n$ and unital completely positive maps $\tilde\theta\colon A \to \mathrm{Mat}_m(\mathbb{C})$ and $\tilde\gamma\colon \mathrm{Mat}_m(\mathbb{C}) \to B$ such that $\tilde\gamma \circ \tilde\theta = \gamma \circ \theta$.

Let p be the support of $\theta(1)$. Then the image of θ is contained in $p\mathrm{Mat}_n(\mathbb{C})p \cong \mathrm{Mat}_m(\mathbb{C})$. Thus replacing $\mathrm{Mat}_n(\mathbb{C})$ by $\mathrm{Mat}_m(\mathbb{C})$ and γ by $\gamma(p \cdot p)$, we may assume that $p = 1$, that is, $x = \theta(1)$ is invertible. Then set $\tilde\theta = x^{-1/2}\theta(\cdot)x^{-1/2}$ and $\tilde\gamma = \gamma(x^{1/2} \cdot x^{1/2})$.

A.13. Let B be a unital C*-subalgebra of a unital C*-algebra A. A **conditional expectation** of A onto B is a unital positive map $E\colon A \to B$ such that $E(bac) = bE(a)c$ for all $b, c \in B$ and $a \in A$. Since $\sum_{i,j} b_i^* E(a_i^* a_j) b_j = E(x^*x)$, where $x = \sum_i a_i b_i$, the map E is completely positive.

Let M be a von Neumann algebra, $N \subset M$ a von Neumann subalgebra, $E\colon M \to N$ a normal conditional expectation. Similarly to the case of normal states, there exists the smallest projection $p \in M$ such that $E(p) = 1$. This projection is called the support of E. Then $E(a) = E(ap) = E(pa)$ for $a \in M$. Let $a \in M$, $a \ge 0$, and assume $s(a) \le p$, where $s(a)$ denotes the support of a. Then $s(a) \le s(E(a))$. Observe that in the classical case, when E is defined by an averaging of a function along atoms of a measurable partition, this statement is obvious.

To prove it in the general case, we shall first check that p commutes with N. If $u \in N$ is unitary, then $E(upu^*) = uE(p)u^* = 1$. Hence $p \le upu^*$. Since this is also true for u^* instead of u, we conclude that $p = upu^*$, so that p commutes with N.

Now set $q = 1 - s(E(a))$. We must show that $qaq = 0$. Since q commutes with p, and $s(a) \le p$, we have $qaq = qpapq = pqaqp$, so that $s(qaq) \le p$. Since E is faithful on pMp, it thus suffices to check that $E(qaq) = 0$. But since $E(qaq) = qE(a)q$, this is obvious.

B

Operator Inequalities

B.1. Let $I \subset \mathbb{R}$ be an interval (of any type). A continuous function $f\colon I \to \mathbb{R}$ is called **operator convex** if
$$f(\lambda x + (1-\lambda)y) \leq \lambda f(x) + (1-\lambda)f(y)$$
for $\lambda \in [0,1]$ and any bounded self-adjoint operators x and y with spectra in I. It is called **operator monotone** if $f(x) \leq f(y)$ for any bounded self-adjoint operators x and y with spectra in I such that $x \leq y$.

For a continuous function $f\colon [0,s) \to \mathbb{R}$ such that $f(0) = 0$ the following conditions are equivalent:

(i) f is operator convex;
(ii) $f(\frac{1}{2}(x+y)) \leq \frac{1}{2}f(x) + \frac{1}{2}f(y)$ for any positive operators x and y of norm less than s;
(iii) $f(v^*xv) \leq v^*f(x)v$ for any v with $\|v\| \leq 1$ and positive x with $\|x\| < s$.

To prove that (ii) implies (iii), set $a = (1-vv^*)^{1/2}$, $b = (1-v^*v)^{1/2}$,
$$X = \begin{pmatrix} x & 0 \\ 0 & 0 \end{pmatrix}, \quad U = \begin{pmatrix} v & a \\ b & -v^* \end{pmatrix}, \quad V = \begin{pmatrix} v & -a \\ b & v^* \end{pmatrix}.$$
Then U and V are unitaries, and
$$U^*XU = \begin{pmatrix} v^*xv & v^*xa \\ axv & axa \end{pmatrix}, \quad V^*XV = \begin{pmatrix} v^*xv & -v^*xa \\ -axv & axa \end{pmatrix},$$
so that
$$\begin{pmatrix} f(v^*xv) & 0 \\ 0 & f(axa) \end{pmatrix} = f\left(\begin{pmatrix} v^*xv & 0 \\ 0 & axa \end{pmatrix}\right) = f\left(\frac{1}{2}U^*XU + \frac{1}{2}V^*XV\right)$$
$$\leq \frac{1}{2}f(U^*XU) + \frac{1}{2}f(V^*XV)$$
$$= \frac{1}{2}U^*f(X)U + \frac{1}{2}V^*f(X)V$$
$$= \begin{pmatrix} v^*f(x)v & 0 \\ 0 & af(x)a \end{pmatrix}.$$

In particular, $f(v^*xv) \leq v^*f(x)v$.

To prove that (iii) implies (i), define

$$X = \begin{pmatrix} x & 0 \\ 0 & y \end{pmatrix}, \quad U = \begin{pmatrix} \lambda^{1/2} & -(1-\lambda)^{1/2} \\ (1-\lambda)^{1/2} & \lambda^{1/2} \end{pmatrix}, \quad P = \begin{pmatrix} 1 & 0 \\ 0 & 0 \end{pmatrix}.$$

Then U is a unitary, and

$$\begin{pmatrix} f(\lambda x + (1-\lambda)y) & 0 \\ 0 & 0 \end{pmatrix} = f(PU^*XUP) \leq Pf(U^*XU)P$$

$$= PU^*f(X)UP = \begin{pmatrix} \lambda f(x) + (1-\lambda)f(y) & 0 \\ 0 & 0 \end{pmatrix}.$$

Thus (i)-(iii) are indeed equivalent.

B.2. Let $\theta: A \to B$ be a contractive positive mapping of unital C*-algebras, $f: [0, s) \to \mathbb{R}$ an operator convex function, $f(0) \leq 0$. Then for any positive $a \in A$ with $\|a\| < s$ we have $f(\theta(a)) \leq \theta(f(a))$.

Since θ is a contraction and $f(0) \leq 0$, we have $f(0)1 \leq \theta(f(0)1)$. Thus replacing f by $f - f(0)$ we may assume that $f(0) = 0$. Replacing further A by the C*-subalgebra generated by a we may assume that A is abelian. Then θ is completely positive by A.3. Hence the result follows from B.1(iii) using the Stinespring dilation of θ.

B.3. We say that an operator convex function f is **strictly operator convex** if $f(\frac{1}{2}(x+y)) = \frac{1}{2}f(x) + \frac{1}{2}f(y)$ implies $x = y$.

The functions $t \mapsto t^2$ on $\mathbb{R}$ and $t \mapsto t^{-1}$ on $(0, +\infty)$ are strictly operator convex.

The first claim follows from the identity

$$\frac{1}{2}x^2 + \frac{1}{2}y^2 - \frac{1}{4}(x+y)^2 = \frac{1}{4}(x-y)^2.$$

To prove the second we have to show that

$$(x+y)^{-1} \leq \frac{1}{4}(x^{-1} + y^{-1})$$

for positive invertible x and y, and equality holds if and only if $x = y$. Setting $a = x^{-1}$ and $b = y^{-1}$, and using

$$x + y = b^{-1}(a+b)a^{-1} = a^{-1}(a+b)b^{-1},$$

we compute

$$(x+y)^{-1} = \frac{1}{2}(a(a+b)^{-1}b + b(a+b)^{-1}a)$$

$$= \frac{1}{4}((a+b)(a+b)^{-1}(a+b) - (a-b)(a+b)^{-1}(a-b))$$

$$\leq \frac{1}{4}(a+b)(a+b)^{-1}(a+b) = \frac{1}{4}(x^{-1} + y^{-1}),$$

and equality holds only if $a = b$, that is, $x = y$.

In fact, any operator convex function on $(-s, s)$ has the form
$$f(t) = \alpha + \beta t + \int_{-s}^{s} \frac{t^2}{s^2 - t\tau} d\mu(\tau),$$
where μ is a positive measure on $[-s, s]$, see e.g. [85], which together with the above result implies that any nonlinear operator convex function on an open interval is strictly operator convex.

The function $t \mapsto -t^{-1}$ on $(0, +\infty)$ is operator monotone. Indeed, if $0 \leq x \leq y$ and x is invertible, then $z = x^{1/2} y^{-1/2}$ satisfies $z^* z \leq 1$. Hence $y^{-1} = x^{-1/2} z z^* x^{-1/2} \leq x^{-1}$.

In fact, a continuous function f on $[0, +\infty)$ such that $f(0) \leq 0$ is operator convex if and only if $t \mapsto t^{-1} f(t)$ is operator monotone on $(0, +\infty)$, see [85].

B.4. The function $t \mapsto t \log t$ on $[0, +\infty)$ is strictly operator convex.

Since $t \mapsto (s + t)^{-1}$ is strictly operator convex for any $s > 0$ by B.3, this follows from the integral representation
$$t \log t = \int_0^\infty \left(\frac{t}{1+s} + \frac{s}{s+t} - 1 \right) ds.$$

Similarly, using the representation
$$\log t = \int_0^\infty \left(\frac{1}{1+s} - \frac{1}{s+t} \right) ds,$$
we see that log is strictly operator concave and operator monotone on $(0, +\infty)$.

B.5. Let $E \colon A \to B$ be a faithful conditional expectation, and $f \colon [0, s) \to \mathbb{R}$ a strictly operator convex function. Then $f(E(a)) \leq E(f(a))$ for any positive $a \in A$ with $\|a\| < s$, and equality holds if and only if $a \in B$.

By B.2 we have to check that if $f(E(a)) = E(f(a))$, then $a \in B$, that is, $E(a) = a$. We have
$$E(f(a)) = f(E(a)) = f\left(E\left(\frac{1}{2} a + \frac{1}{2} E(a) \right) \right)$$
$$\leq E\left(f\left(\frac{1}{2} a + \frac{1}{2} E(a) \right) \right) \leq E\left(\frac{1}{2} f(a) + \frac{1}{2} f(E(a)) \right) = E(f(a)).$$

Since E is faithful, it follows that
$$f\left(\frac{1}{2} a + \frac{1}{2} E(a) \right) = \frac{1}{2} f(a) + \frac{1}{2} f(E(a)),$$
which by strict operator convexity of f implies $a = E(a)$.

B.6. Let A be a C*-algebra, τ a tracial state on A. For $a \in A$ and $p \geq 1$ set $\|a\|_p = \tau(|a|^p)^{1/p}$. Then for $p, q, r \geq 1$ such that $1/r = 1/p + 1/q$ and any $a, b \in A$ we have the **generalized Hölder inequality**

$$\|ab\|_r \leq \|a\|_p \|b\|_q.$$

Replacing A by $\pi_\tau(A)''$ we may assume that A is a von Neumann algebra and τ is a faithful normal trace.

Consider first the case $r = 1$. Let $a = u|a|$ and $b = v|b|$ be the polar decompositions. For $c \in A$ consider the function

$$F(z) = \tau(u|a|^{pz} v|b|^{q(1-z)} c).$$

This function is bounded and continuous on the strip $0 \leq \operatorname{Re} z \leq 1$ and analytic in the interior. For $t \in \mathbb{R}$ we have

$$|F(1+it)| \leq \tau(|a|^p)\|c\| = \|a\|_p^p \|c\|,$$
$$|F(it)| \leq \tau(|b|^q)\|c\| = \|b\|_q^q \|c\|.$$

By the three lines theorem it follows that $|F(z)| \leq \|a\|_p^{p\operatorname{Re} z} \|b\|_q^{q(1-\operatorname{Re} z)} \|c\|$ for any z in the strip. Letting $z = 1/p$ we get

$$|\tau(abc)| \leq \|a\|_p \|b\|_q \|c\|.$$

Since $\|ab\|_1$ is the norm of the functional $\tau(ab \cdot)$, we thus get the required inequality.

For $r > 1$ consider similarly the function

$$F(z) = \tau(u|a|^{zp/r} v|b|^{(1-z)q/r} c).$$

Let r' be such that $1 = 1/r + 1/r'$. Then

$$|F(1+it)| \leq \|cu|a|^{p/r}\|_1 \leq \|u|a|^{p/r}\|_r \|c\|_{r'} = \|a\|_p^{p/r} \|c\|_{r'},$$
$$|F(it)| \leq \||b|^{q/r} c\|_1 \leq \||b|^{q/r}\|_r \|c\|_{r'} = \|b\|_q^{q/r} \|c\|_{r'}.$$

It follows that $|F(s)| \leq \|a\|_p^{sp/r} \|b\|_q^{(1-s)q/r} \|c\|_{r'}$ for any $s \in [0,1]$. For $s = r/p$ we get

$$|\tau(abc)| \leq \|a\|_p \|b\|_q \|c\|_{r'}.$$

Let $ab = w|ab|$ be the polar decomposition. Apply the above inequality to $c = |ab|^{r-1} w^*$. Then we get the required inequality, since $\tau(abc) = \|ab\|_r^r$ and $\|c\|_{r'} = \||ab|^{r-1}\|_{r'} = \|ab\|_r^{r-1}$.

C
Direct Integrals

C.1. Let (X,μ) be a σ-finite measure space. A collection $\{H_x\}_{x\in X}$ of Hilbert spaces together with a subspace $\mathcal{H}$ of $\prod_{x\in X} H_x$ is called a **measurable field of Hilbert spaces** on (X,μ) if the following conditions are satisfied:

(i) for every $\xi = (\xi_x)_{x\in X} \in \mathcal{H}$, the function $X \ni x \mapsto \|\xi_x\|$ is μ-measurable;

(ii) if $\zeta = (\zeta_x)_{x\in X}$ is such that the function $X \ni x \mapsto (\zeta_x, \xi_x)$ is μ-measurable for every $\xi \in \mathcal{H}$, then $\zeta \in \mathcal{H}$;

(iii) there is a sequence $\{\xi^{(n)}\}_{n=1}^{\infty}$ of elements of $\mathcal{H}$ such that $\{\xi_x^{(n)}\}_n$ spans a dense subspace of H_x for μ-a.e. $x \in X$.

Elements of $\mathcal{H}$ are called measurable vector fields. Consider the subspace of $\mathcal{H}$ consisting of elements $\xi = (\xi_x)_x$ such that

$$\int_X \|\xi_x\|^2 d\mu(x) < \infty.$$

Identify two such vector fields if they coincide a.e. We then get a Hilbert space with inner product

$$(\xi, \zeta) = \int_X (\xi_x, \zeta_x) d\mu(x).$$

This Hilbert space is called the **direct integral** of $\{H_x\}_x$ and is denoted by $\int_X^{\oplus} H_x d\mu(x)$.

C.2. Let $\{H'_x\}_x$ and $\{H''_x\}_x$ be two measurable fields of Hilbert spaces on (X,μ). Assume $\dim H'_x = \dim H''_x$ for μ-a.e. $x \in X$. Then the measurable fields are isomorphic in the sense that for a.e. $x \in X$ there exists a unitary operator $U_x \colon H'_x \to H''_x$ such that a vector field $\xi = (\xi_x)_x \in \prod_{x\in X} H'_x$ is measurable if and only if the field $(U_x \xi_x)_x \in \prod_{x\in X} H''_x$ is measurable, cf. [214, Lemma IV.8.12].

In particular, if a measurable field $\{H_x\}_x$ has fibers of constant dimension $n \in \mathbb{N} \cup \{\infty\}$, then it is isomorphic to the constant field with fiber H_0, $\dim H_0 = n$, and then $\int_X^{\oplus} H_x d\mu(x)$ can be identified with $L^2(X, d\mu) \otimes H_0$.

C.3. Let $\{H_x\}_x$ be a measurable field of Hilbert spaces on (X,μ), $H = \int_X^\oplus H_x d\mu(x)$. An operator $a \in B(H)$ is called **decomposable** if there exist operators $a_x \in B(H_x)$ such that for any $\xi \in H$ we have $(a\xi)_x = a_x \xi_x$ for μ-a.e. $x \in X$. We then write $a = \int_X^\oplus a_x d\mu(x)$. If each operator a_x is a scalar then the operator a is called diagonal. The algebra of diagonal operators is identified with $L^\infty(X,\mu)$. A bounded operator is decomposable if and only if it commutes with the algebra of diagonal operators, cf. [214, Corollary IV.8.16].

We write $\int_X^\oplus B(H_x) d\mu(x)$ for the von Neumann algebra of decomposable operators. If φ is a normal state on this algebra, then there exist normal states φ_x on $B(H_x)$ such that for any decomposable operator $a = \int_X^\oplus a_x d\mu(x)$ we have
$$\varphi(a) = \int_X \varphi_x(a_x) d\mu(x),$$
cf. [214, Proposition IV.8.34]. We then write $\varphi = \int_X^\oplus \varphi_x d\mu(x)$.

C.4. Let X be a compact metric space, $\pi \colon C(X) \to B(H)$ a unital representation on a separable Hilbert space. Then there exist a measure μ on X and a measurable field $\{H_x\}_x$ of Hilbert spaces on (X,μ) such that π is unitarily equivalent to the canonical representation of $C(X)$ on $\int_X^\oplus H_x d\mu(x)$ by diagonal operators. Assuming $H_x \neq 0$ the measure class of μ is uniquely determined, and as soon as μ is fixed the measurable field $\{H_x\}_x$ on (X,μ) is unique up to an isomorphism, cf. [214, Theorem IV.8.23]. By C.2 it follows that the pair $([\mu], m)$ is a complete invariant for the representation, where $[\mu]$ denotes the equivalence class of μ, and m is the multiplicity function, $m(x) = \dim H_x$.

To show existence of a measurable field as above it suffices to consider cyclic representations. So assume a unit vector $\xi \in H$ is cyclic. There exists a probability measure μ on X such that $(\pi(f)\xi, \xi) = \int_X f(x) d\mu(x)$ for all $f \in C(X)$. Then the map $C(X) \ni f \mapsto \pi(f)\xi$ extends to a unitary operator $L^2(X, \mu) \to H$ which can be considered as an isomorphism of the direct integral of one-dimensional Hilbert spaces onto H.

C.5. Let $\{H_x\}_x$ be a measurable field of Hilbert spaces on a Lebesgue space (X, μ), $H = \int_X^\oplus H_x d\mu(x)$. Let $\pi \colon A \to B(H)$ be a representation of a separable C*-algebra by decomposable operators. Then for a.e. $x \in X$ there exists a representation π_x of A on H_x such that for any $a = (a_x)_x \in A$ we have $\pi_x(a) = a_x$ a.e., cf. [214, Theorem IV.8.25]. We then write $\pi = \int_X^\oplus \pi_x d\mu(x)$.

C.6. Let $\{H_x\}_x$ be a measurable field of Hilbert spaces on a Lebesgue space (X, μ), and $a = \int_X^\oplus a_x d\mu(x)$ a decomposable self-adjoint operator on $H = \int_X^\oplus H_x d\mu(x)$. Then a_x has pure point spectrum for μ-a.e. $x \in X$ if and only if H decomposes into a direct sum of subspaces which are cyclic for $L^\infty(X, \mu)$ and a-invariant.

Indeed, note that if the representation of $L^\infty(X, \mu)$ on H is cyclic then $L^\infty(X, \mu)$ is maximal abelian in $B(H)$, and hence $a \in L^\infty(X, \mu)$. Since in general we could obtain the decomposition $H = \int_X^\oplus H_x d\mu(x)$ by decomposing H

into cyclic subspaces for $L^\infty(X,\mu)$, see C.4, we conclude that if we can choose those subspaces to be a-invariant then we get that $\{H_x\}_x$ is a direct sum of one-dimensional fields such that a is diagonal on each of them. In particular, a_x has pure point spectrum for a.e. $x \in X$.

To prove the converse statement it suffices to show that any closed $L^\infty(X,\mu)$- and a-invariant subspace $K \subset H$ contains an a-invariant subspace which is cyclic for $L^\infty(X,\mu)$. Let $p \in B(H)$ be the projection onto K. Then p is decomposable, $p = \int_X^\oplus p_x d\mu(x)$, and we have $K = \int_X^\oplus p_x H_x d\mu(x)$. Thus replacing H_x by $p_x H_x$ we may assume that $K = H$. Replacing X by a subset of positive measure we may also assume that the fibers H_x, $x \in X$, have constant dimension, so that by C.2 we have a constant field with fiber H_0. Removing further a subset of zero measure we may assume that X is a complete metric space, for any $v, w \in H_0$ the function $x \mapsto (a_x v, w)$ is Borel, and a_x is a self-adjoint operator with pure point spectrum for any x. Consider

$$\Gamma = \{(x, v, \lambda) \mid \|v\| = 1, \ a_x v = \lambda v\} \subset X \times H_0 \times \mathbb{R}.$$

The set Γ is a Borel subset of $X \times H_0 \times \mathbb{R}$. Consider the projection $\Gamma \to X$ onto the first coordinate. Since a_x has pure point spectrum by assumption, this map surjective. Hence, see e.g. Theorem A.16 in [214, Volume I], there is a measurable cross-section. That is, there exist a measurable function λ and a measurable vector field ξ such that $\|\xi_x\| = 1$ and $a_x \xi_x = \lambda(x) \xi_x$ for any $x \in X$. Then the $L^\infty(X,\mu)$-invariant subspace generated by ξ, which can be identified with $\int_X^\oplus \mathbb{C}\xi_x d\mu(x)$, is a-invariant.

C.7. Let U be a unitary operator on a separable Hilbert space H. Then U defines a representation of $C(\mathbb{T})$ on H. Hence by C.4 there exists a measure ν on $\mathbb{T}$ and a decomposition $H = \int_\mathbb{T}^\oplus H_z d\nu(z)$ such that $U = \int_\mathbb{T}^\oplus z \, d\nu(z)$.

If the multiplicity function m, $m(z) = \dim H_z$, is constant ν-a.e., one says that U has **homogeneous spectrum**.

Consider the decomposition $\nu = \nu_a + \nu_s$ of ν into absolutely continuous (with respect to the Lebesgue measure on $\mathbb{T}$) and singular parts. Then putting

$$H_a = \int_\mathbb{T}^\oplus H_z d\nu_a(z) \ \text{ and } \ H_s = \int_\mathbb{T}^\oplus H_z d\nu_s(z),$$

we have $H = H_a \oplus H_s$. More explicitly, the subspace H_a (resp. H_s) of H consists of all vectors $\xi \in H$ such that the measure ν_ξ on $\mathbb{T}$ such that

$$(U^n \xi, \xi) = \int_\mathbb{T} z^n d\nu_\xi(z) \ \text{ for } \ n \in \mathbb{Z}$$

is absolutely continuous (resp. singular).

If $\nu_s = 0$ and ν_a is equivalent to the Lebesgue measure, one says that U has **Lebesgue spectrum**.

References

1. Aaronson, J.: An introduction to infinite ergodic theory. Mathematical Surveys and Monographs, **50**. American Mathematical Society, Providence, RI (1997).
2. Adler, R. L., Konheim, A. G., McAndrew, M. H.: Topological entropy. Trans. Amer. Math. Soc., **114**, 309–319 (1965).
3. Akiho, N., Hiai, F., Petz, D.: Equilibrium states and their entropy densities in gauge-invariant C*-systems. Preprint math.OA/0409601.
4. Alicki, R., Fannes, M.: Defining quantum dynamical entropy. Lett. Math. Phys., **32**, 75–82 (1994).
5. Alicki, R., Fannes, M.: Quantum dynamical systems. Oxford University Press, Oxford (2001).
6. Anantharaman-Delaroche, C.: Amenability and exactness for dynamical systems and their C^*-algebras. Trans. Amer. Math. Soc., **354**, 4153–4178 (2002).
7. Araki, H.: Gibbs states of a one dimensional quantum lattice. Comm. Math. Phys., **14**, 120–157 (1969).
8. Araki, H.: On uniqueness of KMS states of one-dimensional quantum lattice systems. Comm. Math. Phys., **44**, 1–7 (1975).
9. Araki, H.: Relative entropy of states of von Neumann algebras. Publ. Res. Inst. Math. Sci., **11**, 809–833 (1975/76).
10. Araki, H.: Relative entropy for states of von Neumann algebras. II. Publ. Res. Inst. Math. Sci., **13**, 173–192 (1977/78).
11. Araki, H., Lieb, E.: Entropy inequalities. Comm. Math. Phys., **18**, 160–170 (1970).
12. Avitzour, D.: Free products of C^*-algebras. Trans. Amer. Math. Soc., **271**, 423–435 (1982).
13. Benatti, F. Deterministic chaos in infinite quantum systems. Trieste Notes in Physics. Springer-Verlag, Berlin, 1993.
14. Benatti, F., Hudetz, T., Knauf, A.: Quantum chaos and dynamical entropy. Comm. Math. Phys., **198**, 607–688 (1998).
15. Benatti, F., Narnhofer, H.: Strong asymptotic abelianness for entropic K-systems. Comm. Math. Phys., **136**, 231–250 (1991).
16. Benatti, F., Narnhofer, H.: Strong clustering in type III entropic K-systems. Monatsh. Math., **124**, 287–307 (1997).
17. Benatti, F., Narnhofer, H., Sewell, G. L.: A noncommutative version of the Arnol'd cat map. Lett. Math. Phys., **21**, 157–172 (1991). Errata: Lett. Math. Phys., **22**, 81 (1991).

18. Benatti, F., Narnhofer, H., Uhlmann, A.: Optimal decompositions of quantum states with respect to entropy. Rep. Math. Phys., **38**, 123–141 (1996).
19. Benatti, F., Narnhofer, H., Uhlmann, A.: Optimal decompositions with respect to entropy and symmetries. Lett. Math. Phys., **47**, 237–253 (1999).
20. Berg, K. R.: Convolution of invariant measures, maximal entropy. Math. Systems Theory, **3**, 146–150 (1969).
21. Bezuglyi, S. I., Golodets, V. Ya.: Dynamical entropy for Bogoliubov actions of free abelian groups on the CAR-algebra. Ergodic Theory Dynam. Systems, **17**, 757–782 (1997).
22. Bisch, D.: Entropy of groups and subfactors. J. Funct. Anal., **103**, 190–208 (1992).
23. Bisch, D.: Bimodules, higher relative commutants and the fusion algebra associated to a subfactor. In: Operator algebras and their applications (Waterloo, ON, 1994/1995), 13–63, Fields Inst. Comm., **13**. Amer. Math. Soc., Providence, RI (1997).
24. Blackadar, B.: K-theory for operator algebras. Second edition. Mathematical Sciences Research Institute Publications, **5**. Cambridge University Press, Cambridge (1998).
25. Blanchard, E., Dykema, K. J.: Embeddings of reduced free products of operator algebras. Pacific J. Math., **199**, 1–19 (2001).
26. Blanchard, F., Glasner, E., Host, B.: A variation on the variational principle and applications to entropy pairs. Ergodic Theory Dynam. Systems, **17**, 29–43 (1997).
27. Boca, F., Goldstein, P.: Topological entropy for the canonical endomorphism of Cuntz-Krieger algebras. Bull. London Math. Soc., **32**, 345–352 (2000).
28. Boyko, M., Nessonov, N.: Entropy of Bogoliubov automorphisms on II_1 representations of the group $U(\infty)$. Methods Funct. Anal. Topology, **9**, 317–332 (2003).
29. Bowen, R.: Entropy for group endomorphisms and homogeneous spaces. Trans. Amer. Math. Soc., **153**, 401–414 (1971).
30. Bratteli, O., Robinson, D. W.: Operator algebras and quantum statistical mechanics. 1. Second edition. Texts and Monographs in Physics. Springer-Verlag, New York (1987). 2, ibid. (1997)
31. Breiman, L.: The individual ergodic theorem of information theory. Ann. Math. Statist., **28**, 809–811 (1957).
32. Brown, N. P.: Topological entropy in exact C^*-algebras. Math. Ann., **314**, 347–367 (1999).
33. Brown, N. P.: Topological entropy, embeddings and unitaries in nuclear quasidiagonal C^*-algebras. Proc. Amer. Math. Soc., **128**, 2603–2609 (2000).
34. Brown, N. P.: Characterizing type I C^*-algebras via entropy. C. R. Math. Acad. Sci. Paris, **339**, 827–829 (2004).
35. Brown, N. P., Choda, M.: Approximation entropies in crossed products with an application to free shifts. Pacific J. Math., **198**, 331–346 (2001).
36. Brown, N. P., Dykema, K.; Shlyakhtenko, D.: Topological entropy of free product automorphisms. Acta Math., **189**, 1–35 (2002).
37. Brown, N. P., Germain, E.: Dual entropy in discrete groups with amenable actions. Ergodic Theory Dynam. Systems, **22**, 711–728 (2002).
38. Bures, D., Yin, H. S.: Shifts on the hyperfinite factor of type II_1. J. Operator Theory, **20**, 91–106 (1988).

39. Choda, M.: Shifts on the hyperfinite II$_1$-factor. J. Operator Theory, **17**, 223–235 (1987).
40. Choda, M.: Entropy for $*$-endomorphisms and relative entropy for subalgebras. J. Operator Theory, **25**, 125–140 (1991).
41. Choda, M.: Entropy for canonical shifts. Trans. Amer. Math. Soc., **334**, 827–849 (1992).
42. Choda, M.: Reduced free products of completely positive maps and entropy for free product of automorphisms. Publ. Res. Inst. Math. Sci., **32**, 371–382 (1996).
43. Choda, M.: Entropy of Cuntz's canonical endomorphism. Pacific J. Math., **190**, 235–245 (1999).
44. Choda, M.: A C*-dynamical entropy and applications to canonical endomorphisms. J. Funct. Anal., **173**, 453–480 (2000).
45. Choda, M.: Entropy of crossed products and entropy of free products. J. Operator Theory, **48**, 355–367 (2002).
46. Choda, M.: Dynamical entropy for automorphisms of exact C*-algebras. J. Funct. Anal., **198**, 481–498 (2003).
47. Choda, M.: Entropy of automorphisms arising from dynamical systems through discrete groups with amenable actions. J. Funct. Anal., **217**, 181–191 (2004).
48. Choda, M., Hiai, F.: Entropy for canonical shifts. II. Publ. Res. Inst. Math. Sci., **27**, 461–489 (1991).
49. Connes, A.: Entropie de Kolmogoroff-Sinai et mécanique statistique quantique. C. R. Acad. Sci. Paris Sér. I Math., **301**, 1–6 (1985).
50. Connes, A., Narnhofer, H., Thirring, W.: Dynamical entropy of C*-algebras and von Neumann algebras. Comm. Math. Phys., **112**, 691–719 (1987).
51. Connes, A., Størmer, E.: Entropy of automorphisms in II$_1$ von Neumann algebras. Acta Math., **134**, 289–306 (1975).
52. Connes, A., Størmer, E.: A connection between the classical and the quantum mechanical entropies. In: Operator algebras and group representations, Vol. I (Neptun, 1980), 113–123, Monogr. Stud. Math., **17**. Pitman, Boston, MA (1984).
53. Cornfeld, I. P., Fomin, S. V., Sinaĭ, Ya. G.: Ergodic theory. Grundlehren der Mathematischen Wissenschaften, **245**. Springer-Verlag, New York (1982).
54. David, M.-C.: Paragroupe d'Adrian Ocneanu et algèbre de Kac. Pacific J. Math., **172**, 331–363 (1996).
55. Dinaburg, E. I.: A correlation between topological entropy and metric entropy. (Russian) Dokl. Akad. Nauk SSSR, **190**, 19–22 (1970).
56. Dykema, K. J.: Topological entropy of some automorphisms of reduced amalgamated free product C*-algebras. Ergodic Theory Dynam. Systems, **21**, 1683–1693 (2001).
57. Dykema, K. J.: Exactness of reduced amalgamated free product C*-algebras. Forum Math., **16**, 161–180 (2004).
58. Dykema, K. J., Shlyakhtenko, D.: Exactness of Cuntz-Pimsner C*-algebras. Proc. Edinb. Math. Soc. (2), **44**, 425–444 (2001).
59. Emch, G. G.: Positivity of the K-entropy on non-abelian K-flows. Z. Wahrscheinlichkeitstheorie und Verw. Gebiete, **29**, 241–252 (1974).
60. Emch, G. G.: Nonabelian special K-flows. J. Functional Analysis, **19**, 1–12 (1975).
61. Emch, G. G.: Generalized K-flows. Comm. Math. Phys., **49**, 191–215 (1976).

62. Emch, G. G., Narnhofer, H., Thirring, W., Sewell, G. L.: Anosov actions on noncommutative algebras. J. Math. Phys., **35**, 5582–5599 (1994).
63. Enomoto, M., Choda, M., Watatani, Y.: Generalized Powers' binary shifts on the hyperfinite II$_1$ factor. Math. Japon., **33**, 831–843 (1988).
64. Enomoto, M., Nagisa, M., Watatani, Y., Yoshida, H.: Relative commutant algebras of Powers' binary shifts on the hyperfinite II$_1$ factor. Math. Scand., **68**, 115–130 (1991).
65. Fannes, M.: The entropy density of quasi free states. Comm. Math. Phys., **31**, 279–290 (1973).
66. Fannes, M.: A continuity property of the entropy density for spin lattice systems. Comm. Math. Phys., **31**, 291–294 (1973).
67. Federer, H.: Geometric measure theory. Die Grundlehren der mathematischen Wissenschaften, **153**. Springer-Verlag New York Inc., New York (1969)
68. Furstenberg, H.: Disjointness in ergodic theory, minimal sets, and a problem in Diophantine approximation. Math. Systems Theory, **1**, 1–49 (1967).
69. Gaure, S.: Entropy of binary shifts and its connection with graph theory. PhD Thesis, University of Oslo (2006).
70. Germain, E.: Non commutative entropy computations for cross-product and continuous fields. Preprint (2003).
71. Ghys, É., Langevin, R., Walczak, P.: Entropie géométrique des feuilletages. Acta Math., **160**, 105–142 (1988).
72. Glasner, E.: Ergodic theory via joinings. Mathematical Surveys and Monographs, **101**. American Mathematical Society, Providence, RI (2003).
73. Glasner, E., Weiss, B.: Quasi-factors of zero-entropy systems, J. Amer. Math. Soc., **8**, 665–686 (1995).
74. Gol'dšteĭn, M. Š. Ergodic theorems and entropy of automorphisms of von Neumann algebras. (Russian) Dokl. Akad. Nauk SSSR, **253**, 781–785 (1980).
75. Golodets, V. Ya., Boyko, M. S.: Dynamical entropy for a class of algebraic origin automorphisms. Mat. Fiz. Anal. Geom., **8**, 385–391 (2001).
76. Golodets, V. Ya., Neshveyev, S. V.: Non-Bernoullian quantum K-systems. Comm. Math. Phys., **195**, 213–232 (1998).
77. Golodets, V. Ya., Neshveyev, S. V.: Gibbs states for AF-algebras. J. Math. Phys., **39**, 6329–6344 (1998).
78. Golodets, V. Ya., Neshveyev, S. V.: Dynamical entropy for Bogoliubov actions of torsion-free abelian groups on the CAR-algebra. Ergodic Theory Dynam. Systems, **20**, 1111–1125 (2000).
79. Golodets, V. Ya., Neshveyev, S. V.: Entropy of automorphisms of II$_1$-factors arising from the dynamical systems theory. J. Funct. Anal., **181**, 14–28 (2001).
80. Golodets, V. Ya., Størmer, E.: Entropy of C^*-dynamical systems defined by bitstreams. Ergodic Theory Dynam. Systems, **18**, 859–874 (1998).
81. Golodets, V. Ya., Størmer, E.: Generators and comparison of entropies of automorphisms of finite von Neumann algebras. J. Funct. Anal., **164**, 110–133 (1999).
82. Goodwyn, L. W.: Topological entropy bounds measure-theoretic entropy. Proc. Amer. Math. Soc., **23**, 679–688 (1969).
83. Haagerup, U., Størmer, E.: Maximality of entropy in finite von Neumann algebras. Invent. Math., **132**, 433–455 (1998).
84. Halmos, P. R.: On automorphisms of compact groups. Bull. Amer. Math. Soc., **49**, 619–624 (1943).

85. Hansen, F., Pedersen, G. K.: Jensen's inequality for operators and Löwner's theorem. Math. Ann., **258**, 229–241 (1981/82).
86. Hiai, F.: Minimum index for subfactors and entropy. II. J. Math. Soc. Japan, **43**, 347–379 (1991).
87. Hiai, F.: Minimum index for subfactors and entropy for canonical shifts. In: Current topics in operator algebras (Nara, 1990), 59–70. World Sci. Publishing, River Edge, NJ (1991).
88. Hiai, F.: Entropy for canonical shifts and strong amenability. Internat. J. Math., **6**, 381–396 (1995).
89. Hiai, F., Ohya, M., Tsukada, M.: Sufficiency, KMS condition and relative entropy in von Neumann algebras. Pacific J. Math., **96**, 99–109 (1981).
90. Hiai, F., Petz, D.: The semicircle law, free random variables and entropy. Mathematical Surveys and Monographs, **77**. American Mathematical Society, Providence, RI (2000).
91. Hiai, F., Tsukada, M.: Strong martingale convergence of generalized conditional expectations on von Neumann algebras. Trans. Amer. Math. Soc., **282**, 791–798 (1984).
92. Hudetz, T.: Spacetime dynamical entropy of quantum systems. Lett. Math. Phys., **16**, 151–161 (1988).
93. Hudetz, T.: Quantum topological entropy: first steps of a "pedestrian" approach. In: Quantum probability & related topics, 237–261, QP-PQ, VIII. World Sci. Publishing, River Edge, NJ (1993).
94. Hudetz, T.: Topological entropy for appropriately approximated C^*-algebras. J. Math. Phys., **35**, 4303–4333 (1994).
95. Huruya, T., Tomiyama, J.: Completely bounded maps of C^*-algebras. J. Operator Theory, **10**, 141–152 (1983).
96. Jewett, R. I.: The prevalence of uniquely ergodic systems. J. Math. Mech., **19**, 717–729 (1969/1970).
97. Juzvinskiĭ, S. A.: Calculation of the entropy of a group-endomorphism. (Russian) Sibirsk. Mat. Ž., **8**, 230–239 (1967).
98. Kalikow, S. A.: T, T^{-1} transformation is not loosely Bernoulli. Ann. of Math. (2), **115**, 393–409 (1982).
99. Katok, A.: Lyapunov exponents, entropy and periodic orbits for diffeomorphisms. Inst. Hautes Etudes Sci. Publ. Math., **51**, 137–173 (1980).
100. Katok, A.: Smooth non-Bernoulli K-automorphisms. Invent. Math., **61**, 291–299 (1980). Errata: Invent. Math., **63**, 355 (1981).
101. Katznelson, Y.: Ergodic automorphisms of T^n are Bernoulli shifts. Israel J. Math., **10**, 186–195 (1971).
102. Kawahigashi, Y.: Cohomology of actions of discrete groups on factors of type II_1. Pacific J. Math., **149**, 303–317 (1991).
103. Kawakami, S.: Relative entropy of a fixed point algebra. In: Mappings of operator algebras (Philadelphia, PA, 1988), 233–238, Progr. Math., **84**. Birkhäuser Boston, Boston, MA (1990).
104. Keane, M., Smorodinsky, M.: Bernoulli schemes of the same entropy are finitarily isomorphic. Ann. of Math. (2), **109**, 397–406 (1979).
105. Kerr, D.: Pressure for automorphisms of exact C^*-algebras and a noncommutative variational principle. PhD Thesis, University of Toronto (2001).
106. Kerr, D.: Dimension and dynamical entropy for metrized C^*-algebras. Comm. Math. Phys., **232**, 501–534 (2003).

107. Kerr, D.: Entropy and induced dynamics on state spaces. Geom. Funct. Anal., **14**, 575–594 (2004).
108. Kerr, D., Li, H.: Positive Voiculescu-Brown entropy in noncommutative toral automorphisms. Preprint math.OA/0303090.
109. Kerr, D., Li, H.: Dynamical entropy in Banach spaces. Preprint math.FA/0407386.
110. Kerr, D., Pinzari, C.: Noncommutative pressure and the variational principle in Cuntz-Krieger-type C^*-algebras. J. Funct. Anal., **188**, 156–215 (2002).
111. Kolmogorov, A. N.: A new metric invariant of transient dynamical systems and automorphisms in Lebesgue spaces. (Russian) Dokl. Akad. Nauk SSSR (N.S.), **119**, 861–864 (1958).
112. Kosaki, H.: Relative entropy of states: a variational expression. J. Operator Theory, **16**, 335–348 (1986).
113. Krieger, W.: On unique ergodicity. In: Proceedings of the Sixth Berkeley Symposium on Mathematical Statistics and Probability (Univ. California, Berkeley, Calif., 1970/1971), Vol. II: Probability theory, 327–346. Univ. California Press, Berkeley, Calif. (1972).
114. Kullback, S., Leibler, R. A.: On information and sufficiency. Ann. Math. Statistics, **22**, 79–86 (1951).
115. Lanford, O. E. III, Robinson, D. W.: Statistical mechanics of quantum spin systems III. Comm. Math. Phys., **9**, 327–338 (1968).
116. Lieb, E. H.: Convex trace functions and the Wigner-Yanase-Dyson conjecture. Advances in Math., **11**, 267–288 (1973).
117. Lieb, E. H., Ruskai, M. B.: Proof of the strong subadditivity of quantum-mechanical entropy. J. Mathematical Phys., **14**, 1938–1941 (1973).
118. Lind, D. A.: The structure of skew products with ergodic group automorphisms. Israel J. Math., **28**, 205–248 (1977).
119. Lindblad, G.: Non-Markovian quantum stochastic processes and their entropy. Comm. Math. Phys., **65**, 281–294 (1979).
120. Lindenstrauss, E., Weiss, B.: Mean topological dimension. Israel J. Math., **115**, 1–24 (2000).
121. McMillan, B.: The basic theorems of information theory. Ann. Math. Statistics, **24**, 196–219 (1953).
122. Miles, G., Thomas, R. K.: Generalized torus automorphisms are Bernoullian. In: Studies in probability and ergodic theory, 231–249, Adv. in Math. Suppl. Stud., **2**. Academic Press, New York-London (1978).
123. Moriya, H.: Variational principle and the dynamical entropy of space translation. Rev. Math. Phys., **11**, 1315–1328 (1999).
124. Moulin Ollagnier, J.: Ergodic theory and statistical mechanics. Lecture Notes in Mathematics, **1115**. Springer-Verlag, Berlin (1985).
125. Narnhofer, H.: Free energy and the dynamical entropy of space translations. Rep. Math. Phys., **25**, 345–356 (1988).
126. Narnhofer, H.: Dynamical entropy, quantum K-systems and clustering. In: Quantum probability and applications, V (Heidelberg, 1988), 286–295, Lecture Notes in Math., **1442**. Springer, Berlin (1990).
127. Narnhofer, H.: Quantized Arnol'd cat maps can be entropic K-systems. J. Math. Phys., **33**, 1502–1510 (1992).
128. Narnhofer, H., Størmer, E., Thirring, W.: C^*-dynamical systems for which the tensor product formula for entropy fails. Ergod. Th. & Dynam. Sys., **15**, 961–968 (1995).

129. Narnhofer, H., Thirring, W.: From relative entropy to entropy. Fizika, **17**, 257–265 (1985).
130. Narnhofer, H., Thirring, W.: Dynamical entropy of quasifree automorphisms. Lett. Math. Phys., **14**, 89–96 (1987).
131. Narnhofer, H., Thirring, W.: Quantum K-systems. Comm. Math. Phys., **125**, 565–577 (1989).
132. Narnhofer, H., Thirring, W.: Algebraic K-systems. Lett. Math. Phys., **20**, 231–250 (1990). Errata: Lett. Math. Phys., **22**, 81 (1991).
133. Narnhofer, H., Thirring, W.: Dynamical entropy of quantum systems and their abelian counterpart. In: On Klauder's path: a field trip, 127–145. World Sci. Publishing, River Edge, NJ (1994).
134. Narnhofer, H., Thirring, W.: C^*-dynamical systems that are asymptotically highly anticommutative. Lett. Math. Phys., **35**, 145–154 (1995).
135. Narnhofer, H., Thirring, W.: Equivalence of modular K and Anosov dynamical systems. Rev. Math. Phys., **12**, 445–459 (2000).
136. Narnhofer, H., Thirring, W.: Realization of two-sided quantum K-systems. Rep. Math. Phys., **45**, 239–256 (2000).
137. Neshveev, S. V.: Quantum Markov K-systems. (Russian) Mat. Fiz. Anal. Geom., **5**, 87–94 (1998).
138. Neshveyev, S. V.: On the K-property of quantized Arnold cat maps. J. Math. Phys., **41**, 1961–1965 (2000).
139. Neshveyev, S. V.: Entropy of Bogoliubov automorphisms of CAR and CCR algebras with respect to quasi-free states. Rev. Math. Phys., **13**, 29–50 (2001).
140. Neshveyev, S., Størmer, E.: The variational principle for a class of asymptotically abelian C^*-algebras. Comm. Math. Phys., **215**, 177–196 (2000).
141. Neshveyev, S., Størmer, E.: Entropy of type I algebras. Pacific J. Math., **201**, no. 2, 421–428 (2001).
142. Neshveyev, S., Størmer, E.: The McMillan theorem for a class of asymptotically abelian C^*-algebras. Ergodic Theory Dynam. Systems, **22**, 889–897 (2002).
143. Neshveyev, S., Størmer, E.: Ergodic theory and maximal abelian subalgebras of the hyperfinite factor. J. Funct. Anal., **195**, 239–261 (2002).
144. Neshveyev, S., Størmer, E.: Maximal abelian subalgebras of the hyperfinite factor, entropy and ergodic theory. Acta Math. Sin., **19**, 599–604 (2003).
145. Neumann, J. von: Thermodynamik quantenmechanischer Gesamtheiten. Gött. Nachr., **1**, 273-291 (1927).
146. Ocneanu, A.: Quantized groups, string algebras and Galois theory for algebras. In: Operator algebras and applications, Vol. 2, 119–172, London Math. Soc. Lecture Note Ser., **136**. Cambridge Univ. Press, Cambridge (1988).
147. Ohya, M., Petz, D. Quantum entropy and its use. Texts and Monographs in Physics. Springer-Verlag, Berlin (1993).
148. Okayasu, R.: Perturbation theoretic entropy of the boundary actions of free groups. Proc. Amer. Math. Soc., **134**, 1771–1776 (2006).
149. Oleksenko, V. M.: Properties of the entropy of the actions of the groups $\bigoplus \mathbb{Z}_n$ by automorphisms on the algebra of canonical anticommutative relations. Dopov. Nats. Akad. Nauk Ukr. Mat. Prirodozn. Tekh. Nauki, no. 3, 28–31 (1988).
150. Ornstein, D.: Bernoulli shifts with the same entropy are isomorphic. Advances in Math., **4**, 337–352 (1970).
151. Ornstein, D. S.: An example of a Kolmogorov automorphism that is not a Bernoulli shift. Advances in Math., **10**, 49–62 (1973).

152. Ornstein, D. S., Shields, P. C.: An uncountable family of K-automorphisms. Advances in Math., **10**, 63–88 (1973).
153. Ozawa, N.: Nuclearity of reduced amalgamated free product C*-algebras. Sūrikaisekikenkyūsho Kōkyūroku, **1250**, 49–55 (2002).
154. Park, Y. M.: Dynamical entropy of generalized quantum Markov chains. Lett. Math. Phys., **32**, 63–74 (1994).
155. Park, Y. M., Shin, H. H.: Dynamical entropy of space translations of CAR and CCR algebras with respect to quasi-free states. Comm. Math. Phys., **152**, 497–537 (1993).
156. Pedersen, G. K.: C*-algebras and their automorphism groups. London Mathematical Society Monographs, **14**. Academic Press, Inc., London-New York (1979).
157. Peters, J.: Entropy on discrete abelian groups. Adv. in Math., **33**, 1–13 (1979).
158. Petz, D.: Properties of the relative entropy of states of von Neumann algebras. Acta Math. Hungar., **47**, 65–72 (1986).
159. Petz, D.: First steps towards a Donsker and Varadhan theory in operator algebras. In: Quantum probability and applications, V (Heidelberg, 1988), 311–319, Lecture Notes in Math., **1442**. Springer, Berlin (1990).
160. Petz, D.: Entropy of Markov states. Riv. Mat. Pura Appl., **14**, 33–42 (1994).
161. Phillips, J.: Automorphisms of full II_1 factors, with applications to factors of type III. Duke Math. J., **43**, 375–385 (1976).
162. Pimsner, M., Popa, S.: Entropy and index for subfactors. Ann. Sci. Ecole Norm. Sup. (4), **19**, 57–106 (1986).
163. Pimsner, M., Popa, S.: Iterating the basic construction. Trans. Amer. Math. Soc., **310**, 127–133 (1988).
164. Pinsker, M. S.: Dynamical systems with completely positive or zero entropy. (Russian) Dokl. Akad. Nauk SSSR, **133**, 1025–1026; translated as Soviet Math. Dokl., **1**, 937–938 (1960).
165. Pinzari, C., Watatani, Y., Yonetani, K.: KMS states, entropy and the variational principle in full C*-dynamical systems. Comm. Math. Phys., **213**, 331–379 (2000).
166. Pollicott, M.: Lectures on ergodic theory and Pesin theory on compact manifolds. London Mathematical Society Lecture Note Series, **180**. Cambridge University Press, Cambridge (1993).
167. Pop, C., Smith, R. R.: Crossed products and entropy of automorphisms. J. Funct. Anal., **206**, 210–232 (2004).
168. Popa, S.: Maximal injective subalgebras in factors associated with free groups. Adv. in Math., **50**, 27–48 (1983).
169. Popa, S.: Classification of amenable subfactors of type II. Acta Math., **172**, 163–255 (1994).
170. Powers, R. T.: Representations of uniformly hyperfinite algebras and their associated von Neumann rings. Ann. of Math. (2), **86**, 138–171 (1967).
171. Powers, R. T.: An index theory for semigroups of *-endomorphisms of $B(H)$ and type II_1 factors. Canad. J. Math., **40**, 86–114 (1988).
172. Powers, R. T., Price, G. L.: Binary shifts on the hyperfinite II_1 factor. In: Representation theory of groups and algebras, 453–464, Contemp. Math., **145**. Amer. Math. Soc., Providence, RI (1993).
173. Price, G. L.: Shifts on type II_1 factors. Canad. J. Math., **39**, 492–511 (1987).
174. Price, G. L.: The entropy of rational Powers shifts. Proc. Amer. Math. Soc., **126**, 1715–1720 (1998).

175. Price, G. L.: Shifts on the hyperfinite II_1 factor. J. Funct. Anal., **156**, 121–169 (1998).
176. Pusz, W., Woronowicz, S. L.: Functional calculus for sesquilinear forms and the purification map. Rep. Mathematical Phys., **8**, 159–170 (1975).
177. Reed, M., Simon, B.: Methods of modern mathematical physics. I. Academic Press, New York–London (1972). II, ibid. (1975). III, ibid. (1979). IV, ibid. (1978).
178. Ribenboim, P.: Classical theory of algebraic numbers. Universitext. Springer-Verlag, New York (2001).
179. Rockafellar, R. T.: Convex analysis. Princeton Mathematical Series, **28**. Princeton University Press, Princeton, N.J. (1970).
180. Robinson, D. W.: Statistical mechanics of quantum spin systems. Comm. Math. Phys., **6**, 151–160 (1967).
181. Rohlin, V. A.: On the fundamental ideas of measure theory. (Russian) Mat. Sbornik N.S., **25(67)**, 107–150 (1949).
182. Rohlin, V. A.: Metric properties of endomorphisms of compact commutative groups. (Russian) Izv. Akad. Nauk SSSR Ser. Mat., **28**, 867–874 (1964).
183. Rohlin, V. A., Sinaĭ, Ja. G.: The structure and properties of invariant measurable partitions. (Russian) Dokl. Akad. Nauk SSSR, **141**, 1038–1041 (1961).
184. Rudolph, D. J.: Fundamentals of measurable dynamics. Ergodic theory on Lebesgue spaces. Oxford Science Publications. The Clarendon Press, Oxford University Press, New York (1990).
185. Ruelle, D.: A variational formulation of equilibrium statistical mechanics and the Gibbs phase rule. Comm. Math. Phys., **5**, 324–329 (1967).
186. Ruelle, D.: Statistical mechanics on a compact set with $\mathbb{Z}^\nu$ action satisfying expansiveness and specification. Trans. Amer. Math. Soc., **185**, 237–251 (1973).
187. Sauer, N.: On the density of families of sets, J. Combinatorial Theory Ser. A, **13**, 145–147 (1972).
188. Sauvageot, J.-L., Thouvenot, J.-P.: Une nouvelle définition de l'entropie dynamique des systems non-commutatifs. Comm. Math. Phys., **145**, 411–423 (1992).
189. Sauvageot, J.-L.: Ergodic properties of the action of a matrix in $SL_2(2,\mathbb{Z})$ on a non-commutative 2-torus. Preprint (1995).
190. Sauvageot, J.-L.: The canonical channel for dynamical quantum entropy. In: Quantum probability communications, 321–328, QP-PQ, X. World Sci. Publishing, River Edge, NJ (1998).
191. Sawin, S.: Relative commutants of Hecke algebra subfactors. Amer. J. Math., **116**, 591–604 (1994).
192. Schmidt, K.: Dynamical systems of algebraic origin. Progress in Mathematics, **128**. Birkhäuser Verlag, Basel (1995).
193. Seneta, E.: Nonnegative matrices and Markov chains. Second edition. Springer Series in Statistics. Springer-Verlag, New York (1981).
194. Shale, D., Stinespring, W. F.: States of the Clifford algebra. Ann. of Math. (2), **80**, 365–381 (1964).
195. Shannon, C. E.: A mathematical theory of communication. Bell System Tech. J., **27**, 379–423, 623–656 (1948).
196. Shelah, S.: A combinatorial problem; stability and order for models and theories in infinitary languages. Pacific J. Math., **41**, 247–261 (1972).

197. Sigmund, K.: Affine transformations on the space of probability measures. In: Dynamical systems, Vol. III—Warsaw, pp. 415–427, Asterisque, **51**. Soc. Math. France, Paris (1978).
198. Simon, B.: Trace ideals and their applications. London Mathematical Society Lecture Note Series, **35**. Cambridge University Press, Cambridge-New York (1979).
199. Sinaĭ, Ja.: On the concept of entropy for a dynamic system. (Russian) Dokl. Akad. Nauk SSSR, **124**, 768–771 (1959).
200. Sinaĭ, Ja. G.: On a weak isomorphism of transformations with invariant measure. (Russian) Mat. Sb. (N.S.), **63(105)**, 23–42 (1964).
201. Sinclair, A. M., Smith, R. R.: The completely bounded approximation property for discrete crossed products. Indiana Univ. Math. J., **46**, 1311–1322 (1997).
202. Slawny, J.: On factor representations and the C^*-algebra of canonical commutation relations. Comm. Math. Phys., **24**, 151–170 (1972).
203. Smith R. R.: Completely bounded maps between C^*-algebras. J. London Math. Soc. (2), **27**, 157–166 (1983).
204. Smith, R. R.: Completely contractive factorizations of C^*-algebras. J. Funct. Anal., **64**, 330–337 (1985).
205. Strătilă, Ş.: Modular theory in operator algebras. Editura Academiei Republicii Socialiste Romania, Bucharest; Abacus Press, Tunbridge Wells (1981).
206. Størmer, E.: Entropy of some automorphisms of the II_1-factor of the free group in infinite number of generators. Invent. Math., **110**, 63–73 (1992).
207. Størmer, E.: Entropy of some inner automorphisms of the hyperfinite II_1-factor. Internat. J. Math., **4**, 319–322 (1993).
208. Størmer, E.: Entropy in operator algebras. In: Recent advances in operator algebras (Orleans, 1992), Asterisque **232**, 211–230. Soc. Math. France, Paris (1995).
209. Størmer, E.: States and shifts on infinite free products of C^*-algebras. In: Free probability theory (Waterloo, ON, 1995), 281–291, Fields Inst. Comm., **12**. Amer. Math. Soc., Providence, RI (1997).
210. Størmer, E.: Entropy of endomorphisms and relative entropy in finite von Neumann algebras. J. Funct. Anal., **171**, 34–52 (2000).
211. Størmer, E.: A survey of noncommutative dynamical entropy. In: Classification of nuclear C^*-algebras. Entropy in operator algebras, 147–198, Encyclopaedia Math. Sci., **126**, Springer, Berlin, 2002.
212. Størmer, E., Voiculescu, D.: Entropy of Bogoliubov automorphisms of the canonical anticommutation relations. Comm. Math. Phys., **133**, 521–542 (1990).
213. Svendsen, A. L.: Endomorphisms and automorphisms from subfactors illustrating non-commutative entropy. Math. Scand., **97**, 266–280 (2005).
214. Takesaki, M.: Theory of operator algebras. I. Springer-Verlag, Berlin (2002). II, ibid. (2003). III, ibid. (2003).
215. Thomsen, K.: Topological entropy for endomorphisms of local C^*-algebras. Comm. Math. Phys., **164**, 181–193 (1994).
216. Tomczak-Jaegermann, N.: The moduli of smoothness and convexity and the Rademacher averages of trace classes $S_p (1 \leq p < \infty)$. Studia Math., **50**, 163–182 (1974).
217. Uhlmann, A.: Relative entropy and the Wigner-Yanase-Dyson-Lieb concavity in an interpolation theory. Comm. Math. Phys., **54**, 21–32 (1977).

218. Umegaki, H.: Conditional expectation in an operator algebra. IV. Entropy and information. Kōdai Math. Sem. Rep., **14**, 59–85 (1962).
219. Vapnik, V. N., Červonenkis A. Ja.: The uniform convergence of frequencies of the appearance of events to their probabilities. (Russian) Teor. Verojatnost. i Primenen., **16**, 264–279 (1971).
220. Vik, S.: Fock representation of the binary shift algebra. Math. Scand., **88**, 257–278 (2001).
221. Voiculescu, D.: Symmetries of some reduced free product C^*-algebras. In: Operator algebras and their connections with topology and ergodic theory (Buşteni, 1983), 556–588, Lecture Notes in Math., **1132**. Springer, Berlin (1985).
222. Voiculescu, D.: Almost inductive limit automorphisms and embeddings into AF-algebras. Ergodic Theory Dynam. Systems, **6**, 475–484 (1986).
223. Voiculescu, D.: Entropy of dynamical systems and perturbations of operators. Ergodic Theory Dynam. Systems, **11**, 779–786 (1991).
224. Voiculescu, D.: Entropy of dynamical systems and perturbations of operators. II. Houston J. Math., **17**, 651–661 (1991).
225. Voiculescu, D.: Entropy-invariants of dynamical systems and perturbations of operators. In: Mathematical physics, X (Leipzig, 1991), 303–307. Springer, Berlin (1992).
226. Voiculescu, D.: The analogues of entropy and of Fisher's information measure in free probability theory. II. Invent. Math., **118**, 411–440 (1994).
227. Voiculescu, D.: Dynamical approximation entropies and topological entropy in operator algebras. Comm. Math. Phys., **170**, 249–281 (1995).
228. Voiculescu, D. V., Dykema, K. J., Nica, A.: Free random variables. A noncommutative probability approach to free products with applications to random matrices, operator algebras and harmonic analysis on free groups. CRM Monograph Series, **1**. American Mathematical Society, Providence, RI (1992).
229. Walczak, P.: Dynamics of foliations, groups and pseudogroups. Instytut Matematyczny Polskiej Akademii Nauk. Monografie Matematyczne (New Series), **64**. Birkhäuser Verlag, Basel (2004).
230. Walters, P.: A variational principle for the pressure of continuous transformations. Amer. J. Math., **97**, 937–971 (1975).
231. Walters P.: An introduction to ergodic theory. Graduate Texts in Math., **79**. Springer-Verlag (1982).
232. Wassermann, S.: Exact C^*-algebras and related topics. Lecture Notes Series, **19**. Seoul National University, Research Institute of Mathematics, Global Analysis Research Center, Seoul (1994).
233. Watatani, Y., Wierzbicki, J.: Commuting squares and relative entropy for two subfactors. J. Funct. Anal., **133**, 329–341 (1995).
234. Weyl, H.: Über die Gleichverteilung von Zahlen modulo Eins. Math. Ann., **77**, 313–352 (1916).
235. Wigner, E. P., Yanase, M. M.: Information contents of distributions. Proc. Nat. Acad. Sci. U.S.A., **49**, 910–918 (1963).
236. Yin, H. S.: Entropy of certain noncommutative shifts. Rocky Mountain J. Math., **20**, 651–656 (1990).
237. Zeller-Meier, G.: Produits croises d'une C^*-algebre par un groupe d'automorphismes. J. Math. Pures Appl. (9), **47**, 101–239 (1968).
238. Zygmund, A.: Trigonometric series. Vols. I, II. Third edition. Cambridge Mathematical Library. Cambridge University Press, Cambridge (2002).

List of Symbols

$A \hat{*} B$	the full free product of the C*-algebras A and B, p. 251
$(A, \varphi) * (B, \psi)$	the reduced free product of the C*-algebras A and B with respect to the states φ and ψ, p. 251
$(H_1, \xi_1) * (H_2, \xi_2)$	p. 252
$\mathcal{A}(H)$	the CAR-algebra over H, p. 227
α_U	the Bogoliubov automorphism defined by the unitary operator U, p. 228
$\eta(t)$	$-t \log t$
$\int_X^\oplus \pi_x d\mu(x)$	the direct integral of the representations π_x, p. 276
$\int_X^\oplus \varphi_x d\mu(x)$	the direct integral of the states φ_x, p. 276
$\int_X^\oplus a_x d\mu(x)$	the direct integral of the operators a_x, p. 276
$\int_X^\oplus H_x d\mu(x)$	the direct integral of the Hilbert spaces H_x, p. 275
ω_A	the quasi-free state defined by the operator A, p. 228
Φ_f, Ψ_f	p. 123
$\rho_\omega(g, h)$	$\omega(g, h)\overline{\omega(h, g)}$
σ_X	the binary shift corresponding to the bitstream X, p. 217
τ_X	the canonical tracial state on $A(X)$ or $M(X)$, p. 217
$\theta_1 * \theta_2$	the free product of completely positive maps, p. 252
$\|x\|_2$	$\tau(x^*x)^{1/2}$
$\|x\|_\varphi$	$\varphi(x^*x)^{1/2}$
$\|\gamma\|_\varphi$	$\sup_{\|x\|\leq 1} \|\gamma(x)\|_\varphi$
ξ^-	$\vee_{k=-\infty}^{-1} T^k \xi$
$\xi^\pm$	$\vee_{k=-\infty}^{\infty} T^k \xi$
$A(X)$	the C*-algebra associated with the bitstream X, p. 211

List of Symbols

$A \otimes B$	the minimal tensor product of the C*-algebras A and B
$A \rtimes_{\beta,\omega} G$	the full crossed product of A by G with respect to the 2-cocycle ω, p. 121
$A \rtimes_{r,\beta,\omega} G$	the reduced crossed product of A by G with respect to the 2-cocycle ω, p. 121
$C^*(G,\omega)$	the full twisted group C*-algebra of G with respect to the 2-cocycle ω, p. 185
$C_r^*(G,\omega)$	the reduced twisted group C*-algebra of G with respect to the 2-cocycle ω, p. 130
$W^*(G,\omega)$	the twisted group von Neumann algebra of G with respect to the 2-cocycle ω, p. 186
E_ξ	the conditional expectation onto $L^\infty(X/\xi, \mu)$, p. 4
E_N	the trace preserving conditional expectation onto N
$h(\mathcal{U};T)$	the topological entropy of the homeomorphism T with respect to the open cover $\mathcal{U}$, p. 9
$h(F;T), h(T)$	p. 188
$H_\lambda(\gamma_1,\ldots,\gamma_n;\xi_1,\ldots,\xi_n)$	the mutual entropy of the channels $\gamma_1,\ldots,\gamma_n$ with respect the decomposition of the state defined by the coupling λ and partitions $\xi_1,\ldots,\xi_n$, p. 43
$H_\mu(\xi)$	the entropy of the partition ξ with respect to the measure μ, p. 3
$h_\mu(\xi;T)$	the entropy of the transformation T with respect to the partition ξ and the measure μ, p. 6
$H_\mu(\xi|\zeta)$	the conditional entropy of ξ with respect to the partition ζ and the measure μ, p. 4
$h_\mu(T)$	the entropy of the transformation T with respect to the measure μ, p. 6
$h_\varphi(\alpha)$	the dynamical entropy of α with respect to φ, p. 48
$h_\varphi(\gamma;\alpha)$	the entropy of the automorphism α with respect to the channel γ and the state φ, p. 48
$H_\varphi(\gamma_1,\ldots,\gamma_n)$	the mutual entropy of the channels $\gamma_1,\ldots,\gamma_n$ with respect to the state φ, p. 34
$H_\varphi(\gamma_1,\ldots,\gamma_n;\{\varphi_{i_1\ldots i_n}\})$	the mutual entropy of the channels $\gamma_1,\ldots,\gamma_n$ with respect to the decomposition $\varphi = \sum_{i_1,\ldots,i_n} \varphi_{i_1\ldots i_n}$, p. 34
$H_\varphi(P|Q)$	the relative entropy of the algebras P and Q with respect to the state φ, p. 157
$h_\varphi^{ST}(\alpha)$	the Sauvageot-Thouvenot entropy of α with respect to φ, p. 77
H_a, H_s	p. 277
$H_{\lambda,\gamma}(\xi)$	the entropy defect of the decomposition of the state $\lambda \circ (\gamma(\cdot) \otimes 1)$ defined by the coupling λ and the partition ξ, p. 78

$H_{\lambda,\gamma}(\xi\|\zeta)$	the entropy defect of the decomposition of the state $\lambda \circ (\gamma \otimes i_\zeta)$ defined by the coupling λ and the partition ξ, p. 79
$h_{top}(T)$	the topological entropy of the homeomorphism T, p. 9
$ht(\alpha)$	the topological entropy of α, p. 96
$ht(\Omega, \delta; \alpha)$	p. 96
$ht(\Omega; \alpha)$	the topological entropy of the automorphism α with respect to the finite set Ω, p. 96
i_ξ	the embedding mapping $L^\infty(X/\xi) \hookrightarrow L^\infty(X)$
$M(X)$	the von Neumann algebra associated with the bitstream X, p. 219
$M \bar{\otimes} N$	the von Neumann algebra tensor product
$N(\mathcal{U})$	the cardinality of a minimal open subcover of the cover $\mathcal{U}$, p. 9
$P(H, \Omega, \delta)$	p. 133
$P_\alpha(H)$	the pressure of α at H, p. 134
$P_\alpha(H, \Omega)$	the pressure of α at H with respect to Ω, p. 134
$P_\alpha(H, \Omega, \delta)$	p. 134
$S(\varphi)$	the von Neumann entropy of the positive linear functional φ, p. 21
$S(\varphi, \psi)$	the relative entropy of the positive linear functionals φ and ψ on a finite dimensional C*-algebra, p. 15
$S(\varphi, \psi)$	the relative entropy of the positive linear functionals φ and ψ, p. 26
$T_2(\mathcal{X})$	the type 2 constant of the Banach space $\mathcal{X}$, p. 111
U_a, U_s	the parts of U with absolutely continuous and singular spectral measures, respectively, p. 229
$X \subset_\delta Y$	for each $x \in X$ there exists $y \in Y$ such that $\|x - y\| < \delta$
$\dim_M(H)$	the dimension of the Hilbert space H relative to the II_1-factor $M \subset B(H)$, p. 164
$\text{rcp}_A(\Omega, \delta)$	the rank of a completely positive approximation of $\Omega \subset A$ to within $\delta > 0$, p. 93
Tr_A	the canonical (nonnormalized) trace on the finite dimensional C*-algebra A, p. 15

Index

2-cocycle, 121

Abelian model, 36
Algebra
   exact, 94
   nuclear, 28
   of canonical anticommutation
      relations, 227
Amenable action, 126
Approximating net, 49
Asymptotically abelian system with
   locality, 138
Asymptotically highly anticommutative
   system, 91

Basic construction, 171
Bernoulli shift, 7
   noncommutative, 51
   on the hyperfinite II_1-factor, 52
Binary shift, 217
Bogoliubov automorphism, 228

C^*-dynamical system, 48
   with singular spectrum, 87
Canonical shift, 173
CAR-algebra, 227
   even, 232
Centralizer of a state, 21, 35
Channel, 34
Commuting square, 159
   condition, 168
Complete order embedding, 103
Completely positive map, 16, 265
Conditional expectation, 16, 269

Coupling, 43
   canonical, 43
   stationary, 77
Crossed product
   full, 121
   reduced, 121

Decomposable operator, 276
Density operator, 15
Dimension of a Hilbert space relative to
   a II_1-factor, 164
Direct integral of Hilbert spaces, 275
Dynamical system, 5
   ergodic, 7
   topological, 9
      uniquely ergodic, 11
   uniformly mixing, 69

Entropic K-system, 69
Entropy
   completely positive approximation,
      104
   conditional, 4
   dynamical, of an automorphism, 48
   mean, 152
   mutual, of channels, 33, 34
   of a measure preserving transforma-
      tion, 6
   of a partition, 3
   of an abelian model, 36
   of an automorphism, with respect to
      a channel, 48
   perturbation-theoretic, 105
   relative

of measures, 13
of positive linear functionals, 15, 26
of subalgebras, 157
Sauvageot-Thouvenot, 77
topological
  of a homeomorphism, 9
  of an automorphism, 96
von Neumann, 21
Entropy defect, 35
Exact group, 126

Factor system, 7
Free product
  full, 251
  reduced, 251
Free shift, 254

Generating
  partition, 7
  sequence, 167
Gibbs state, 144
Golden-Thompson inequality, 65

Hölder's inequality, generalized, 274
Horn's inequality, generalized, 112

Independent algebras, 66
Index of a subfactor, 164
Irrational rotation algebra, 206

Jewett-Krieger theorem, 11
Jones tower, 171

Kolmogorov-Sinai theorem, 7

Local algebras, 138
Local operators, 138

Measurable field of Hilbert spaces, 275
Mirroring, 173
Modular
  conjugation, 35
  group, 35
  operator, 35

Nuclear homomorphism, 94

Operator convex function, 271
Operator monotone function, 271
Operator system, 103
Ornstein's theorem, 7
Orthogonal decomposition of a state, 35

Peierls-Bogoliubov inequality, 30
Pimsner-Popa inequality, 165
Pinsker's formula, 107
Pressure, 134

Quasi-free state, 228

Rank
  of a completely positive approximation, 93
  of an algebra, 21

Schwarz map, 16, 266
Separated set, 10
Set
  nonperiodic, 213
  periodic, 213
Shannon-McMillan-Breiman theorem, 11
Sinai's theorem, 8
Singular spectrum, 87
Skew-symmetric bicharacter, 185
Spanning set, 9
State
  equilibrium, 137
  Gibbs, 144
  KMS, 144
  quasi-free, 228
Stinespring's dilation, 266
Strictly operator convex function, 272
Subfactor
  extremal, 167
  irreducible, 165
  of finite depth, 178
Subgradient, 137

Thermodynamic inequality, 24
Topological model, 5
Tunnel of subfactors, 171

Unitary operator
  with homogeneous spectrum, 277
  with Lebesgue spectrum, 277

Variational principle, 10
  for asymptotically abelian systems with locality, 138

W^*-dynamical system, 48
  entropic K-system, 69
  with completely positive entropy, 223

Printing: Krips bv, Meppel
Binding: Stürtz, Würzburg